The European Scene
A Geographic Perspective
SECOND EDITION

JAMES R. McDONALD

Prentice Hall
Upper Saddle River, New Jersey 07458

Library of Congress Cataloging-in-Publication Data

McDonald, James R.
   The European Scene : a geographic perspective / James R. McDonald
— 2nd ed.
     p. cm.
   Includes bibliographical references and index.
   ISBN 0–13–368614–0
   1. Europe—Geography.   2. Europe—Economic conditions.   I. Title.
D900.M35   1997   96–43933
914—dc20   CIP

Acquisition Editor: Dan Kaveney
Production and Page Composition: Joan Eurell
Manufacturing Manager: Trudy Pisciotti
Executive Managing Editor: Kathleen Schiaparelli
Assistant Managing Editor: Shari Toron
Art Director: Jayne Conte
Cover Image: © 1990 Bassignac/Gamma-Liaison
Cover Photo Research: Teri Stratford

© 1997, 1992 by Prentice-Hall, Inc.
Simon & Schuster / A Viacom Company
Upper Saddle River, NJ 07458

All rights reserved. No part of this book may be reproduced, in any form or by any means, without permission in writing from the publisher.

Printed in the United States of America.

10 9 8 7 6 5 4 3 2 1

ISBN 0-13-368614-0

Prentice Hall International (UK) Limited, *London*
Prentice Hall of Australia Pty. Limited, *Sydney*
Prentice Hall of Canada, Inc., *Toronto*
Prentice Hall Hispanoamericana, S.A., *Mexico*
Prentice Hall of India Private Limited, *New Delhi*
Prentice Hall of Japan, Inc., *Tokyo*
Simon & Schuster Asia Pte. Limited, *Singapore*
Editora Prentice Hall do Brasil, Ltda., *Rio de Janeiro*

Contents

	Map List	v
	Preface	vii
	Introduction	1

ONE THE NATURAL SETTING

1	The Physical Geography of Europe	11
2	European Weather and Climate, Vegetation, and Soils	29

TWO HUMAN GEOGRAPHY AND THE DEVELOPMENT OF THE CULTURAL LANDSCAPE

3	Europe in the World Population Picture	51
4	Restless Europe: The Role of Migration	61
5	European Language Patterns and Religious Influences	73
6	The Settlement and Urbanization of Europe	97

THREE PATTERNS AND IMPLICATIONS OF EUROPEAN ECONOMIC ACTIVITY

7	Preagricultural Europe: Plant and Animal Domestication and its Diffusion into Europe	111
8	Classic European Economies: Alpine and Mediterranean	119
9	European Agriculture, Forestry, and Fisheries	139
10	European Resources	153
11	Industrial Location in Europe: The Evolution of the European Industrial Model	171
12	European Tourism	191
13	The European Environment	203
14	Supranational Institutions in Europe	217

Four National and Regional Profiles

15	Northern Europe	229
16	The British Isles	245
17	France	255
18	The Benelux States	269
19	Germany	279
20	The Alpine States	289
21	Italy, Greece, and the Mediterranean Islands	301
22	Iberia	321
23	Eastern Europe and the Baltic States	331
	Conclusion: Europe and the New Century	361
	Index	367

Map List

I.2	The nations of Europe — 1996.
I.4	The Ural line traditionally divides Europe from Asia.
1.1	European population density.
1.2	Europe: Geologic framework.
1.6	Europe: Simplified landform regions.
1.7	Extent of Pleistocene glaciation in Europe.
1.10	Europe's principal waterways.
2.1	Europe: Generalized climatic types.
2.4	European drought: Does it reflect the "greenhouse effect?"
2.7	Europe: Generalized vegetation regions.
2.8	Europe: Generalized soil types.
5.3	European language regions.
5.4	Switzerland's major language regions.
5.5	Belgium's language structure.
5.7	Europe: Religious strength.
5.9	The spread of Islamic influence into Europe.
6.5	Europe: Millionaire cities.
6.8	Paris: Successive wall lines.
7.3	Diffusion routes of plants and ideas: Middle East to Europe.
8.1	Europe: Areas above 1,000 feet in altitude.
8.6	Europe: Northern limits of selected crops.
9.2	Europe: Major land–use types.
9.5	Europe: Major production regions of selected crops.
10.1	Europe: Energy resources.
10.7	Europe: Location of selected metallic minerals.
10.9	Resource conflict in Western Europe.
11.2	Europe: Selected manufacturing centers and the industrial heartland.
11.6	The English Midlands were home to the Industrial Revolution.

11.7 The Ruhr is Europe's most concentrated industrial district.
13.2 Europe: Concentration of low pH values.
14.1 Membership of the European Union (dates of accession shown).
15.1 Norway.
15.2 Sweden.
15.3 Denmark and its strategic location at the entrance to the Baltic.
15.4 Finland.
15.5 Iceland.
15.6 The Baltic nations.
16.1 The United Kingdom.
16.2 The Irish Republic
16.4 Greater London: One of Europe's largest metropolitan areas.
17.1 France.
17.4 Greater Paris.
17.7 France: Nuclear electricity plants, 1996.
17.9 Monaco: A ministate on the Mediterranean coast.
18.1 Belgium.
18.2 The Netherlands.
18.3 Luxembourg.
18.5 Major Dutch reclamation projects.
19.1 Germany.
19.4 Berlin, a city reunited after 45 years of division.
20.1 Switzerland.
20.2 Austria.
20.7 Liechtenstein: Alpine ministate.
21.1 Italy.
21.5 Rome rediscovers its ancient importance as modern Italy's capital.
21.7 The ministate of San Marino.
21.8 Greece.
21.9 Malta and Cyprus.
22.1 Spain.
22.2 Portugal.
22.8 Andorra: A ministate in the Pyrenees.
23.1 Poland.
23.2 The Czech and Slovak republics.
23.5 The Danube links many of the nations of Eastern Europe.
23.8 Hungary.
23.10 Romania.
23.11 Bulgaria.
23.14 The Yugoslav states: Political divisions.
23.16 The Yugoslav states: Ethnic complexities.
23.17 Albania.
23.18 European nations from the former Soviet Union: Belarus, Moldova, Ukraine.

Preface
to the Second Edition

The first edition of this book appeared some five years ago, which means that the writing "went to bed," except for very minor changes, in June 1991. Five years is a long time in the world, and an even longer time when one is attempting to produce a book that is reasonably contemporary. Even as I compose these introductory words, I am aware that at least several months will pass until the work is in circulation: months during which the details of Europe will change in countless ways—some perhaps spectacular—that I am powerless to consider. To some extent one can guess. In 1991, as the Soviet Union fell apart, it was not too imaginative to consider the Baltic states as soon-to-be-independent northern European nations. On the other hand, despite speculation regarding tensions and potential political stress, who could have forecast that Yugoslavia would collapse into Europe's most brutal war since 1945? In this edition, I again take some chances, looking cautiously at what the future might hold for the Yugoslav states and the European Union; and at other large uncertainties that loom over the European scene. There is the certain knowledge, however, that many things will change beyond the confines of any reasonable prediction.

Yet not everything is different. One of the continuing themes of the book is "continuity and change" in Europe, and there is at least as much of the one as the other. I have spent extensive time in Europe in each of these intervening years, particularly on the back roads and byways, and have come to appreciate even more the underlying stability of much of the continent. Europe's natural environment has evolved over millions of years, and short-term changes are necessarily subtle. Climatic patterns have also developed over a long timeframe, although certain recent changes have been striking. Additionally, human modifications of the landscape, as well as the complexity of cultural institutions and economic interests, have deep roots in Europe's past. Thus, some chapters of the first edition have been only modestly rewritten to reflect the changes of five years, while others have been extensively revised. All maps have been corrected, updated, and improved to better reflect the realities of 1996-97; new tables and diagrams, as well as many new photographs, have been added; and all data have been updated to provide the most recent available numbers. Unless otherwise credited, all photographs are by the author.

Since publication of the first edition, I have been naturally gratified by the considerable number of instructors who have found the book suitable to their courses, many of whom have communicated this to me. The original idea was to produce a "user-friendly" book that would be accessible to an undergraduate audience while allowing maximum flexibility to instructors in designing their own schedules; this was apparently accomplished in the first edition, and the same philosophy has been continued in the second.

I would like to thank the many reviewers who provided constructive comments on the first edition, pointing out numerous flaws and suggesting possible (and in many cases obvious) improvements. Two people who read the book thoroughly, Dr. Sean M. McDonald as an instructor at the University of Northern Colorado and Ms. Agneta H. Hagman as a student at the University of Washington, were especially helpful. Many other reviewers provided guidance for the second edition, and I am also grateful for their insights. For the errors which will undoubtedly have crept into this edition, I have unfortunately no one but myself to blame. The genealogy of my European interests was noted in the preface to the first edition, and need not be repeated here; but I must especially note the encouragement and enthusiasm of my editors at Prentice Hall: Dan Kaveney and, earlier, Ray Henderson, who have been unfailingly positive about this project; and also the skilled Prentice Hall production team, led by Joan Eurell, which has taken many burdens off the author. Thanks also to John Sanderson, of Horizon Design, for his skilled cartographic work. Finally, I am deeply indebted, now and forever, to my wife and European traveling companion, Dr. Sharon C. McDonald.

James R. McDonald

For Chris and Sean

The Next Generation of Europhiles

Introduction

EUROPE: THE CONTINENT OF CHANGE

If Europe has had a consistent image over the years since 1945, it has been as a continent of dramatic and often unpredictable change. The heavy weight of history, the landscapes touched by and reflective of the past, the people and ideas that have in a very real sense created the modern world: all this formed the essence of a politically and economically fragmented continent entering the last half of the twentieth century. Yet, as the 1990s have unfolded, this image has altered dramatically and unexpectedly once again. From being a footnote to contemporary events, Europe has reinforced the trends of its own history and of recent global events by becoming yet again a continent of sudden and profound change, challenging its own traditions as well as the established world order. Two major developments, each with important and multiple implications, have sparked this new era of revolution in Europe: one is concerned with destroying and reconstructing an old order, the other with building something entirely new.

What riveted the world's attention during the period 1989–91 and made Europe a sort of day-to-day exercise in current events, was the sudden, total, and almost universally unforeseen collapse of nearly all of the communist regimes in Eastern Europe. Progressive failure of state-directed command economies to meet national objectives and consumer demands, combined with the winds of glasnost and perestroika blowing from a Soviet Union increasingly unwilling or unable to militarily enforce its writ across the region, changed the political environment beyond recognition in the course of a few brief months. Not surprisingly, these changes, by both their suddenness and their depth, have had a ripple effect of implications for the rest of Europe and indeed for the entire world.

Later, the Soviet Union itself collapsed into a collection of fifteen very diverse independent countries, many of which consider themselves quite European, others that look elsewhere for inspiration. The inevitable turmoil engendered by the sudden end of a classic monolithic system has also had profound implications for Europe. Nor have the post-Soviet 1990s been an era of unrestricted peace and cooperation for Europe. The tragic evolution of the situation in what was—in 1990—a peaceful Yugoslavia has brought home to all of Europe the fragility of its current general tranquility and has raised substantial questions as to the future political evolution of the continent.

Second, and again in sharp counterpoint to the national divisiveness and jealously guarded frontiers of a half century ago, Europe appears to be embarking seriously on a course of economic and, eventually perhaps, political union unprecedented in its entire history. The European Union, from its modest beginnings in 1957, has become a powerful institution challenging at once the established nation-state structure so characteristic of historic Europe and the international image of the continent as economically powerful but hopelessly fragmented. The years through the first decade of the twenty-first century loom as critical for the Union, as Europe takes decisive steps to achieve a degree of integration that may enable it to recapture its historic position as the motor of the world's economy and the arbiter of its political order.

From these two great events, and from their interaction, other significant changes in European realities and ideas flow. Old military and economic alliances break down, or are forced to critically reexamine their rationales and future roles. Nationalities and ethnic minorities, long stifled by repressive or centralizing governments, awaken to demand a place in the new European sun, often fostering tensions and conflict as they rejoice in unaccustomed freedoms. The environmental disaster that much of Eastern Europe has become in the name of economic progress is exposed to the light, and becomes a concern of the utmost priority for the entire continent from its international organizations to its local political units and even to its individual citizens.

Figure I.1 Destruction of the infamous Berlin Wall in 1989-90 emphasized the dramatic changes sweeping Eastern Europe.

Introduction

As the twentieth century draws to a close, Europe is thus in many ways the most exciting continent of all to consider. Innumerable aspects of its political, cultural, and economic geography are changing beyond recognition. Yet, at the same time, there is a certain continuity to the European scene that is both reassuring and promising. The natural landscapes of the continent, with their strong component of human management, remain remarkably timeless. The vitality and creativity of Europe's peoples, that have produced the dynamism and variety that are the hallmarks of the continent, persist undiminished. The economic strength of Europe continues to have a decisive impact on the entire world.

Continuity and change, perhaps the words that best describe Europe in the 1990s, are also those that summarize the geographic approach to space and time. A geography of Europe, then, should attempt to encompass as many relevant variables of the modern scene as possible. But how should such a work be organized? What are the possible frameworks that should be considered?

SYSTEMATIC VERSUS REGIONAL APPROACHES

Geography studies the physical, cultural, and economic phenomena of the earth in their spatial interactions. Hardly anything is arranged either perfectly symmetrically or entirely randomly across the surface of the globe, and geography seeks to describe and explain the patterns of distribution—often complex—and to suggest ways of seeing that might help to interpret the past, justify the present, and possibly predict the future. It seeks to offer a way of understanding the bewildering diversity all around us and, as an essentially ecological science, emphasizes the inevitable interrelationships among all the world's systems, both natural and human.

In setting about this admittedly sizeable task, geographers have traditionally used two distinctive approaches: the systematic and the regional. Both are useful; both are still widely employed. This book is an attempt to synthesize the two in the setting of Europe. As such, some explanation might be in order.

Classical regional studies such as, for example, many books dealing with Europe, tend to arrange material almost entirely by country or by major subregion (e.g., "Northern Europe"). This approach is based on the familiar geographic concept of the "region" as the ultimate unit of synthesis of all geographic information: that is, a unique space which can be differentiated from any other, and described in terms of its particular characteristics of landforms, climate, cultures, historical evolution, economic practices, ways of life, or political outlooks. Using such an approach for Europe, we might therefore (after some brief general remarks) discuss the region on the basis of its forty-four or so individual nations, proceeding either alphabetically (an extreme case!) from Albania to Yugoslavia; or more logically by subregion (Germany, Switzerland, and Austria, for instance, discussed sequentially as "central Europe").

This strategy, however, has some major drawbacks, many of which have become more serious with the passage of time. Chief among these is the fact that Europe—in size a small continent—has many characteristics that inevitably cut across national boundaries: basic patterns of climate, for example, or problems of air and water pol-

Figure I.2 The nations of Europe — 1996.

lution, or the area in which a certain language is spoken. Over the course of the past half century, this trend has become much more marked, as European nations have developed important political and economic institutions in which many of them are grouped (the European Union is a case in point), and which are having the gradual effect of reducing the significance of national boundaries, thus blurring their effectiveness as regional limits. Many crucial decisions in Europe today are being made on the basis of multinational frameworks of one kind or another, and the significance of these can only imperfectly be explained in any traditional context of regional subdivision. Moreover, rapid advances in communications, transportation, and other aspects of the late twentieth century technological revolution have tended somewhat to "homogenize" Europe in terms of information, movement, and economic patterns and problems; this trend has been accelerated by the momentous political changes of the past decade.

The internationalization of modern cities and the growing importance of interurban networks, which literally bypass many conventional distinctions among nations, have also played a key role in this process, as have the vast international and internal human migrations which have recently created truly cosmopolitan societies in many urban districts of Europe. The result of these interacting trends has been to render the traditional regional approach to describing the continent somewhat dated. Nonetheless, it is clearly not possible to ignore entirely those national and subregional units which are such a familiar part of the European scene, and which continue to play a key role—in reality as well as in the imagination—in the day-to-day functioning of the continent. We must surely consider these at some point.

In the other basic strategy for considering geographic information—the systematic approach—the distribution patterns of all basic phenomena which must be considered in order to gain some understanding of the total region under discussion (again, Europe in the present case) are observed and noted wherever they occur, rather than discussed country by country. Landforms and their significance, population characteristics, and resource availability are thus considered on a continental, rather than national, basis. This approach has a number of advantages. First of all, it enables the "big picture" of various topics to emerge more clearly, which is difficult to accomplish when the relevant information is scattered at several points in the text. Second, it permits the logical grouping of distinctive units which would otherwise be fragmented. For example, it makes greater sense to discuss the characteristics of the Alps as a single, complex mountainous region than to divide them among separate descriptions under Austria, Switzerland, Germany, France, Italy, and Slovenia. Finally, it diminishes the often artificial importance of political boundaries as geographical determinants, and enables a more modern picture of international interactions to emerge.

In the Europe offered here, therefore, the basic approach to the presentation of geographic information will be a systematic one, with specific national or local examples used at many points to emphasize the patterns that are developed. The latter part of the book, however, consists of a number of national and subregional profiles, in which basic information about the major European nations is summarized and their main national characteristics identified. In this way, it is hoped that the two traditions of geography—the systematic and the regional—can be combined so as to enhance understanding of this most fascinating of continents.

WHY EUROPE?

Of all of the world's major regions, Europe is in many ways the least likely candidate for a featured role. It is, for a start, the smallest by far of the continents (if the former Soviet Union is excluded), and consists largely of a collection of islands and peninsulas attached to the western extremities of the vast Eurasian landmass. In fact, the ex-Soviet Union and Europe combined are comparable in size to the other major continents. It is only 750 miles (1200 kilometers) from the Baltic Sea to the Black Sea, less than 250 miles (400 kilometers) from the Mediterranean Sea across France to the Atlantic Ocean. Enjoying the elaborate system of superhighways, a motorist can easily travel in four or even five countries in the course of a single day!

If its small size fails to interest us, we might well ask if a strong resource base is what makes Europe important (perhaps on the Middle Eastern model). Here again, the answer is negative. Europe is a continent deficient in almost all of the major industrial minerals, in energy commodities, in most of the raw materials its industries process, and even in certain agricultural products. The continent as a whole is an enormous net importer of these basic items, and only in a few countries do their production and export play an important economic role.

If, however, in attempting to understand the importance of Europe, we turn to its population, the picture changes considerably. It is, in effect, the people of Europe—their numbers, their skills, their inventiveness, and above all their impact on the rest of the world—that have given Europe the success, the prosperity, the self-confidence, and the key role in world affairs that it still enjoys. From classical times through the Middle Ages and the Age of Exploration, down to the most recent periods of history, Europeans have left their mark on the world to a degree out of all proportion to the size and resources of their continent. This is a pattern which still persists: many Americans know the names of (and often something about) European countries with populations of only 5 or 10 million—Switzerland, Sweden, the Irish Republic, for example—yet around the world other nations with populations ten times as large remain totally, and embarrassingly, unknown.

Europe not only discovered the New World in the fifteenth and sixteenth centuries, it exercised colonial control over much of the globe at one time or another until the mid twentieth century. During the same period it also exported millions of its unwanted or undervalued citizens to populate large areas of the world. Moreover, wherever Europeans and European influence went, substantial cultural change—much of it permanent—also occurred. English and French became the languages of international communication over widespread areas of Africa and Asia (modern India could probably not function as a multiethnic nation without the "glue" of English). European ideas and philosophies often replaced older values, and European systems of legal and social organization are the general rule.

Finally, it has been the European economic model which has largely become the wonder of the world, to be envied, resented, and emulated by nearly all nations. The European system of agriculture (both crops and technologies) spread wherever Europeans settled. Elsewhere, Europeans established plantations to satisfy their own requirements; these are still a crucial element of the economies of numerous tropical and subtropical countries. It was also Europeans who controlled and profited most from long-distance trading to Asia and the Americas. Most important, the European model of industry, as articulated in the industrial revolution and all of its continuing implications, has set the standard for the world.

Thus, Europe has a significance in the world that goes far beyond its size and internal resources. No part of the globe has altogether escaped its influence, and by taking a closer look at its geographic patterns we can, to a degree true of no other continent, begin to understand something of how the world thinks and acts.

Introduction

Figure I.3 Area and population data for European countries, 1990.

	Area (1,000 sq. mi.)*	Population (millions)*†	Current Annual Rate of Growth (%)
Albania	10.6	3.3	1.7
Andorra‡	(185)	(45,000)	1.2
Austria	31.9	8.1	0.1
Belarus§	80.2	10.3	(−0.3)
Belgium	11.8	10.2	0.1
Bosnia-Herzegovina	19.7	3.6	0.6
Bulgaria	42.7	8.4	(−0.4)
Croatia	21.8	4.4	0.0
Cyprus	3.6	0.7	0.9
Czech Republic	30.6	10.3	(−0.1)
Denmark	16.4	5.2	0.2
Estonia§	17.4	1.5	(−0.5)
Finland	117.6	5.1	0.3
France	212.4	58.4	0.3
Germany	134.9	81.7	(−0.1)
Greece	50.5	10.5	0.1
Hungary	35.7	10.2	(−0.3)
Iceland	38.9	0.3	1.0
Ireland	26.6	3.6	0.5
Italy	113.5	57.3	0.0
Latvia§	24.9	2.5	(−0.7)
Liechtenstein‡	(60)	(30,000)	0.5
Lithuania§	25.2	3.7	(−0.1)
Luxembourg	(990)	0.4	0.4
Macedonia	9.9	2.1	0.8
Malta	(120)	0.4	0.6
Moldova§	14.2	4.3	0.3
Monaco‡	(0.50)	(25,000)	0.0
Netherlands	13.1	15.5	0.4
Norway	118.5	4.4	0.4
Poland	117.6	38.6	0.2
Portugal	35.5	9.9	0.1
Romania	88.9	22.6	(−0.2)
Russia§	6,592.8	147.7	(−0.5)
San Marino‡	(20)	(30,000)	0.3
Slovakia	18.8	5.4	0.3
Slovenia	7.8	2.0	0.0
Spain	192.8	39.3	0.1
Sweden	158.9	8.9	0.1
Switzerland	15.4	7.1	0.3
Ukraine§	233.1	51.1	(−0.5)
United Kingdom	93.3	58.8	0.2
Vatican City‡	(0.1)	(1,000)	0.0
Yugoslavia#	26.9	10.2	0.3

*Figures in parentheses are actual numbers.

†Population Reference Bureau 1996 estimates [except for ‡].

‡1986 United Nations estimates.

§Former Republics of the ex-USSR; now considered independent European nations.

#"Yugoslavia" now technically consists of the former federal republics of Serbia and Montenegro, plus the autonomous regions of Kosovo and Vojvodina.

THE LIMITS OF EUROPE—SOME ARBITRARY DECISIONS

A final problem should be resolved by way of introduction: what exactly constitutes "Europe"? What are its boundaries? How much should be included in this book? There is, of course, a certain continental core or heartland which would be included under any definition, but toward the edges some of the distinctions become fuzzier, and the "Europeanness" of some regions less obvious. There are at least three decisions that must be made.

To the east, Europe blends almost imperceptibly into Asia. The flat, featureless landscapes of northern Germany or Poland continue unchanged for hundreds of miles into Russia, while climatic change and many cultural patterns are transitional rather than abrupt. Only the north–south trending Ural Mountains offer a distinctive geographic contrast across this space; not surprisingly, they have traditionally been used as the dividing line between European and Asiatic Russia. At the southern end of the Urals, this line is less convincingly extended along the Ural River and the north face of the Caucasus Mountains to the Black Sea and hence to the Mediterranean.

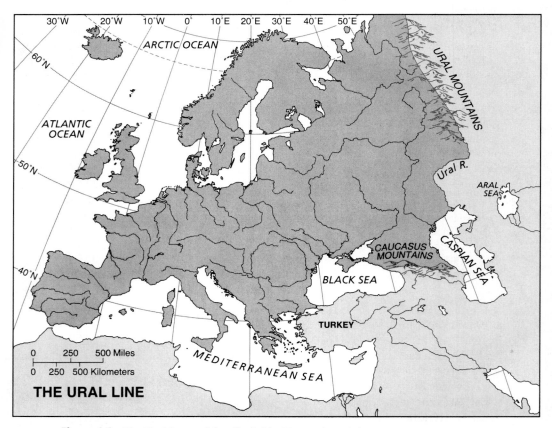

Figure I.4 The Ural line traditionally divides Europe from Asia.

Introduction

Should we accept this limi[t and not go] west of it in our study? There i[s ...] gained independence upon the [...] many links of history, politics, a[nd ...] their physical landscapes. Set a[gainst ...] plexity of the Russian Federatio[n ...] in countless ways to create the d[...]

Moscow's connections wit[h ...] to Central Asia or the far easter[n ...] Thus, it is probably a better idea [...] cept for those areas where its inc[...] the political life of the Eastern E[uropean ...] ary of Russia—certainly another h[...] now ends, although three new ex-S[oviet ...] Latvia, Lithuania), with their strongly [...] along the western border of the former S[oviet Union ...] va, are certainly worth inclusion.

Another area of uncertainty concerns [...] one of the world's most consistent physical, h[istorical ... dis-] cussion of the Mediterranean economy, Chapter 8). Some twenty modern nations lie in or around the classical sea of antiquity, and for all of them the sea represents an important component of national life. It is certainly reasonable to consider the region as a whole, rather than simply its northern, European fringe: Spain has considerably more in common with Morocco than with Sweden, yet we arbitrarily exclude North Africa and the Middle East. As a partial compromise, the essential unity of the Mediterranean lands will be emphasized whenever the region is discussed.

A final problem of definition involves Turkey, like Russia, a country culturally and economically between east and west, with a traditional physical division as well. The Urals line between Europe and Asia passes from the Black Sea to the Mediterranean Sea along an internationalized channel known generally as the Turkish Straits. These comprise the Bosphorous, Sea of Marmara, and Dardanelles. On the western, or European, side of this water, a small piece of modern Turkey (a part of the region of Thrace) is a reminder of the influence (up to 1918) of the Ottoman Empire, when Turkey ruled a sizeable share of southeastern Europe.

This problem is further complicated by the fact that the great city of Istanbul (formerly Constantinople and Byzantium before that) lies on the European shore. Since this was one of the largest and most influential cities of Europe for many centuries, it is difficult to consider it "non-European" today. As is the Soviet Union, however, Turkey is probably best discussed as a national unit in some larger West Asian context, rather than abstracting a small area to consider as part of Europe.

Thus, in each of our boundary problems, we have taken a fairly "minimalist" approach, rejecting rather than adding marginal regions for our definition of "Europe." This is probably just as well, since there is much to consider in even the smaller version of the continent; but it should be emphasized that these judgements are purely arbitrary. If nothing else, this exercise proves again the substantial difficulty inherent in one of geography's most familiar exercises: that of drawing lines on the map.

Introduction

thought, it is sobering to reflect that the dramatic political... nationalisms since 1989 have made even the number of ...ubject to change. From the 1989 list, which contained thirty-... one, East Germany, has now vanished into a reunited German ...er hand, internal strains have resulted in a violent breakup of Yu-... instead of one country there are now (for the moment) five. Czecho-... also divided into two nations. Elsewhere, the Baltic republics (Estonia, ...thuania), having won their independence from a fragmenting Soviet Union, ...ed to the European list (where indeed they were found from 1919 to 1940). ...r new nations formed from the former USSR (Ukraine, Belarus, Moldova) are ...ded to the total (which is based on the latest United Nations groupings), as is Russia itself. Europe evolves in many ways, not least politically.

ADDITIONAL READING

CLOUT, H., et.al. *Western Europe: Geographical Perspectives*. London and New York: Longman, 1985.

HOFFMAN, G., ed. *Europe in the 1990s: A Geographic Analysis*, 6th ed. New York: Wiley, 1989.

ILBERY, B., *Western Europe: A Systematic Human Geography*, 2nd ed. Oxford and New York: Oxford University Press, 1986.

JORDAN, T., *The European Culture Area: A Systematic Geography*, 3rd ed. New York: Harper & Collins, 1996.

PART ONE
The Natural Setting

1
Physical Geography of Europe

Physical geography—the patterns of natural phenomena on the earth's surface—has much to do with the distribution of humankind and the prospects of economic success. The elements of physical geography generally include geologic processes, landforms, weather and climate, natural vegetation, and soils; and all of these, alone or in combination, have a major impact on human activities. In fact, geographic thought during the 1920s and 1930s tended toward the view that nearly all human and economic patterns could be explained on the basis of their relation to the natural world alone. Although this concept ("environmental determinism") has lost favor, since it fails to take into account those historical, cultural, and political factors which also condition human distribution and activities, it nonetheless expresses a very real relationship: it is extremely difficult to understand the observed human landscapes in any of the world's regions without having some appreciation of those factors of the natural environment which are exerting their influence.

The growth of environmental interest in recent years, and the gradual realization by increasing numbers of the world's peoples that we are, after all, still obligated to relate to the natural world in many important ways, have led to a renewed interest in understanding the patterns and processes of the natural world. Since Europe, perhaps as much as any of the world's major regions, is conditioned by the reality of its natural setting, it might be useful to spell out these relationships in a general way.

THE INFLUENCE OF GEOLOGICAL AND PHYSICAL FACTORS

It is easy enough to say that Europeans are strongly influenced by their natural environment, but what exactly does this mean? Some of the ways in which this relationship might be identified are as follows:

1. **Population distribution.** Looking at a map of population density in Europe (or, for that matter, any other continent), it is clear that people are not evenly arranged across the landscape, and there is at least a strong suspicion that the patterns observed are not random: various factors are acting to attract more people to one place, fewer to another, and among these factors are some physical realities (Figure 1.1). Put

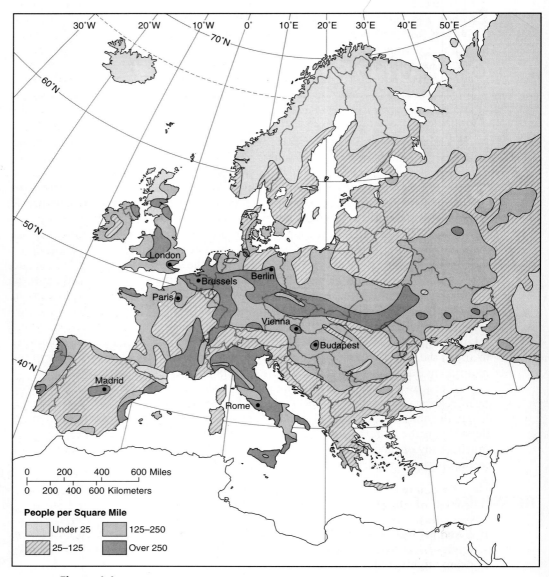

Figure 1.1 European population density.

simply, there is a negative correlation between elevation and population: fewer people live in the mountains than on the plains or in the valleys. Yet why is this true? If the image of the mountains in our time is a positive one, why don't more people live there? The answer is that humanity is of necessity an economic species, and that the mountains offer relatively few opportunities for economic progress, or even survival.

2. **Economic activity.** The highlands of the world are sparsely populated primarily because of the difficulty of earning a living in this environment. First of all, mountainous areas are characterized primarily not by sheer elevation, but by the high percentage of their surface in steep slope. There is little flat land in the mountains, and this means that there is relatively little scope for agriculture, the most basic of human economies. Farming steep slopes without extensive (and expensive) terracing is virtually impossible, since soil is removed by gravity almost as soon as it forms. Second, the highland climate is also an inhibiting factor in terms of agriculture and other traditional economies. Highland areas are characterized by longer, colder winters, and therefore have shorter growing seasons than lowlands; thus they have no advantage in competing with better-favored regions in terms of agricultural production. Finally, the steep slopes and generally rugged character of mountains pose severe problems of isolation. In both economic and cultural terms, the highlands are "far away," and their attraction for human settlement is thus sharply restricted. Obviously, people have been inventive in devising ways of coping with these problems (see "Alpine economy," Chapter 8); but the population map shows clearly that most people choose to cope with the mountains by avoiding them.

3. **Climatic influences.** Landform distribution is also important in explaining the distribution of people and their activities because of its impact on climate over a larger area than on the mountains alone. By blocking or channeling winds and the movement of air masses which dominate climatic conditions at the earth's surface, mountains have a profound effect on rainfall and temperature patterns over vast areas of most continents, and thus on the potential for human settlement and activity. In Europe, for example, the major mountain barriers (Alps, Pyrenees) have a generally east–west trend, and are thus parallel to the main flow of air masses (see "Weather and Climate," Chapter 2). Thus, these mountains pose no barrier to the movement of moisture and relatively mild ocean temperatures across much of the European landmass, while at the same time they create striking differences between Central Europe and the Mediterranean region to the south. In contrast, the mountains of Northern Europe, mainly those dividing Norway from Sweden, run roughly north–south, or at right angles to the movement of air masses. Thus, the difference in climatic conditions between the mild, wet coast of Norway and the colder, dryer interior of Sweden is striking, especially during the winter months.

4. **Settlement sites.** Landforms are also significant in determining the location of thousands of villages and towns in Europe, many of which have now grown into quite large cities. In this regard, it is important to remember that Europe is a continent upon which humanity has been making its mark for a very long time, and human decisions about where to locate settlements were thus often made at a time when the technological ability to cope with environmental problems was limited and any slight advantage offered by the natural setting took on a considerable importance. Natural

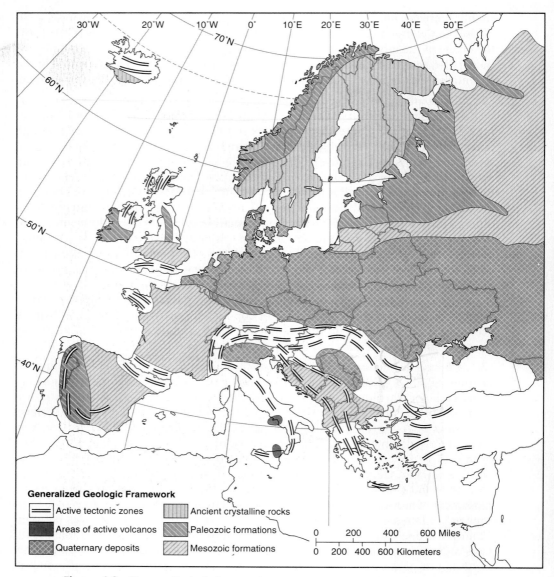

Figure 1.2 Europe: Geologic framework.

passes through mountainous or hilly terrain, for example, always attracted human movement; thus developed communities for control, taxation, or the provision of services. Places where rivers could be easily crossed or forded, or where they flowed to the sea, also were primary early location sites. The Romans, among others, as they laid out their famous straight roads across the empire, were always on the lookout for the lower pass or the easier crossing to aid their well-organized but unsophisticated (in the modern sense) engineering technology. Since routes once laid out and places once es-

Figure 1.3 The Alps demonstrate both the beauty and the problems of Europe's mountains: Gruyères, Switzerland.

tablished tend to have a life and an inertia of their own, many modern cities and towns can trace their origins back to antiquity or the Middle Ages; their locations can only be explained with reference to the advantages of site offered by the environment.

    5.   Boundaries and political concerns; cultural separation. Elements of the natural environment often play a major role in determining the boundaries between nations, and what might be called the "quality" of these boundaries may play a crucial role in the political history of the world's various regions. In Europe, a densely populated region with a long history of political turbulence and aggressive nationalism, the quality of boundaries and the importance of "natural" frontiers have played an important role in determining national boundaries throughout history. Positive natural boundaries, those fixed along coastlines or the highest crests of mountain barriers where few national interests come together, have generally been very successful. The 22-mile- (35-kilometer) wide channel

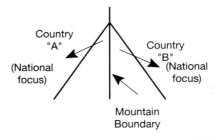

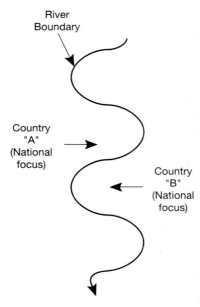

Figure 1.4 "Good" vs. "Bad" boundaries. In the top figure, a national boundary is drawn along a mountain crest. Both country "A" and country "B" normally direct their attention away from the mountains toward the more productive lowlands; thus the boundary tends to be stable and uncontroversial. In the bottom figure, a river is used as the national boundary. Both country "A" and country "B" would like to exercise greater control over the productive valley; thus the frontier can be insecure and a focus of international stress.

separating Great Britain from France (until recent completion of the channel tunnel) has been extremely effective in protecting Britain and permitting its independent development (see regional discussion of Great Britain, Chapter 16), while Switzerland owes at least part of its long record of independence and neutrality to the unchanging, easily defended frontiers traced along the highest elevations of the Alps and Jura mountains.

On the other hand, boundaries laid out along river valleys have generally been less successful. River valleys tend to concentrate population and economic potential, rather than dividing them, and when strong national interests clash across such a weak frontier, political stress and even warfare are a strong possibility. The Rhine River as a boundary between France and Germany is a classic example. Even worse is the case

Physical Geography of Europe

in which national boundaries are traced across flat, undifferentiated plains, or along small rivers with no physical distinctions on either side. Here, power politics and obscure historical motivation become most important, and local populations tend to suffer accordingly. Surely no nation in Europe has a more troubled history then Poland, and a fair number of the problems that have beset the Poles over the centuries can be traced to the fact that the boundaries of the Polish nation to both east and west (or toward their traditionally more powerful neighbors) have been founded on minor, indefensible rivers, or have simply been expressed as lines on a map.

Additionally, the formidable obstacles posed by mountain barriers to the movement of peoples and the circulation of ideas have been instrumental in determining many cultural patterns, such as those of language distribution. France and Spain, for example, share a common frontier of some 300 miles (480 kilometers), but the Pyrenees mountains, which form this border, are among the most difficult of all European mountain ranges to cross or penetrate. The result is that they form a significant cultural boundary between peoples who might otherwise be much more similar. Other examples of this process in Europe would include the strong sense of independence among the Swiss states, or cantons, stemming from cultural isolation in the Alpine valleys.

6. Resource distribution. Nearly all of the various mineral resources that have been valued by humans at various times in history are formed by certain geologic processes. This means of course that they are unevenly distributed (see Chapter 10, "European resources"), and that there is a strong correlation between the location of these resources and the nature of the geologic environment. Coal, petroleum, and natural gas, for instance, are products derived from organic materials (plant and animal remains). Thus, to find them, it is necessary to search areas of the earth's surface where rock formations are of an age associated with advanced organic forms. Since the rocks of Northern Europe, as an example, are so old as to predate any major life forms, there is no point in prospecting here for these materials.

In contrast, most metallic minerals, from gold and silver through the less glamourous industrial minerals, are generally formed in association with the same geologic processes that produce mountains. Therefore, if one is seeking these minerals, there is no advantage in looking in the same places as one would look to find oil. In Europe, as elsewhere, the highlands have thus been the major source of metals throughout history, and mining continues to be a major support of the mountain economy in many countries (see Chapter 8, "Alpine economy").

7. Physical danger. A final general way in which the natural environment influences human activities and decisions is through the direct threat of physical danger. The southern margins of Europe—the Mediterranean basin and its extensions into the Middle-East—are some of the most geologically active regions of the earth's crust (another being the rim of the Pacific Ocean), and are subject to volcanic and tectonic events which annually claim lives and cause substantial property damage. One basic principle of geology is that the forces which shaped the earth as we see it today are the same as those still operating to produce change. This principle is well illustrated in the Mediterranean, where there are several active volcanos, including Mt. Etna on Sicily, which erupted as recently as 1983, and Mt. Vesuvius, near Naples in southern Italy, whose famous eruption in 79 AD buried the Roman towns of Pompeii and Herculaneum under tons of ash.

More serious even than the threat of volcanic eruptions is the danger from earthquakes across the Mediterranean basin. No year passes without report of some devastating quake occurring in southern Europe, North Africa, or the Middle East. The disasters in Armenia (1988) and Iran (1990) were recent reminders of the instability of this region. Essentially, the Mediterranean is the contact zone between two of the world's great crustal plates, the Eurasian and African, and as these push against each other the resultant shocks and tremors have a profound effect on the human cultures at the surface. Citizens of the geologically stable parts of northern Europe have little to fear from geologic processes, but to the people of the south, the possibility of violence from the natural environment is a daily threat.

It should also be noted that the island of Iceland, culturally and economically a part of Europe, is another geologic "hot spot." At the junction of two of the major seams in the earth's crust, Iceland is the scene of continuous volcanic activity, with new eruptions occurring almost annually, and entire new mountains sometimes being formed. On the bright side, the Icelanders have made maximum use of their natural setting to generate a large share of their energy requirements by geothermal means. Hot water and heat for the island's cities are largely assured by tapping into the heated formations just below the earth's surface.

River Systems

Another aspect of the natural environment of Europe that has had a profound influence on the distribution of population and on human economic activity is the system of rivers. It is always useful to remember that Europe was settled at an early point in human history, when little technology was available, and the abundant rivers and streams that characterize nearly all of the continent (except for some of the driest portions of the Mediterranean region) thus had a crucial importance to early peoples and

Figure 1.5 Volcanic activity is a frequent occurrence in Mediterranean Europe: Mt. Etna on the Italian island of Sicily.

have continued to play a central role in the evolution of cultural and economic patterns. The importance of waterways on the European scene lies in three major areas, and while these are not unique to Europe, the relationships are perhaps better developed here than on other continents.

   1. Transportation. Prior to the nineteenth century, waterways provided by far the safest and most reliable means of transport across much of the world. Roads were primitive, difficult, and often dangerous, and in Europe it was most often the waterways that stitched together the populations and economies of the various emerging nations. As a result, location on a waterway navigable at least for shallow-draft vessels was a prime requisite for development, and thousands of towns and cities of modern Europe owe their location, on what may today seem a very modest stream, to this requirement; it was the way in which they were linked to the outside world.

   2. Water supply. The streams and rivers of Europe also provided the most obvious and direct source of water supply for early populations. Although the Romans built impressive aqueducts to provide water to various cities in the relatively dry Mediterranean world, most Europeans took the simpler solution of locating their settlements directly on the water courses. So long as the cities remained small in population, the pattern of water supply/sewage disposal was quite realistic; thus, many modern cities have simply grown where they began: on the water.

   3. Power. Finally, the combination of generally well-balanced seasonal runoff and a large number of hilly or mountainous areas enabled most parts of Europe to establish a system of water power sites that provided the energy for much early industry until the rise of the steam engine (see Chapter 10, "European resources"). Although the mills and their ponds are by now often a relict of the past, the cities and towns thus sited may remain in their original waterside locations.

THE PHYSICAL GEOGRAPHIC REGIONS OF EUROPE

The Highlands

In considering the major physical divisions of Europe, it is convenient to think of a series of generally parallel east–west zones, and to realize that, geologically speaking, the major highland areas of northern Europe are older, those of the south younger. The highlands can be divided into four groups, based on age:

   1. The Fenno-Scandian Shield. This part of northern Finland and Sweden, with extension into the Kola peninsula of Russia, is one of the oldest rock formations exposed anywhere on the earth's surface. Formed at least 600 million years ago, these crystalline rocks have been smoothed by the forces of erosion to the point where few surfaces now exceed 2000 feet (610 meters) in elevation.

   2. The Caledonian formations. About 300 million years ago, the Caledonian orogeny, or period of mountain building, created a new series of mountains in Europe, of which the most important are the Scandinavian highlands that occupy most of Norway and divide that country from Sweden. Rugged, with peaks up to 5000 feet (1525 meters), and severely eroded by later glaciation, these mountains are among the

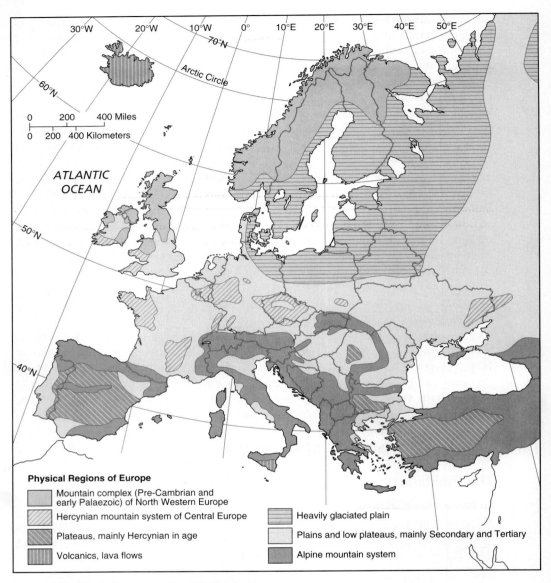

Figure 1.6 Europe: Simplified landform regions.

most desolate and sparsely inhabited regions of Europe. The Scottish Highlands ("Caledonia" is an ancient name for Scotland), the Southern Uplands of Scotland, Wales, the Lake District of England, and the Irish mountains are also of this age.

    3. The Hercynian formations. Some 100 million years ago, another major orogeny produced the complex systems of hills and low mountains that characterize much of Central Europe. Ranging in elevation up to 6000 feet (1830 meters), but more commonly in the 2000–3000 feet (610–915 meters) range, and generally forested, these areas are perhaps best typified by the "Mittelgebirge" ranges of central and southern Germany and the Czech Republic, including the famous Black Forest ("Schwartzwald"). The Ardennes and Vosges of northern France and Belgium, the Central Massif and the hills of Brittany and Normandy (also in France), the Pennines and Cornwall in Britain, are all of this age.

Similar in age but quite different in structure is the high plateau that makes up most of the Iberian peninsula. A generally flat formation up to 8000 feet (2440 meters) in elevation, dissected by several major rivers, the Iberian plateau is a unique European landform region.

4. **The Alpine formations.** Finally, to the south, the Alpine orogeny created the youngest and most spectacular formations in Europe. Some 20 million years old, the mountains of the Alpine period are young enough to have been less eroded than ranges farther north, and the continent's highest elevations are accordingly found here. Mont Blanc, in the French Alps near the common border with Italy and Switzerland, is the highest peak in Europe. At 15,771 feet (4807 meters), it is not among the world leaders by any means, but it is impressive—covered with permanent snow and often shrouded in clouds—and is one of several peaks that exceed 10,000 feet (3500 meters).

The Alps are in reality a complex series of mountain chains displaying great geologic and topographic variety. Beginning with the Maritime Alps that run northward from the Mediterranean between France and Italy, the Alps then turn eastward, the major ranges being found in Switzerland, northern Italy, and Austria. Farther along, the Alpine formations are continued in the various mountain masses of the Yugoslavian states, Bulgaria, and Greece, connecting beyond this with the mountains of the Middle East. The Carpathian ranges of Eastern Europe are also of Alpine age.

West of the Alps, the Pyrenees divide France from Spain. A single block-faulted range with the steep face to the north, the Pyrenees are in many respects a more difficult and imposing barrier than the Alps: there are still comparatively few roads across the border except along the narrow coastal strips. Although not as high as the Alps (the highest peak is Pico de Añeto, Spain, at 11,168 feet—3404 meters), the Pyrenees and their extension along the north coast of Spain as the Cantabrians have a strong impact on climatic patterns and human activities.

Later Modifications

As previously noted, geology teaches that the processes which formed the landscapes we see before us are the same processes which continue to modify them today. This principle is clearly illustrated in the physical geography of Europe. Mention has already been made of the continuing tectonic activity which particularly characterizes Iceland and the Mediterranean region (see "physical danger" section of this chapter).

An even more recent process which has modified the landscapes of a major part of Europe has been recent or Pleistocene glaciation. As in northern North America, northern Europe was invaded by continental ice sheets on four occasions during this period, the most recent advance melting back only about twelve thousand years ago (no more than eight thousand years in many parts of northern Scandinavia). This means that the ice-modified landscapes have had almost no time (geologically speaking) to return to their preglacial condition. The main center of ice accumulation was over the Fenno-Scandian shield, and as the ice moved southward, it severely eroded the landscapes of northern Europe and eventually covered all of the continent north of the Baltic Sea. The hundreds of lakes which are today such a feature of Finland and Sweden are reminders of this recent period of ice scouring and drainage disruption.

To the west, the great ice sheet nearly covered the Scandinavian mountains, gouging the valleys on the Norwegian coast hundreds of feet below sea level (unlike streams, ice can continue to cut valleys well below the level of lakes or the sea so long as pressure from behind is maintained). When the ice eventually receded, sea water invaded these deep valleys, penetrating inland in some cases for more than 100 miles (160 kilometers) and creating the famous fjords which so characterize the modern Norwegian coastline (see box below). South of the modern Baltic Sea, which roughly coincides with the southernmost advance of the most recent Pleistocene ice sheet, floodwaters from the melting ice deposited vast amounts of relatively infertile sand and gravel that had been carried by the glaciers, creating what is today a surprisingly unproductive, sparsely populated landscape in northern Germany and Poland.

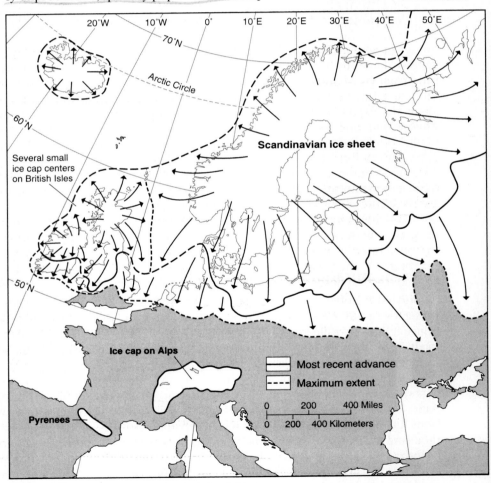

Figure 1.7 Extent of Pleistocene glaciation in Europe.

PERCEPTION: NORWAY AND ITS FJORDED COASTLINE

No European nation has been more profoundly influenced by the circumstances of its natural environment than Norway. Confined largely to the narrow western slope of the Scandinavian mountains, between Sweden and the Atlantic, the country is some 1100 miles (1770 kilometers) long, but rarely more than 100 miles (160 kilometers) wide, except in its extreme southern section. Aligned essentially northeast to southwest, the landscape that is now Norway was thus nearly at right angles to the glacial ice pouring over the mountains and into the Atlantic. The ice cut steep-sided, U-shaped valleys along the courses of the pre-existing rivers, and, since moving ice is entirely capable of eroding below sea level, deepened these valleys by hundreds of feet. Eventually, as the earth's climate warmed, the rate of melting exceeded the rate of ice advance, causing the glaciers to "retreat" and allowing sea water to fill the deepened valleys.

The fjords are truly spectacular and are one of modern Europe's most impressive landscapes. Some, such as the Sognefjord or the Hardangerfjord, with their many arms and extensions, extend over 100 miles (160 kilometers) inland from the open ocean. The world's largest ships can navigate most of them with ease, anchoring if required within a few feet of the shore. Yet at the same time they have posed substantial problems to the cultural and economic development of Norway. Although fine natural harbors, their steep sides and the generally rugged character of their glaciated hinterlands mean that there is little flat land and few land–water interactions associated with them. The only bits of land capable of supporting farming and settlement are at the heads of the fjords, where the ice-carved valleys are no longer below sea level. These areas are known as "viks," and it was from them that the "vikings" sailed away from overpopulation and limited resources to terrorize and colonize other parts of Europe and even North America.

The fjords also pose problems in that they can only be crossed by boat, by air, or by circuitous routes around their interior ends. Since Norway is a linear country, transport and communication connections among its various regions are vital, yet these are often rendered physically impossible or prohibitively expensive by the deep, dissecting fjorded valleys. An entirely overland trip between south and north (without using time-consuming and weight-restricted ferries) is only possible by detouring through Sweden. Thus, the natural landscape threatens national unity. Maintaining population and viable economies in the various regions semi-isolated by the fjords is one of modern Norway's most pressing and perplexing problems.

While the continental glaciers were active in northern Europe, cooler climate and abundant precipitation in the Alps gave rise to another large expansion of mountain or alpine glaciers which substantially modified these recent mountains and their surrounding region. Moving downslope from permanent ice and snow fields at higher elevations, these glaciers carved steep-sided, flat-bottomed U-shaped valleys, eroded many mountain peaks into jagged "horns" (the Matterhorn in Switzerland is the most famous), and generally created the spectacular mountain scenery that we associate with the Alps today. Numerous waterfalls, and the well-known lakes of Switzerland, Austria, and northern Italy are beautiful leftovers from the glacial period, while glaciers may still be seen today at higher elevations. What might be termed a "fringe benefit" provided to modern civilization in this region by the ice age is the greater access well into the interior provided by the glaciated valleys. Much settlement and most transportation patterns are concentrated in these valleys.

Elsewhere, other mountains of Alpine age, being lower in elevation, more southerly in latitude, and/or nearer the coast, did not accumulate ice to nearly the same extent as was true in the Alps themselves, and thus have been much less modified by the processes of glacial erosion. The Pyrenees are a good example of this. Although there are some glacial features at higher altitudes, much of the chain was very little influenced. One reason why access into many parts of the Pyrenees today is so difficult is because there are no broad glacial valleys, such as were created in the Alps.

Is the ice age at an end? Or is the earth only in another interglacial phase? These intriguing questions are prompted by the fact that the period between the retreat of the last ice sheet and today (some 10,000 years for an average figure) is shorter than the period between each of the other advances of Pleistocene ice. It may be, therefore, that a fifth advance of the ice (and possibly others) is in the earth's fairly near future. There is much speculation today about climatic change and man's role in causing it (see Chapter 2, "European weather and climate"); and while most discussion centers around a probable earth warming in the years ahead, the fact that a cooling of the earth's atmosphere by only 2 to 3 degrees Celsius on average (for example through the "nuclear winter" scenario) would be sufficient to trigger renewed ice accumulation in northern Europe certainly gives added interest to the scientific debate.

Figure 1.8 The Geiranger Fjord is typical of the beautiful but rugged glaciated landscapes of Norway.

European Lowlands

In this discussion of Europe's physical geography, emphasis has thus far been placed almost entirely on the uplands, that is, on the mountains and rough hills that, as we have seen, inhibit settlement and human economic activity. This emphasizes the reality that much of Europe is, in fact, composed of this sort of unpromising terrain, and also reminds us of the critical difference between a tourist and a resident: the spectacular mountains of Switzerland or the haunting beauty of the Scottish highlands form landscapes that we might very well enjoy visiting, but they are not home to many people. This is true around the world; most of the human population is in fact crowded onto only a small percentage of the total landmass. The reasons are clear: agriculture, transportation, city-building—all the vital support activities of modern cultures—are best carried out where the climate is agreeable and where the land is flat to gently rolling, fertile, and well-drained. The scarcity of such areas becomes increasingly apparent as we wander through the world, observing the intensifying land-use conflicts as more people and activities compete for a fixed amount of "good" space. The European scene gives us an ideal vantage point from which to appreciate this reality.

In fact, there is only one large, contiguous area of essentially flat land in Europe. This is the region that borders the North Sea, with an extension along the Baltic, that is known as the North Sea Plain or Great European Plain. Including northern France, eastern England, Belgium, the Netherlands, northern Germany and Poland, and parts of Denmark and southern Sweden, and broadening eastward into the great plains of European Russia and Ukraine, this vast extent of lowland stands out on any map as a focus of population, cities, agriculture, transportation lines, industrial development, and countless other measures of human success and progress. It is by far the most important region of Europe today, and as the continent moves into a new century, this is the region that still leads the way. From the point of view of the tourist, the landscapes here are dull and without great interest; from the geographer's perspective, this is the pulsing heart of Europe. In this context, it is worth remembering that, with few exceptions, prosperity in any region has historically been founded most often on successful agriculture: farming generates the profit that is invested in other activities. This puts the world's rich agricultural regions—including much of the North Sea Plain—into a proper perspective, and emphasizes a rather ominous bottom line: there are very few such regions.

Apart from the North Sea Plain and its extensions to the east, the only other major area of flat lowland in Europe is the central valley of the Danube River and its tributaries in eastern Europe, known as the Alföld. This region, which includes most of Hungary and parts of Serbia, Croatia, Bosnia, Romania, and Austria, has been the breadbasket of southeastern Europe for centuries. Although the area suffers in many places from flooding and soil salinity, it remains clearly the focus of activity in this part of Europe. Perhaps it is not surprising that the Magyar (Hungarian) people, who have occupied much of the Alföld since the tenth century, have been able to maintain their rich cultural traditions in a region of almost constant political turmoil.

Beyond the North Sea Plain and the Alföld, areas of flat, productive land in Europe consist mainly of river valleys and often narrow strips of coastal plain. The valley of the Po in northern Italy is the most significant of these (the traditional heartland of the Italian nation), but in nearly all countries the relatively small extent of quality land represented by these features has assumed an unusual degree of importance.

Figure 1.9 The Rhine is one of Europe's most important and most famous rivers: From the Lorelei Rock, Germany.

Coastlines

The coastline of Europe generally mirrors the larger landform regions of the interior in the sense that much of it is rugged and mountainous. Most of the Mediterranean coast, Scandinavia, and Finland, and a great deal of the British shoreline are of this type, while only the North Sea (north of Normandy in France), the south shore of the Baltic, southwestern France, and a few parts of the Mediterranean coast have smooth shorelines. Many coastal plains, moreover, such as those related to such rivers as the Po, Rhône, Rhine, and Danube, are generally marsh or shallow wetland, and have required enormous human effort and ingenuity to be made productive. As a result of these conditions, coastal regions have often been isolated from their natural hinterlands, and the development of the harbors and ports upon which so much of Europe's prosperity has depended has been a difficult proposition. Not surprisingly, the relatively few good harbors with easy connections to a rich hinterland (such as, for example, Marseille, France) have been important places since remote antiquity.

Rivers and Streams

Finally, any brief review of the physical geography of Europe should reinforce again the importance of the region's major river systems. To an "old" continent such as Europe, that is, to one in which humans have been present in signficant numbers for a very long time, running water has an exceptional importance, and the location of places, establishment of transportation patterns, and siting of economic activity—

Physical Geography of Europe

often still present today—can be explained in a surprising number of cases by relationship to running water. Mention has already been made of the classic trilogy of water location advantages: water supply, transportation, and power.

Many rivers will be identified and discussed at different points in this text. Mention might be made here, however, of a few which have special prominence on the European scene. Rising in Switzerland, the Rhine connects the great industrial regions of eastern France and western Germany to the North Sea by way of the Netherlands. Also from Switzerland, the Rhône flows southward through France to the Mediterranean. The Danube, which rises not far from the Rhine in Germany, is Europe's only major eastward flowing river, linking together nearly all the nations of southeastern Europe. The Seine in France and the Thames in Britain are important mainly because they serve to link Europe's two greatest cities (Paris and London) to the sea, while the Po is the classic river of northern Italy. In the western part of the former Soviet Union, most major rivers flow north to south; and the Dneiper, Dneister, and especially the Volga have played critical roles in the human and economic history of Russia, Ukraine, Belarus, and Moldova. These and numerous other rivers serve today to stitch together the fabric of Europe, providing water, power, and transportation, as they have done since the earliest times.

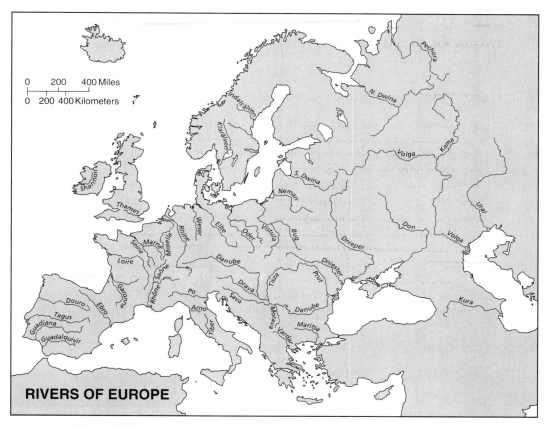

Figure 1.10 Europe's principal waterways.

ADDITIONAL READING

AGER, D.V. *The Geology of Europe.* London and New York: McGraw Hill, 1980.
BECKINSALE, M., and R. BECKINSALE. *Southern Europe.* New York: Holmes and Meier, 1975.
EMBLETON, C. *Geomorphology of Europe.* New York: Wiley, 1984.
SHACKLETON, M. *Europe.* London: Longmans Green, 1959.
WATTS, M. T. *Reading the Landscape of Europe.* New York: Harper & Row, 1971.

DISCUSSION QUESTIONS

1. What is the origin of the Norwegian fjords? How have these features influenced development of the Norwegian nation?
2. In a geographic sense, what is likely to constitute a "good" boundary between nations? Consider some European examples.
3. In Europe, iron ore but no coal is found in northern Sweden. What might this tell us about the relationship between geologic processes and modern resources?
4. Earthquakes are common occurrences across southern Europe, but are rare to the north. Why is this?
5. France and Spain share a long common boundary—300 miles (485 kilometers)—yet have developed quite different languages and cultures. What physical aspect of their boundary helps to account for this?

2
European Weather and Climate, Vegetation, and Soils

WEATHER AND CLIMATE

Of all the factors of the natural environment that influence the course of human events and serve to distinguish one of the world's regions from another, weather (the day-to-day occurrence of atmospheric phenomena) and climate (the long-range pattern characteristic of a given area) are the most significant. Although by no means absolutely deterministic, weather and climate can normally provide us with a great deal of understanding of countless variations in the earth's cultural and economic landscapes.

As a striking example, climate is the absolute bottom line for agriculture; no amount of skilled management or technological inputs can go very far to mitigate the effects on farming of conditions that are colder or dryer than usual. Weather management, and even reliable long-range prediction, continue to baffle science. The relationship between agriculture and climate is thus worldwide and much in the news. Drought in Africa, unpredictable rain-bearing winds in India, frost in Florida, cool summers in Ukraine: all have direct economic—and perhaps even more serious—impacts on human populations. As the world's numbers of people continue to increase rapidly, placing added pressure on all food production systems, the climate question becomes more acutely posed, the margin for coping with extreme events more narrow.

Besides agriculture, the relationships of weather and climate to tourism and energy use are very close, while individual weather events (hurricanes, tornados) serve to remind us again of the power of natural phenomena and man's only marginal ability to control them. In each of the world's regions, weather and climate are important keys to understanding. Europe is certainly no exception.

Factors Influencing Weather and Climate

Before reviewing the European climates and noting their impacts on the human landscapes, perhaps it would be useful to briefly review the basic elements of weather and climate. What factors determine the pattern of climate that will be found in a given place? Is there a sense of order to the distribution of climates? Very broadly, six factors may be recognized:

1. **Latitude.** One of the most basic controls of weather and climate is simply the latitude of a given place (latitude is the distance north or south of the equator, measured in degrees). Since the earth is a sphere, more sun energy is received near the equator, where the noon sun is at a high angle all year, than at the poles, where the sun never rises far above the horizon. Since it is solar energy that drives the earth's dynamic atmosphere (and is in fact the ultimate source of all the earth's energy), one major result of these latitudinal differences is thus the formation of three consistent pressure belts based on the nearly constant heating and cooling. In a band extending roughly 5 degrees either side of the equator, high temperatures create a low pressure zone—the equatorial low—which is a major factor in creating the humid, rainy conditions prevalent at low latitudes. In contrast, cooling near the poles creates in each hemisphere a polar high, from about 80 degrees poleward, where high pressure and the resultant dry conditions are the basic climatic realities. Between these extremes, places in intermediate locations can to some extent calculate their likely climate as a function of latitude. This is not infallible, however; Europe's high-latitude location would seem to argue a very cold dry climate, yet such is not generally the case. Other factors are at work.

2. **Earth rotation.** If the earth was a stationary body suspended in space, atmospheric circulation would be as simple as the climate pattern. Heated air would rise from the equator, flow evenly toward the poles, cool and sink, and return along the earth's surface in a single vast convectional pattern. Things are not, however, so simple. The rotational movement of the earth on its axis creates dynamic pressure zones and wind systems which affect most world climates and have a notable impact on Europe. Although these patterns are complex, they may be simplified for our purposes. As the earth spins, rotational motion causes air rising aloft at the equator to pile up above about 30 to 40 degrees in each hemisphere. This column of air then descends of its own weight, creating two marked zones of high pressure: the subtropical highs. Between these belts and the polar highs, two compensating areas of generalized low pressure (the subpolar lows) tend to form between 60 and 70 degrees. Thus, to the three thermal pressure zones, four dynamic ones are added, each having a profound effect on local climates.

Moreover, the earth's rotation introduces another complication which, as we shall see, is of vital importance to Europe. Air continues to flow from high pressure to low pressure, of course, even with the introduction of new pressure zones; but the rotation of the earth as a spherical body means that air currents (or in fact anything in motion) are deflected to the right in the northern hemisphere, to the left in the southern hemisphere. This is known as the Coriolis effect, and while it is studied as an interesting scientific curiosity, its principal effect from a geographic point of view is to create six general zones of air movement between the pressure belts: two easterly "trade wind" belts between 5 and 30 degrees, two areas of relatively steady westerly flow between 40 and 60 degrees, and two polar easterly zones in the general area of 70 to 80 degrees. Note the exceptional importance to Europe of being in the path of the westerly airflow.

3. **Seasonal change.** The earth as a physical body has one more trick up its sleeve which creates areas of climatic complexity, including a major region of Europe. As a simple rotating sphere, the earth's wind and pressure belts would be as described above. However, the earth is not vertical on its axis with respect to incoming solar energy, but is instead inclined some 23.5 degrees. This means that, as the earth revolves around the sun during the year, insolation varies substantially even at the same latitude, producing the seasons which are such a feature of yearly climatic change, notably in the mid-latitudes. Geographically, what this means is that all of the wind and pressure belts we have described migrate: 5 to 10 degrees northward during the northern hemisphere summer; 5 to 10 degrees southward during the northern hemisphere winter (which is of course the southern hemisphere summer). While this is of only academic interest in some regions, in those areas which come under the influence of two strikingly different influences during the year (such as southern Europe or central Africa), it can be an absolutely critical climatic factor.

4. **Land–water contrast: the continental effect.** The factors discussed thus far in this section would account for the patterns of climate if the earth were a homogeneous body, composed of the same material and of equal elevation everywhere. Such, however, is far from the case. In particular, the earth is covered about two thirds by water, while the landmass accounts for only one third. Since land and water behave very differently in terms of heat gain and loss during the year, the location of places with reference to the oceans is important in determining their climates. Land surfaces absorb more solar energy during the summer than do water bodies, and similarly lose more energy during the winter. Thus, coastal areas of most continents will have a much smaller range of temperatures during the year—milder winters and cooler summers—than will places farther from the water at the same latitude. With its numerous surrounding seas and embayments, Europe is the continent most likely to be influenced by this maritime effect, which brings moderate conditions to unusually high latitudes.

5. **Ocean currents.** Not only is the presence or absence of nearby water bodies critical to the climate of any region, but the temperature of the water offshore is also an important factor. Driven by the same sorts of dynamic forces that move the atmosphere, the world's oceans are also in movement in a series of complex drifts and currents that may alter from time to time with dramatic effects. Generally speaking, currents moving from tropical areas to higher latitudes are warm currents; those

originating at high latitudes and moving toward the equator are cold currents. In both cases, the result may be water temperature quite different from what one might expect at any given latitude.

In this context, Europe is significantly affected by a warm current that has its origins in the Atlantic Ocean near the equator, is propelled westward by the trade winds (blowing east to west in both hemispheres), and is deflected into the northern hemisphere by the fact that the "hump" of northeastern Brazil is south of the equator. Moving through the Caribbean and along the east coast of North America at a steady rate of about 8 miles per hour, this warm water is the famous Gulf Stream. Deflected to the right and picked up by the westerly winds, it crosses the Atlantic as the North Atlantic Drift, bringing surprisingly warm water to the high latitudes of the European coasts.

6. **Mountain barriers.** A final factor influencing local weather and climate is the arrangement of landforms—particularly mountain ranges—on the earth's surface. Although conditions at high altitudes in the atmosphere influence the earth's weather patterns significantly, much local daily weather and many regional climatic conditions are determined by low-altitude phenomena. Thus, although the hills and mountains of the earth are insignificant in elevation when compared to the height of the atmosphere, they are nonetheless capable of exercising a powerful influence on climatic patterns.

In Europe, for example, the major mountain barriers (those of Alpine age) generally trend east to west, or parallel to the major flow of air masses in the westerly wind belt. This being the case, relatively mild maritime air is carried far into the interior of the continent (contrast this to the Pacific Northwest region of North America, where mountain ranges running at right angles to the prevailing winds create substantial differences between coastal and interior climates). At the same time, the Alps, Pyrenees, and other ranges, by forming a substantial barrier between north and south in Europe, accentuate the differences between the climates of the north and those of the Mediterranean, making it possible for the traveler to experience vastly different climatic environments within a few hours. In Scandinavia, on the other hand, the major mountain barrier runs northeast to southwest, and the consequent differences in climate are striking. Along the Norwegian coast, conditions are moderate throughout the year to an astonishing latitude, while across the mountains in the interior of Sweden, much colder and dryer conditions obtain.

The preceding list of causative climatic factors is general for all regions of the world. Where important modifications appear (and there are many), they can almost always be explained within this same general framework. If we now look more closely at the specific climatic types of Europe, we can clearly see that the continent's climates may be explained to a large extent by: 1) location within the westerly wind flow throughout the year for all of northern and central Europe; 2) transitional location between the westerlies (winter) and the subtropical high (summer) of the Mediterranean region; 3) exceptionally mild water off all the European coasts as a result of the Gulf Stream/North Atlantic Drift; and 4) the east to west orientation of the major European mountain barriers. On this basis, it is possible to recognize three distinctive types of climatic region in Europe, and to identify a few others of less significance.

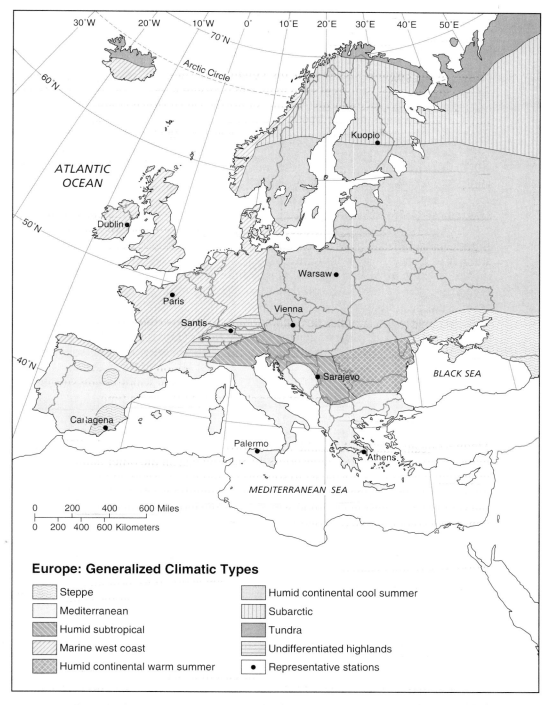

Figure 2.1 Europe: Generalized climatic types.

European Climatic Types

1. **The Marine West Coast climate.** Certainly the type of climate most often associated with Europe, the Marine West Coast, or maritime, climate covers most of the North Sea Plain, including nearly all of Great Britain and Ireland, France except for its Alpine and Mediterranean fringes, the Benelux states, western Germany, and southern Scandinavia, extending northward along the Atlantic coast to remarkably high latitudes. As its name implies, this climate is found wherever in the world a westerly air flow brings maritime conditions of moisture and temperature stability across a continent. The climate features unusually mild winters—no month has an average temperature below freezing—correspondingly cool summers and, as a result, a very small range of monthly average temperatures throughout the year. London, for example, has only a 24-degree (Fahrenheit) range between the average temperatures of its warmest and coldest month.

   A constant supply of moisture also means that this climate is characterized by abundant, well-distributed rainfall, with a high percentage of cloud cover and a large number of days on which some rain is recorded. The green landscapes of countries such as Ireland are legendary, and it is the foolish tourist who visits this part of Europe at any season without sweaters and rain gear! Because the water off the European coasts is comparatively warm, and because the major mountain ranges pose few obstacles to air mass movement into the interior, much of Europe has a remarkably moderate climate, although its general latitude range (London and Paris are at 48 to 52 degrees north) is that of the upper Great Lakes region up to Hudson's Bay in North America.

2. **The Continental climates.** With increasing latitude, increasing distance from the mild waters of the Atlantic, or increasing elevation in Europe, climatic variation from month to month also begins to increase. In particular, colder winters begin to be the rule, with at least one month and frequently more falling below freezing on average, and summers becoming slightly warmer. The result is the climatic type known as Continental, in which landmass characteristics begin to take precedence over marine ones. In Europe, this is the climatic pattern found from central Germany through much of Poland, the Czech Republic, and Slovakia (and on into Belarus, Ukraine, and western Russia); it is also found in most of Scandinavia away from the Atlantic coast and in the Baltic States and Finland, except for the extreme north, where climatic conditions become even more severe. As under the Marine West Coast climate, precipitation is well balanced in the Continental climates (with colder winters, the moisture requirements are somewhat lower, so slightly diminished precipitation amounts pose no problem). On the southern or western edges of the Continental region, warm summer temperatures may be found; to the north and east, cooler summer conditions may prevail, although these are not generally critical distinctions.

3. **The Mediterranean climate.** The southern region of Europe is the type area for one of the world's most unique climates: the Mediterranean. This is the climate of the west coast of continents at latitude 30 to 35 degrees (southern California and bits of Chile, South Africa, and Australia are also climatically "Mediterranean"), and results from the transitional nature of these latitudes. During the winter months, the area is under the influence of the westerly wind belt: storms regularly enter the

Figure 2.2 Palms and bananas in a garden on Lake Geneva reflect the surprisingly mild climate of Western Europe: Lausanne, Switzerland.

Mediterranean basin, as they do across the area of Marine West Coast climate, and rainfall is adequate and well distributed. During the summer months, however, the region gradually comes under the influence of the subtropical high-pressure zone as this migrates northward, and the same conditions of descending air and high pressure that create the Sahara desert bring several months of nearly absolute drought to the Mediterranean basin. The result is the only major type of dry-summer climate in the world; this again may either have a hot summer away from the coast, or a milder summer near the influence of water.

Of additional interest in the Mediterranean basin is the role of the major European mountain ranges in accentuating the differences between this region and central Europe. Running generally east to west, the Alps, Pyrenees, and other ranges tend at least partially to block the movement of storms into the Mediterranean during the winter, as well as to restrict drought conditions to the south during the summer. This results in a surprisingly sharp demarcation of the true Mediterranean climate, which is found in most of Spain and Portugal south of the Pyrenees–Cantabrian system, southeastern France, central and southern Italy, Greece and the coastal zones of Croatia, Montenegro, and Albania; and the islands of the Mediterranean, as well as adjacent parts of North Africa and the Middle East. Not surprisingly, this climate supports a very distinctive assortment of natural vegetation, and has posed substantial challenges to agriculture and other human activities (see "Mediterranean economy," Chapter 8).

4. **Minor climatic types.** In addition to these three major types of climate, several other varieties occupy smaller parts of the continent. In southeastern Europe, particularly Hungary, Romania, Bulgaria, Moldova and Serbia, more southerly latitude combined with greater distance from the Atlantic leads to hot, humid summers and a region of humid subtropical climate. This is often a climatic type found on the east coast of continents at latitudes about 30 to 35 degrees (thus: a nonmaritime condition), and is characteristic of such regions as the southeastern United States, southern China, and central Argentina.

In Spain, the natural Mediterranean conditions in the summer combine with the elevation of the plateau and the blocking effect of the Pyrenees–Cantabrians and—in the south—the Sierra Nevada, which serve to restrict winter rainfall as well, to produce small areas of semiarid or steppe climate. No other regions of continually dry climate are found until, moving eastward, one reaches the southern republics of the former Soviet Union (notably southern Ukraine) or the interior of the Middle East.

At higher elevations in more northerly latitudes, as in the Scandinavian mountains or along the Arctic coast, climatic conditions deteriorate—particularly in the sense of longer, colder winters—to the point where trees cannot survive and the sparse scrub vegetation gives its name to the climate as well: Tundra. Finally, mountainous areas, with their great variety of elevations, slopes, and exposure angles, have a complex assortment of microclimates which are difficult to classify at large scale. These are often simply generalized as Highland climates; the Alps are the best example of a Highland climate in Europe.

As a final thought, it is also possible to understand a great deal about Europe by observing the climatic types and weather phenomena that the continent does not have. First of all, Europe has no tropical climates which, with their constantly high levels of heat and humidity and resultant year-round disease and insect pest problems, have always proven a very difficult environment for humans. Second, with minor exceptions, Europe has none of the dry climates (steppe or desert) which are again such blank areas on the world population map because of the great difficulty posed to human economic systems by constant aridity. Finally, Europe, with the exception of its limited Arctic or highland fringes and the extensive ice cap of Greenland, suffers relatively little from the extreme winter conditions that render vast tracts of the earth such as Siberia and the Canadian Arctic virtually uninhabitable on any regular basis.

Figure 2.3 Europe: Representative climatic stations.

Climate Type	Station	Elev (ft)	Temp/Prec	J	F	M	A	M	J	J	A	S	O	N	D	Annual
Marine West Coast	Dublin, Ireland	155	T(°F)	41	41	43	46	51	56	59	59	55	50	44	41	
			P (in)	2.7	2.2	2.0	1.9	2.3	2.0	2.8	3.0	2.8	2.7	2.7	2.6	30
	Paris, France	164	T(°F)	36	38	43	50	55	62	65	64	58	50	42	37	
			P (in)	1.4	1.4	1.5	1.6	1.9	2.2	2.0	1.9	1.9	2.3	1.9	1.7	22
Continental	Warsaw, Poland	394	T(°F)	25	27	34	45	57	62	65	64	56	47	36	28	
			P (in)	1.2	1.1	1.3	1.5	1.9	2.6	3.0	3.0	1.9	1.7	1.4	1.4	22
	Vienna, Austria	636	T(°F)	27	31	38	48	56	64	66	65	58	48	37	30	
			P (in)	1.5	1.3	1.8	2.1	2.8	2.8	3.1	2.7	2.0	1.9	1.8	1.8	26
Mediterranean	Palermo, Italy	233	T(°F)	52	51	54	58	65	71	75	78	72	67	59	53	
			P (in)	4.4	3.8	2.6	2.1	1.4	.6	.3	.6	1.8	3.7	4.2	5.2	31
	Athens, Greece	351	T(°F)	48	49	53	59	68	76	81	81	74	67	58	51	
			P (in)	2.2	1.5	1.4	.8	.8	.6	.2	.4	.6	1.7	2.8	2.7	16
Humid Subtropical	Sarajevo, Bosnia	1,870	T(°F)	41	47	55	62	69	73	72	66	58	49	43	43	
			P (in)	2.4	2.2	3.1	2.4	2.8	3.1	2.4	2.5	2.9	4.1	3.1	2.8	34
Subarctic	Kuopio, Finland	814	T(°F)	14	13	21	34	46	55	61	56	49	38	27	18	
			P (in)	2.0	1.4	1.5	1.2	1.7	2.4	2.8	3.0	2.4	2.0	1.8	1.6	24
Highland	Säntis, Switzerland	8,202	T(°F)	16	17	17	23	30	36	41	40	37	29	23	17	
			P (in)	5.5	7.2	6.5	8.1	7.6	9.5	12.0	10.5	8.3	7.0	4.8	5.9	96
Steppe	Cartagena, Spain	43	T(°F)	51	53	55	59	63	70	75	76	72	65	58	52	
			P (in)	1.8	1.4	1.6	1.1	1.0	.8	.1	.2	1.4	1.7	1.9	2.1	15

Source: Adapted from H.L. Nelson, *Climatic Data for Representative Stations of the World*.

Another advantage to Europe in this sense is the general absence of really violent weather. Mid-latitude location and the general moderation of the surrounding seas keep Europe free from such scourges as hurricanes (a phenomenon of low-latitude tropical areas) and tornados (generally restricted to North America). Even summer thunderstorms are a fairly rare occurrence over most of the continent. Severe storms and high tides in the North Sea can be a substantial threat during the winter months, while occasional outbreaks of arctic air from Siberia bring clear skies and bitterly cold air to much of Europe for several days each winter. The contrast in pressure and temperature conditions north and south of the Alps also creates numerous well-known local wind systems, some of which can be locally damaging. Among the best known of these are the mistral, which brings cold, dry air from northern Europe to the Mediterranean, especially through the Rhone valley; the sirocco, a hot, dry wind from the Sahara that sears the valleys and passes of the Mediterranean coast in summer; and the Föhn, warming air which descends the south side of the Alps (usually in the spring), causing rapid snowmelt and bringing the danger of avalanches.

The high-latitude location of Europe provides another advantage in the form of long hours of daylight during the summer months. A delight for tourists (seeing the "midnight sun" in northern Norway is a classic excursion), this factor provides a more practical benefit in the form of expanded inputs of sun energy during the growing season, and thus better agricultural productivity. Ironically, the warm, rainy tropical world, with no winter season, often has agricultural values below those of Europe because each day of the year is almost exactly as long as any other.

In summary, the climatic pattern of Europe can be easily understood given the continent's position with reference to the major causes of climatic variation outlined at the beginning of this section. By their moderation in both temperature and rainfall characteristics, these climates have proven hospitable to human settlement and basic economic activities, especially agriculture. While many factors of history, technology, political organization, and human ingenuity have played a major role in advancing Europe to its present position among the world's regions, the fact that its climates put few insurmountable obstacles in the way of human progress must certainly be taken into account.

PERCEPTION: THE EUROPEAN GREENHOUSE EFFECT

Over the past fifteen years, a mass of compelling scientific evidence has led to the inevitable conclusion that the temperature of the earth's atmosphere is increasing, and that, moreover, the rate of this increase is more rapid than any similar event of which traces can be found in the geologic record. Briefly summarized, the reason for this dramatic rise in atmospheric temperature (perhaps 0.5 degrees Celsius over the century following 1861, conservatively increasing to as much as 0.3 degrees Celsius per decade at the present time) lies in the greater volume of "greenhouse gases," especially carbon dioxide, being released into the atmosphere by various natural processes and human activities, most notably the growing volume of combustion of fossil fuels by industry and the destruction by burning of large areas of tropical forest. As these gases accumulate in the atmosphere, they inhibit the escape of radiant heat energy from the earth (while apparently doing little to block the entry of solar radiation), thus causing the temperature of the earth, encased in its atmospheric blanket, to rise; hence the "greenhouse" analogy.

If the greenhouse is a reality—and by now the general pattern seems incontrovertible although numerous complexities and uncertainties make precise calculations of levels and rates impossible—the implications for the world are profound. The oceans would warm and expand as glaciers melt, raising sea level from 1 to 4.5 feet (.3 to 1.5 meters) by the year 2050 (depending on the exact magnitude of the warming trend). Agriculture might expand into areas where growing seasons are now too short, while vanishing from other marginal regions where higher temperatures would increase evaporation, accelerating the processes of drought. Many ecosystems would be endangered by relatively sudden temperature change that plant species could not cope with through migration.

In Europe, all of these impacts would be felt, and would result in profound changes to both natural and human landscapes. Rising sea level would threaten thousands of coastal communities, including such important cities as Venice, Stockholm, and even London; while most of the Netherlands (roughly 40% of the country is land reclaimed from below the sea) would be endangered. The melting of glaciers and small ice fields in Scandinavia and especially the Alpine nations (already clearly occurring) would change the mountains' image and imperil much of the lucrative winter sports industry. The 1994–95 European winter games scheduled for southern Spain had to be cancelled as a result of insufficient snow cover in the Sierra Nevada.

Agriculture might be able to expand to the north throughout the continent, while hotter, dryer conditions would restrict it in the south; thus the mix of crops grown and the relative importance of agriculture to various national economies would change considerably.

Without doubt, the most affected region would be the Mediterranean. Already seasonally transitional between the cool, moist marine climate of northwestern Europe and the hot, dry deserts of North Africa, the European shores of the Mediterranean would find themselves facing even more months of drought and higher temperatures. This would lead to water shortages in a region of comparatively dense population; and while the tourism industry might welcome additional sunshine, many traditional activities such as agriculture would come close to vanishing altogether.

Is the greenhouse here? It is human nature to see in every hot, dry summer and every cold, snowy winter evidence of a trend; but climatic change is a long-term process compounded of many variables, such that conditions in a specific year mean little. Nonetheless, it is difficult to ignore certain patterns in Europe suggestive of a rapidly altering climatic balance. During the period December 1988 to February 1989, southern Europe received less than half its normal precipitation (Figure 2.4), while hot, dry conditions more normally associated with August occurred across much of Europe in May, 1990. For many stations, in fact, the 1980s were overall the warmest decade since reasonably comparable records have been kept, while 1990 appears (worldwide) to have been the warmest year ever recorded. Measurements taken in 1995 reveal that the thickness of the Antarctic ice cap is increasing, which seems at first glance to contradict the global warming theory, but may in fact reflect the extra available moisture that a warming atmosphere may be providing.

On the other hand, other types of climatic change may slow or even reverse the warming trend. The eruption of Mt. Pinatubo in the Philippines in 1991 put such a volume of volcanic ash and other particulate matter into the atmosphere that a slight global cooling trend was noted into 1994.

Figure 2.4 European drought: Does it reflect the "greenhouse effect?"

Nonetheless, despite such short-term interruptions, the greenhouse process appears to remain a long-term reality. The greenhouse effect may be less imminent than the more sensational reports imply, but nearly all European countries are now taking the possibility entirely seriously.

ADDITIONAL READING

FLOHN, H., and R. FANTECHI, eds., *The Climate of Europe: Past, Present, and Future.* Hingham, MA: Reidel, 1984.

HARE, F.K., *The Restless Atmosphere.* London: Hutchinson, 1966.

RIEHL, H. *Introduction to the Atmosphere.* New York: McGraw-Hill, 1972.

JEFTIC, L. et al, eds., *Climatic Change and the Mediterranean.* London: Edward Arnold, 1992.

TREWARTHA, G., *The Earth's Problem Climates.* London: Methuen, 1961.

WALLEN, C., ed., *Climates of Northern and Western Europe.* Volume 5 of *World Survey of Climatology.* New York: Elsevier, 1970.

DISCUSSION QUESTIONS

1. The average January temperature at Marquette, Michigan is 16 degrees Farenheit (–9 degrees Celsius). At Brest, France, the January average is 45 degrees Farenheit. (7 degrees Celsius). Both stations are at about the same latitude (46 to 48 degrees north.) How can you explain the difference in January average temperatures?
2. Spokane, Washington and Basel, Switzerland are both located about 250 miles (400 kilometers) from the west coast of a continent at about 48 degrees north latitude, and are about at the same elevation. Yet Basel is considerably milder in the winter, cooler in the summer, and wetter than Spokane. How can you account for this?
3. Assume that the "greenhouse effect" is a reality. Speculate on the changes this might produce in Europe by the year 2050.
4. Parts of the Iberian plateau in Spain are the only truly semiarid landscapes of Europe west of Ukraine. Why are these areas unusally dry?
5. To obtain 'guaranteed' sunshine for a summer vacation in Europe, you would probably visit the Mediterranean region. Why?

NATURAL VEGETATION

Understanding the broad pattern of landform distribution and climatic types that characterize Europe provides a good picture of the natural environments that might be encountered there. Two other elements, upon which humans have depended and which they have in turn substantially modified, however, should also be considered to fill in some of the detail of the landscape. These elements are the natural vegetation that was present before human intervention and which may occasionally survive or (more frequently) have been replaced by other forms; and soils, upon which the most critical of all human economic activities—agriculture—is founded. Since both of these elements owe much of their pattern of distribution to climatic changes from place to place, it is appropriate to discuss them at this point.

When early humans began to expand their numbers in Europe to a significant degree—perhaps some 60,000 years ago—they found a generally forested continent which provided both challenge and opportunity. Over the centuries, as technology improved, as the need to clear more land for farming became universal, and as the demand for wood as fuel and construction material steadily increased, most of the original forest cover was removed. In fact, by the middle of the twentieth century, Europe could correctly be identified as the continent with the smallest percentage of its land in forest or other natural vegetation types. Original cover could only be found in the more remote districts of northern Europe or the mountainous areas, or where some accident of history might have kept a small tract of woods safe from the axe (see the discussion of forestry, Chapter 9).

In more recent years, preservation and recovery of forested areas, as well as other types of natural vegetation, have become a national priority for most nations, with the result that Europe is by now probably adding to its overall vegetative cover at a more rapid rate than any other continent. Reforestation of large areas is widespread, providing both an eventual cash crop (the generally mild climates promote rapid tree growth), and such other advantages as recreation space, wildlife habitat, and watershed protection. Little of this admirable effort is designed to accurately reproduce the original cover, however. Conifer forests, more "useful" and faster growing, have often replaced the broadleaf forests of earlier times. European pride in a newly reforested landscape has been reflected in recent concern about the environmental consequences of acid rain and other pollutants on the woodlands of every region (see environmental discussion, Chapter 13).

The types of natural vegetation cover originally found in Europe, and which may still be observed at least partially, are closely related to the climate patterns. This is of course not surprising, since the climatic distributions we have noted were suggested by a botanist, Wladimir Köppen, who based his work on changes in the plant communities. Thus, just as there are three major climatic regions in Europe, so are there also three distinctive types of natural vegetation:

1. **The broadleaf deciduous forest (temperate deciduous forest).** Under the Marine West Coast climate, with its generally mild temperatures and plentiful precipitation, a broadleaf forest developed based on species familiar to the middle latitudes of all continents, such as oak, elm, maple, and beech. With the marked seasonal change evidenced in the changing length of day, the trees lose their leaves for a short dormant season, but in the mild environment of the North Sea Plain spring comes early, and leaves and blossoms begin to reappear as early as February in some favored locations. Very little of this original forest remains. Only where small pieces were preserved, as in royal hunting preserves (for example, the "New Forest" in Southern England dating from the time of William the Conqueror), can large trees of this type be seen today.

2. **The needleleaf evergreen forest (the coniferous forest).** As one moves from the Marine West Coast climate into the Continental varieties, that is, by moving farther east, farther north, or higher up the slopes of mountain ranges, the vegetation cover also changes. As the climate becomes colder and thus drier (since frozen water is useless to

Figure 2.5 Extensive reforestation has recently occurred in most European countries: Beech forest in Luxembourg.

plants, cold winters are therefore the same as a drought), plants better adapted to withstand dry conditions begin to dominate. With their small leaves, deep roots, and thick bark, most evergreen (coniferous) species (pine, spruce, fir, etc.) are thus successful in this environment, and so come gradually to dominate the forest composition. There is a broad transition zone, of course, as the climate gradually deteriorates from west to east, between the open broadleaf deciduous forests of western Europe and the solid conifer forest, or taiga, of northern Europe and the vast reaches of Siberia. It is, incidentally for the same reason, drought, that coniferous species are widely distributed around the coastlines of Europe and other continents. The sandy soils common to coastal environments allow rainfall to percolate rapidly, often before vegetation has a chance to utilize it fully. The result, even in areas of abundant rainfall, may effectively be a drought condition for the plant community, in which evergreens, with their greater adaptability to such an environment, again become dominant.

3. The Mediterranean scrub forest. With its unique climatic regime, featuring particularly the dry summer season, the Mediterranean region of southern Europe (and indeed all of the world's areas of this distinctive climate) poses unusual problems for its plant community. On the one hand, mild temperatures and moderate rainfall during the winter months permit vegetation to grow normally in what is usually a dormant season. On the other hand, the summer drought, with associated higher temperatures, means that the plants must be specially adapted to survive. The result of these contradictions is a plant commmunity unlike any other in the world. All of the native plants of the Mediterranean share certain adaptive features: small size, small leaves, deep roots to obtain any available moisture, thick bark to protect the stem against excessive water loss, and a broad canopy to shade the plant and reduce evaporation. The most distinctive of these unique vegetation communities is a collection of scrubby plants, averaging perhaps a meter (3 feet) in height, which forms a continuous ground cover across much of the Mediterranean basin, and which is known by a distinctive name in each of the bordering cultures: Maquis in southern France, Macchia in Spain (and Chaparral in the Mediterranean climate region of Southern California!). Larger trees of the region share these same characteristics. These would include the olive (see discussion of the Mediterranean economy, Chapter 8), several types of small pines, and a variety of live oaks, including the famous cork oak, whose thick, protective bark is stripped off and utilized by man without permanent damage to the tree.

Figure 2.6 "Maquis" vegetation of the Mediterranean region: Greece.

A particular problem in assessing Mediterranean vegetation arises from considering how much of the original cover has been lost as a result of human intervention, and at the same time how much of the present landscape is covered by introduced species. As late as the beginning of the Classical period—perhaps 3000 years ago—the Mediterranean, reflecting a long period of relatively cooler and wetter postglacial climate, was much more densely forested, as the writings of many prominent Greek authors of the time suggest. This had become a very fragile environment, however, and with humanity's expanding numbers and the increasing ability of technology to modify the landscape, the Mediterranean had been largely deforested (and in fact environmentally severely damaged (see Environmental discussion, Chapter 13) by the early centuries of the Christian era.

At the same time, the spread of new technologies also included the introduction of irrigation across the region. These techniques were generally introduced as part of the Islamic impact of the eighth and ninth centuries AD, and dramatically altered the vegetative face of the entire Mediterranean basin. The mild temperatures, abundant sunshine, and at least moderate amounts of good soil meant that, with the addition of permanent water supplies where major rivers or aqueducts could be used to provide year-round irrigation, the Mediterranean became something closely resembling the proverbial Garden of Eden. Many of the plants with which we most closely associate the Mediterranean world today, therefore, are comparatively recent introductions that would not survive without the certainty of irrigation during the dry summers. Surely the best known of these plants are the citrus fruits: native to Southeast Asia, oranges, lemons, and other citrus plants have become legendary in the Mediterranean (a city in Spain gave its name to the Valencia orange), yet they depend totally on irrigation for survival. Moreover, a truly bewildering variety of other exotic plants from around the world is found in the modern Mediterranean landscape wherever water can be consistently provided.

In addition to these three major regions of natural vegetation in Europe, a number of minor types of vegetative complexes may be recognized, most of which are again entirely consistent with climatic constraints. For example, on the extreme Arctic fringes of Europe (as well as other circumpolar regions of North America and Russia), and at the higher elevations of Europe's major mountain ranges, very short growing seasons and extremely cold winters mean that trees cannot survive, and the basic vegetation becomes a complex of small shrubs and specially adapted flowers known collectively as the tundra. This hardy yet delicate vegetation provides the basic ground cover throughout the North.

Related to the tundra complex is the scrub vegetation found in many of the coastal environments of northern and western Europe. Here, the constant westerly wind, often of considerable velocity, evaporates moisture rapidly so that, even in regions of heavy rainfall, the plant community lives in a virtual drought. Moreover, the wind exercises a powerful mechanical weathering effect, literally grinding away the growing shoots of plants except where these are sheltered in valleys or on the reverse slopes of hills. Among the plants that survive in the often barren landscapes of highland Scotland and Wales, Brittany, and other exposed regions are such well-known species as heather, gorse, and broom.

Europe has very little truly dry climate, and thus shows few areas of dry-land vegetation. In the semiarid climates of central and southern Spain, some areas of short grass or steppe vegetation are found, but this is of limited extent. In fact, of all the major types of vegetation found in the mid-latitudes of the world, Europe is most underrepresented by the grasslands. Only in the valley of the middle Danube (or Alföld) in Hungary and adjacent regions of Eastern Europe is there any considerable extent of grassland, and because of soil problems (see following discussion), this is not nearly as productive a grassland region as Ukraine, the North American Midwest, or the Argentine Pampas. The famous steppes of Ukraine and southern Russia, based on a semiarid climate, are among the world's best-known grasslands, and have become a vital region of grain (especially wheat) production. Elsewhere, extensive grasslands are only to be found in the deltas of a few of Europe's major rivers, notably the Rhône, whose grassy delta region in southern France, the Camargue, is a distinctive and colorful region, and the Danube, which has built a complex delta into the Black Sea.

SOILS

Along with climate, soils are the most important determinant of agricultural productivity, and the pattern of their distribution may thus be closely correlated with such factors as food production and farm income. Surveying the world, it is both interesting and sobering to realize just how little of the earth's surface is blessed with rich soils. Europe, unfortunately deficient (except for Ukraine) in the world's most productive soil types, nonetheless enjoys a range of soils that, with careful management and intelligent utilization, can be reasonably productive. One of the reasons why such innovations in agricultural practice as crop rotation were perfected in Europe was the constant need both to conserve and to improve the existing soil base.

Soils are the long-term products of a wide range of natural factors. Climatic conditions and natural vegetation cover play perhaps the most important roles (the correlation among climate, vegetation, and soil patterns is usually a very strong one); but such factors as parent material, slope, and drainage efficiency also play their parts. The result is a very complex pattern, especially at local scale, where small changes in the natural environment may produce dozens of different soil types within a single square mile. Moreover, soil classification is both complicated and subject to constant revision. The most commonly accepted classification at the present time—that of the United States Department of Agriculture—is known as the Comprehensive Soil Classification System (CSCS), or "seventh approximation," inevitably suggesting that we are somewhere between the sixth and the eighth and perhaps correctly emphasizing the fact that there may in reality be no universally applicable and enduring system.

At the broad scale of this book, it may be best to simplify the distribution patterns of soils by recognizing that there are basically three complex processes by which soils are formed: those of the humid middle-latitudes, the tropical world, and the deserts. Each of these processes varies considerably in intensity and in the resultant pattern of specific soils, and their definitions tend to be broad rather than narrowly

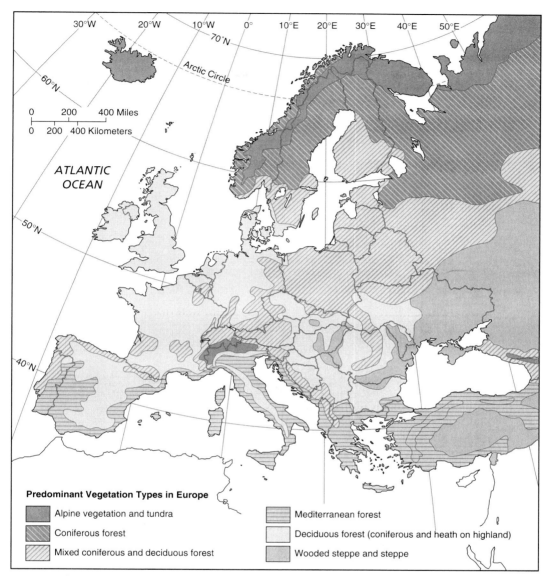

Figure 2.7 Europe: Generalized vegetation regions.

scientific. Nonetheless, their descriptive characteristics help to explain the nature of many of the world's most distinctive landscapes, and to provide some solid clues as to the distribution of human populations and agricultural activity.

In the tundra regions of extreme northern Europe, the harsh climatic conditions and lack of significant vegetation produce a poorly developed group of soils known in the CSCS as inceptisols. Given its mid- to high-latitude location and generally well-distributed rainfall, however, Europe's soils are largely the result of the podzolic process. Under the solid coniferous forest (or taiga) of northern Europe and Russia, the true podzol (the term "podzol" derives from the Russian term for "ashen soil") soils that form are acidic, heavily leached in the upper layers, tightly recemented in the lower layers, and almost totally unmanageable and unproductive. These are known in the CSCS classification as spodisols. Over much of Europe, however, conditions are luckily not so extreme. Milder temperatures, shorter winters, and a broadleaf or mixed forest cover result in less acid ground water and thus soils with at least some potential. Recognizable levels of leaching above and redeposition below produce what are called grey-brown podzolic soils, and it is these, often heavy and difficult to work, upon which European agriculture is based. These may also be referred to as alfisols. Substantial areas of histosols (bog and moorland soils), and entisols (immature soils on alluvial or sand deposits) are also found in Europe, while soils of the mountainous districts can be extremely complex as a result of slope variation and local microclimates.

There are, clearly, many important areas of difference; the pattern is by no means uniform. There is a strong correlation, in fact, between small regions of unusually rich soils and economic (and even political) success. For example, some parts of central Europe are covered by a thin layer of fine wind-blown material dating from the retreat of the most recent ice sheet. This material, known by the German term loess (also called limon in France), adds considerably to the natural fertility of the soil and is responsible for such rich agricultural regions as the Paris basin and the Börde region of central Germany (see regional profiles). In a similar way, the deep, rich sediments of many river valleys (such as the Po of northern Italy) have given them an unusual significance in the development of several countries. Conversely, where soils are produced by the podzolic process on such acidic rocks as granite, the result is a relatively poor agriculture on what are known as "cold" soils. Many limestones, on the other hand, will produce very rich soils.

In the Mediterranean region of southern Europe, the pattern of soil types is even more complicated than elsewhere. While cool temperatures and reasonably abundant rainfall produce podzolic conditions during the winter months, the dry, hot summers complicate the picture considerably. As surface evaporation largely exceeds rainfall, moisture moves upward through the soil by capillary action (in effect, the process that operates year-round in the world's desert regions), thus redistributing minerals from bottom to top and creating very productive soils that have a tendency to maintain themselves well. If modern agriculture is sometimes less important in the Mediterranean than might be expected, it is the fault of the dry growing season climate and a lack of flat land, rather than of the soil conditions.

Also in the Mediterranean zone, widespread volcanic activity produces pockets of quality soil as the often mineral-rich lava weathers over time. Although it seems foolhardy, villages are built and farming takes place surprisingly near the crater of Mt. Etna on Sicily, for example. Farmers are willing to risk the danger and the periodic need to abandon their lands and homes because of the better harvests they can expect from the volcanic soils.

As a final note, it should be remembered that much of Europe, because of its general absence of grasslands, is also without large extents of those superb soils—the world's richest—which support the bountiful agriculture of Ukraine or Iowa. These are

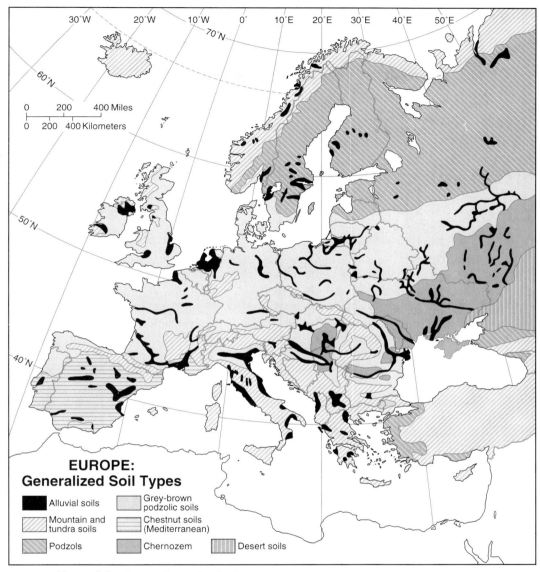

Figure 2.8 Europe: Generalized soil types.

known traditionally as chernozems ("black earth" in Russian), or as mollisols in the CSCS classification. Except for Ukraine, the only other fairly large original European grassland region—the Alföld, or central basin of the Danube River—is certainly productive, but much of its value is compromised by poor drainage and a high water table, which produce widespread areas of waterlogged or alkaline soils.

ADDITIONAL READING

DETWYLER, T.R., *Man's Impact on Environment.* New York: McGraw-Hill, 1971.
KUBIENA, W.L., *The Soils of Europe.* London: Murby, 1953.
TANSLEY, A.G., *The British Isles and Their Vegetation.* London: Cambridge University, 1939.
THIRGOOD, J.V., *Man and the Mediterranean Forest.* New York and London: Academic Press, 1981.
THOMAS, W.L., ed., *Man's Role in Changing the Face of the Earth.* Chicago: University of Chicago, 1956.

DISCUSSION QUESTIONS

1. Describe the podzolic soil formation process. How do the grey-brown podzolic soils common in Europe differ from the true podzols?
2. What is loess? In which parts of Europe has it served to increase soil fertility?
3. Describe the maquis, or Mediterranean scrub forest. Why has this unusual plant community developed in southern Europe?
4. Reforestation is now common practice in most European countries, yet many ecologists and environmentalists are not entirely pleased with the results. Why?
5. Europe's prehuman natural vegetation cover included very few areas of grassland. Why is this significant?

PART TWO

Human Geography and the Development of the Cultural Landscape

3
Europe in the World Population Picture

Of all the major trends that influence the relative development and prospects of the world's various regions, the characteristics of population are perhaps the most significant. Thoughtful observers have noted that whatever problem any individual may put at the top of his or her list—hunger, poverty, international tensions, environmental stress, etc.—rapidly increasing population numbers make conditions worse. It is, in fact, possible to estimate with some precision the relative economic situation of any country or population based on demographic statistics alone; only those countries with massive reserves of exportable resources escape this rule to some extent. The population crisis (as many see it) has two components: rapid overall population growth, especially in the less developed world, and explosive urbanization, also concentrated in less developed countries. Both of these trends are creating conditions absolutely without parallel in the history of the world; and without historical models for guidance, we can only guess at their implications for the future.

The growth of Europe's population over the centuries gives us some useful clues as to what is happening today in other parts of the world. Before analyzing this evolution in more detail, however, it might be helpful to take a brief look at the world population picture as of 1996, and to assess Europe's position in the overall picture. (See Figure 3.1: World Population Summary).

In Figure 3.1, the major world regions (small populations such as Oceania are not considered here) are divided into two groups: developed and less developed economies. It is important to note, by the way, that the term "development" is a loaded one, and must be used with considerable care. Many of the world's regions that are rich and highly developed in a cultural sense are poor only economically, and the term "underdeveloped" can often be insulting. Development terms will be used in this discussion only in an economic sense, and even there may unfortunately reflect a certain bias as to what constitutes true "development." Nonetheless, the figures are revealing in terms of the similarities in population characteristics displayed by regions of comparable economic development level.

The first three regions largely constitute the economically developed or "rich" world. Their population characteristics are remarkably similar, featuring low birth and death rates, correspondingly moderate rates of population growth, and a long period required to double population (if, indeed, this will ever come about). Note that even among these regions, Europe has by far come the closest to stabilizing its population at current levels. The annual increase in North America is slightly higher as a result of immigration (some 400,000 legal immigrants enter the United States annually); while a somewhat higher birthrate in the former Soviet Union reflects the fact that about half of this population is comprised of non-European groups which still have higher fertility rates. Another common feature of the population of these three major regions is the relatively moderate percentage of the total population under fifteen years of age (22 to 25 percent). This fact will be discussed later in this section. Finally, in terms of economic success (as measured by gross national product per capita— "GNP"), these three regions are clearly far ahead of the world average despite the current economic adjustment problems affecting eastern Europe and the successor states of the Soviet Union.

FIGURE 3.1 World Population Summary—1996

Continent	Pop. (Millions)	Birth Rate/ Death Rate (Per 1,000)	Annual Increase %	Doubling Time (Yr)	Under 15 (%)	G.N.P. ($/Cap)
DEVELOPED REGIONS *(caution: includes some poor nations)*						
Europe*	580	11/11	(–0.1)	—	19	12,310
N. America	295	15/9	0.6	114	22	25,220
Russia	148	9/15	–0.5	—	21	2,650
LESS DEVELOPED REGIONS *(caution: includes some rich nations)*						
L. America	486	26/7	1.9	36	35	3,290
Asia	3,501	24/8	1.6	43	32	2,150
Africa	732	41/13	2.8	25	44	660
World	**5,771**	**24/9**	**1.5**	**45**	**32**	**4,740**
NATIONAL EXAMPLES						
Hungary	10.2	11/14	(–0.3)	—	18	3,840
Nicaragua	4.6	33/6	2.7	26	44	330
China	1,217.6	17/7	1.1	66	27	530
Ethiopia	57.2	46/16	3.1	23	49	130
Kenya	28.2	40/13	2.7	25	48	260

Source: Population Reference Bureau.

*Excludes Russia; includes other European former USSR states.

Europe in the World Population Picture

The second group of major regions generally constitutes the economic "have-nots" of the world (although note that such a broad-brush treatment groups such relatively wealthy nations as Japan, Hong Kong, South Africa, and the Middle Eastern oil states with their much less fortunate neighbors). Here again, the population characteristics show substantial similarity within the group; but they are radically different from those of the developed regions. Note especially that although death rates are little different from those of Europe, birthrates are two to four times as great. Thus, it is clear that it is the continued relatively high birthrates in these regions that is largely responsible for their extraordinary rate of annual increase, as well as a doubling time that seems uncomfortably short. Moreover, signs of the future can be read in the fact that 37 to 45 percent of the population of these regions is less than 15 years of age today. Finally, despite the few rich nations grouped geographically in this category, it is clear from the GNP data how vast a gap exists between the developed and less developed world in any economic sense.

The world summary figures on the next line (Figure 3.1) are of casual interest perhaps as cocktail party conversation, but nonetheless carry a few significant messages. First of all, it is clear that world population is increasing at the staggering rate of about one billion people each twelve years. Second, although the world rate of population growth has declined from 2.0 percent to 1.5 percent over the past twenty years—a remarkable fact in itself—this still represents an unsettling rate of growth. Third, a world doubling rate of only forty-five years means simply that most of today's geography students, who can expect with reasonable confidence to be alive at that point in time, will witness a total population substantially exceeding 10 billion. This is a figure, by the way, based not on some eventual increase in the birthrate, but on the reasonably moderate projections involving families of people already inhabiting the earth. Finally, the relatively high figure (32 percent) representing the world's younger population explains why total numbers are bound to grow, no matter what the evolution of the birthrate; it also highlights the enormous problems facing developing countries where the provision of services and jobs for such numbers is highly problematic. It is a sobering thought to most geography students to realize that, as they move into their mid-twenties, they enter into the older half of the world's population!

In the final section of the table, a few countries have been selected to demonstrate various models of population change. Hungary is an example of the general pattern of European population stability, in this case actually showing a population loss as the death rate in an aging population exceeds the very low birthrate. This loss of demographic vitality is currently a major concern for the Hungarians (as well as for leaders of several other European nations), as declining numbers of younger citizens raise substantial questions of labor supply, military preparedness, and long-term trends.

Nicaragua is an example of a small nation with runaway population growth. Although numbers are small (4.6 million), a doubling time of only twenty-six years means that an already poor country without extensive resources will almost certainly become even poorer with the passage of time.

China is selected for the table largely because of its enormous 22-percent share of the total world population; clearly, whatever China does about population shows up quickly in the world tables. What China has been doing, often through the use of fairly draconian methods, is reducing its rate of population growth by an unprecedented one-tenth of one percent per year. It has been this dramatic reduction that has enabled the world figure of annual growth to decline sharply in recent years. Nonetheless, it requires very little mathematics to calculate that, even with a growth rate well below the world average, China is still adding people at a rate of over 11 million per year.

Finally, Ethiopia is chosen as an example of one of the world's poorest countries. It still has one of the the higher death rates of any modern nation, but this is declining, and the high birthrate means that numbers are being added even here at an unprecedented rate. With a per capita GNP of only $130, it is legitimate to wonder what will become of the 750,000 new children born in Ethiopia each year.

There are numerous lessons to be drawn from the study of this table, or indeed from any of the mass of similar data documenting the current rapid increase of the world's population. We suggest here only a few of the more striking ones. For example, there is a clear correlation between population growth and economic well-being, even at this scale. A consistent economic growth rate of about 3 percent annually is considered more than satisfactory in the modern world, yet observe the differences as this is applied to different countries. In Europe, with almost no population increase, any economic growth is applied directly to improving the well-being of the current population. In developing countries with population growth rates of 3 percent or better (there are many), a 3 percent rate of economic growth simply serves to feed new mouths at the same depressed level; there is no real progress. Another sobering fact is that, as the world's population grows at an extravagant rate, most of the addition occurs in those nations that can least afford it. For example, between 1993 and 1994, some 87 million people were added to the world's numbers. Of this total, only 9.4 million increased the numbers of the developed world, (as in Figure 3.1, including also Japan and Oceania), while over 78 million were added to those countries of the less developed world, which are in many cases already having difficulty coping with existing numbers. As these growing populations put an increasing brake on real economic growth, the result will inevitably be a poorer rather than a richer world.

PERCEPTION: A DECLINING EUROPE?

Since Europe's population is essentially stable in numbers, it continues to represent an increasingly smaller part of the world total. In 1900, Europe contained 25 percent of the world's people; by 1994, this figure had fallen to just under 13 percent (including Russia and the other European parts of the former Soviet Union). It is clear to any thoughtful person that world population must eventually stabilize at some point and by some means. Many projections now hypothesize that Europe and the other developed regions will have reached a true zero-growth pattern by the middle of the twenty-first century, and that the less developed world will stabilize by about 2100. If this turns out to be the case (and some critics consider it overly optimistic), total world population will probably be somewhere between ten billion (low estimate) and twelve billion, or two to two and one-half times the present number. In that case, Europe's population, which of course will almost certainly stabilize earlier than that of any other region, will be an even smaller percentage of the total. In fact, today's developed nations, including Japan, Oceania, and some other minor additions as well as North America, Europe, and Russia, will contain scarcely more than 10 percent of all the world's people.

How has world population evolved in this way? Since Europe is the continent that has gone the farthest toward population stability, it seems reasonable to consider the processes responsible for this result, perhaps with a view to asking if today's less developed regions might find a useful example to follow. The heart of the matter is what is called the demographic transition, which Europe has largely completed and which less developed regions are only beginning to traverse (Figure 3.2).

For most of human history, both birthrates and death rates were very high, with deaths actually exceeding births in years of famine or plague. Some 25 percent of Europe's population died in the fourteenth century from bubonic plague alone. The result was naturally a total population that grew very slowly. Europe had about 40 million inhabitants at the beginning of the Christian Era. Ten centuries later, the number had probably only barely exceeded 80 million: a rate of growth averaging only some 0.1 percent annually (even before the ravages of the black death a few centuries later). It is worth noting that most of Europe has today returned to that same rate of growth: 0.1 percent annually. It is conventional to begin Europe's demographic transition about the year 1650, when great changes were beginning to alter every facet of the continent's life, certainly including population.

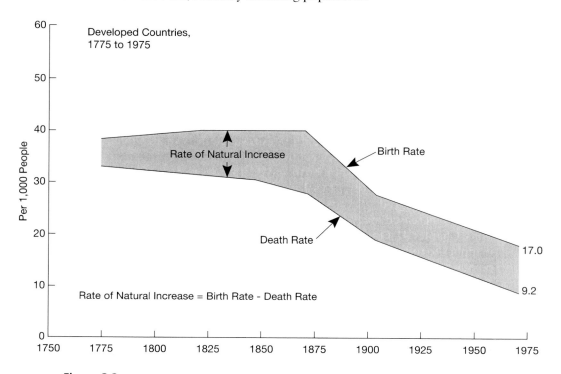

Figure 3.2 The demographic transition in Europe.

In the later years of the seventeenth century, and with increasing speed during the eighteenth, the European death rate began to decline. At the same time, however, the birthrate remained nearly as high as ever, the result being a sharp increase in the population numbers. Death rates decline before birthrates (this part of the model seems consistent in all parts of the world) because death rates are largely technologically controlled. Thus, as Europe began to witness dramatic improvements in medical care, public health, educational opportunities, and agricultural productivity, the traditional shadow of early death was pushed farther into the background; life expectancy went up and mortality declined year by year. Birthrates remained high for a much longer period, however, because they are largely culturally controlled and thus not subject to the sort of technological "fixes" that dramatically reduce deaths. The value of large families was so ingrained in the European mentality that several generations of radical change in the social and economic realities of the continent were to pass before traditional values were reexamined by large numbers of people.

In those centuries (most of history) in which nearly everyone made a direct living off the land, larger families meant more useful farm labor and thus greater prosperity. In the absence of organized systems of social security, children were also the only hope of old-age care for parents, while the necessity of ensuring survival of a male heir was a powerful incentive in most European societies. During centuries of high death rates, consequently, large families meant both present prosperity and future security; it is, therefore, small wonder that this concept was hard to change.

With the increasing modernization of Europe, however, and especially with the onset of the industrial revolution and the consequent beginnings of the shift from a largely rural to a primarily urban population, large families became a burden rather than an asset. Space was inevitably scarce in urban housing, and children could rarely contribute significantly to the economic well-being of the family, as they had on the farm. Moreover, as education became widespread, and eventually compulsory, the expense of raising children increased sharply. As these profound changes began to alter the fabric of society in the nineteenth century, birthrates began to decline more sharply so that, by about the 1930s, a new population equilibrium had been established in which low birth and death rates resulted in as small a rate of growth as had the much higher rates of three centuries earlier.

During the period when birth and death rates were farthest apart, that is, from the late 1700s to the early 1900s, Europe witnessed a true population explosion. This problem was solved, of course, by the mass emigration of millions of Europeans, as numbers increased much faster than economic opportunities (see European Migrations, Chapter 4). One of the great imponderables of the demographic transition as it applies to today's less developed regions (which, as we have seen, are at the crucial point where death rates have declined sharply while birthrates remain high) is whether the process will run a normal course, given the fact that it is today impossible to export large numbers of people.

Another useful way of considering population data, and a way that is revealing of the nature of Europe's inhabitants, is to construct *population pyramids,* diagrams in which the percentage of the population in each age group is shown as a bar (male and female population considered separately): the longer the bar, the greater the percentage of total population. European nations generally have pyramids which are, in effect, nearly rectangular. This means that each generation is approximately the same size, the obvious result of a near-zero-growth rate of increase. This situation is most advantageous to the Europeans in that it permits governments and agencies to plan ahead with substantial confidence. The level of services that must be provided for the young, the number of jobs that will be required, and the demands of old age can generally be accommodated because they are nearly the same for each generation. The developing nations generally have much more broadly based pyramids, reflecting the very large percentages of their population in the younger age groups (recall that 44 percent of Africa's population is under fifteen years of age; this figure is as high as 49 percent in some countries). The problems of providing education, services, and jobs for such numbers are obviously staggering, while the large size of these generations is the reason that world population will continue to increase rapidly well into the future even if the rate of growth begins to decline more sharply.

In the United States, where the European-type population profile that had developed by 1940 was dramatically upset by the "baby boom" of the 1950s and 1960s, it is possible to gain at least a modest impression of the difficulties facing those nations whose generations are so unevenly divided. The American problems of education, employment, and old age security that this group is creating as it moves through the age table are vexing but not insoluble in a rich country. Far more serious problems of imbalance in much poorer countries make the future for these large generations an uncertain prospect at best.

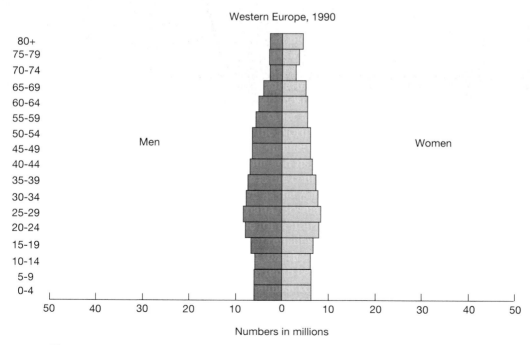

Figure 3.3 Population pyramid: Western Europe.

THE DISTRIBUTION OF EUROPEAN POPULATION

Looking specifically at the distribution of population within Europe, a number of significant points stand out (see Figure 1.3: "Area and population data for European countries: 1996"). Initially, note that the population of Europe is finely divided among a relatively large number of countries, such that the population of each country is relatively small. Of the forty-four countries listed, only six (Russia, Ukraine, Germany, France, Italy, and Great Britain) have populations greater than 50 million; only Russia and Germany exceed 60 million (and, given current trends, their totals may be declining). Russia is now the world's sixth largest country (the former Soviet Union ranked third), while the reunification of Germany has created a nation of about 81 million people: by far Europe's second largest, yet no better than eleventh in the world. Only three others exceed 20 million (Spain, Poland, Romania), while over half the nations listed (twenty-four) have fewer than 10 million inhabitants. It often comes as something of a shock to realize that many familiar European nations, of great significance in the world historically, economically, and culturally, have such small populations. Switzerland (7 million) and Sweden (8.8 million) are good examples.

Moreover, the growth rate figures in the table demonstrate clearly the fact that Europe, of all the continents, is closest to achieving a zero-growth model of population change. In 1996, nine countries (Germany, Hungary, Estonia, Latvia, Bulgaria, Croatia, Russia, Romania, Ukraine) saw their birthrates slip slightly below death rates, resulting in a modest population loss. Seven others recorded no change at all or grew so slightly (0.1 percent) as to be statistically insignificant. In fact, the only European country exceeding the world rate of population growth (1.5 percent annually) is tiny Albania (1.8 percent), a remote, relatively underdeveloped nation (by European standards) on the Adriatic Sea. Beyond this, only Iceland and Cyprus have growth rates as high as 1 percent annually.

Two additional points should be made regarding Europe's population. First, although advantageous in the long run, stabilized or negative growth can present certain problems. If younger generations are slightly smaller, for example, questions of labor supply and the maintenance of traditional activities are raised. Germany, for example, has been forced to take the unpopular step of extending the period of obligatory military service since the numbers of those eligible has been declining. Additionally, increasing longevity in a stable population may increase the numbers of elderly to a level where they become something of a burden for smaller younger generations. The most rapidly growing segment of the European population (in percent) today is that over eighty-five years of age. Smaller working-age populations are forced to increase social security contributions to maintain the elderly, while health care costs are also likely to escalate. Nonetheless, it is clear that the small family and a low-growth mode are popular with most Europeans. Despite substantial government incentives in many countries to boost birthrates and thus the rate of population growth (family allowances, free education, tax advantages, etc.), stability in numbers remains the rule.

Second, it is clear that Europe has relatively few population problems when compared with much of the world. This is not to say, however, that there are no problems at all. Population numbers must always be considered in relation to the economic support base available to maintain them. There is no such thing, in other words, as absolute overpopulation. Europe, as we shall see, is a continent whose overall resource base (considering food, energy, minerals, etc.) is inadequate to support its existing numbers at current levels of national and personal well-being. Only by importing vast quantities of resources from elsewhere in the world can this success be achieved. Any disruption, through political disputes, wars, or the increasing demands of rapidly growing "third world" countries, could mean that Europe would be faced with an overpopulation problem of important dimensions in the sense that numbers—even though growing only slowly—might exceed resources.

ADDITIONAL READING

BEAUJEU-GARNIER, J., *Geography of Population.* London: Longman, 1966.

FREJKA, T., *The Future of Population Growth: Alternative Paths to Equilibrium.* New York: Wiley, 1973.

GLASS, D.V. and D.E.C. EVERSLEY, eds., *Population in History: Essays in Historical Demography.* London: Edward Arnold, 1965.

INTERNATIONAL LABOUR OFFICE, *From Pyramid to Pillar: Population Change and Social Security in Europe.* Geneva: I.L.O., 1989.

MCINTOSH, C.A., *Population Policy in Western Europe.* Armonk NY: M.E. Sharpe, 1983.

DISCUSSION QUESTIONS

1. What is the demographic transition? How did it apply in Europe to create the present pattern of population? What lessons does it hold for the less developed countries today?
2. Why have birth rates historically been harder to reduce than death rates?
3. Europe today is one of the "greying" continents. What does this mean? What are the economic and social implications of this for Europe in the next century?
4. What is a population pyramid, and what can it tell us about the probable future development and problems of any given country? Describe a typical European pyramid.
5. The reunification of Germany has created a nation of some 81 million people. How does this compare with the populations of other major European countries?

4
Restless Europe
The Role of Migration

With the exception of North America, the continent most strongly marked by immigration, no part of the world has been so profoundly influenced by migratory currents in recent centuries as has Europe. Not only have Europeans colonized many parts of the world, thus adding vastly to the influence of European ideas and economic systems, but within the continent itself, migratory trends have rearranged the population in several remarkable ways. It is convenient to consider the phenomena of migration involving Europe in three rather distinct categories: emigration, or movement of population away from the continent to other parts of the world; intracontinental migration, or movement of population from one country to another within Europe; and internal migration, or movements within a particular nation. Finally, in the 1990s, Europe has again become a continent of immigration, as political refugees from other parts of Europe and the world, as well as economic refugees seeking no more than the chance for a better life, have flooded into the more prosperous nations.

EMIGRATION

Until the middle ages, Europe was a region of immigration; peoples from the east arrived in successive waves and often pushed former arrivals into remote corners of the continent. The Celtic peoples that are found today in Scotland, Ireland, Wales, and Brittany were early immigrants who were forced to retreat to the far western fringes of Europe by pressure from newcomers from the east. Historically interesting as these ancient movements are, however, they are of less importance in explaining the face of modern Europe and its impact on the world than are more recent events.

Beginning with the early colonial experiences of the sixteenth and seventeenth centuries, and culminating in the mass exodus of Europeans as a result of the rapid population growth and economic upheavals of the nineteenth and early twentieth centuries, Europe was a continent of emigration on a gigantic scale. The numbers alone are impressive: between about 1830, when Europe's population was some 250 million, and 1930, when the great depression and more stringent American immigration laws combined to slow the exodus, about 60 million people left Europe, 40 million of them permanently. Since these emigrants represented largely the younger, economically active members of society, and since by removing themselves from Europe they also removed their eventual children as well, the long-term impact on both Europe and the regions of immigration was even more profound than their numbers alone might suggest.

To understand these migrations on such a vast scale, it is necessary to consider the conditions in Europe and the receiving regions which led to them. What could have provoked such a volume of human movement? It should be remembered in this regard that most migrations have historically been acts of desperation. In the late twentieth century, residents of advanced nations in Europe (or North America) take an almost casual attitude toward migration, moving easily and often in response to job or educational opportunities, or simply to the urge to see something new. It is sometimes difficult for us to understand that even in the earlier years of this century, most people tended to remain closely tied to their native town or village all of their lives. A weekly trip to a nearby market town, a rare overnight journey to some regional center, perhaps a once-in-a-lifetime trip to the national capital: this was the travel calendar for most people, for whom the prospect of moving permanently to some totally new setting, far from the known and familiar, was a frightening prospect and only rarely even came to mind. There is a basic conservatism in most people that weighs heavily against such a dramatic step.

Against this background, the tide of European emigration appears even more unlikely, but there were at least five kinds of pressures acting to break down the natural reserves of the European masses. Two of these were what are called "push" factors, that is, those tending to force people away from the familiar. Two others were "pull" factors, that is, those tending to attract potential migrants to some different place. A final factor was essentially a catalyst, or facilitating technology.

Figure 4.1 European migration to the United States, 1830–1930
Selected Countries and Regions (Thousands of Immigrants)

Country	1830–50	1851–70	1871–90	1891–1910	1911–30
Great Britain	344	1,031	1,356	797	672
Ireland	991	1,350	1,092	727	367
Scandinavia	17	151	890	877	402
Germany	589	1,739	2,171	847	556
Italy	5	21	363	2,698	1,565
All European Countries	2,100	4,518	7,009	11,695	6,845

Source: U.S. Census Bureau

1. **Population and economic pressures.** The first push factor was the rapidly growing European population and its steadily increasing pressure on the continent's economic support base. Overpopulation is, of course, strictly a relative term, only to be considered in relation to the economic opportunities available to the human group; but as Europe entered the critical phase of its demographic transition in the nineteenth century, the rate of population growth largely outstripped the rate of economic growth. Eventually, as Europe moved from a rural to an urban society and from an agricultural to an industrial economy, plenty of economic opportunity came to be provided, but this was a lengthy, complex process that provided little consolation for individuals caught in the middle. Millions of average Europeans saw the year-by-year conditions of their daily lives steadily worsening, and while this by itself would not have triggered any spectacular wave of emigration, the uncertainties it provoked began to create a sort of "migratory mentality."

2. **Political instability.** An additional push factor was what might be called the general political instability of the times. Europe has, of course, a long history of warfare and turmoil; but the nineteenth century seemed to mark a new high in the sort of internal and international struggles that create a mood of uncertainty and fear in the population, as well as disrupting its economic progress. Beginning with the restructuring of Europe during and after the Napoleonic age, passing through the revolutions of 1848, the reunifications of Italy and Germany, the Franco-Prussian war of 1870–71, and a seemingly endless round of conflicts in most parts of the continent, and culminating in the cataclysm of World War I, the threats of military service, requisition, new taxes, new masters, and economic collapse hung heavily over thousands of villages, adding to the feeling of insecurity and heightening the migratory mentality still further.

Push factors by themselves, however, are rarely sufficient to cause migration on a truly impressive scale. No matter how bad things are, where can one go? Far from Europe, however, other events were taking place that helped to balance the equation by creating pull or attractive factors.

3. **The demand for agricultural labor.** Most significant among these factors was the rapidly increasing demand for agricultural labor that sprang up in many "Europeanized" parts of the world, most notably in North America. As the Americans pushed westward beyond the Appalachian frontier, especially in the years following the Civil War, the demand for settlers to provide stability, justify the rapidly expanding railroad network, and provide a stable economic base in the Midwest and Great Plains, stimulated Congress to pass such measures as the Homestead Acts, which generally provided virtually free land (in many cases a quarter-section or 160 acres) to any settler who would live on the land, farm it, and improve it over a period of time (often five years).

Many Americans from poor-farm regions such as New England seized this opportunity, but it was also literally the chance of a lifetime for millions of Europeans. Most Europeans feeling the winds of change at home were, of course, farmers, and were hemmed in by a combination of very small landholdings, heavy debt and tax burdens, and in many cases a social structure (including absentee landlordism and ethnic/religious discrimination) that had barely changed since the height of the feudal era. Small wonder that the prospect of a free, sizeable piece of what was already gaining a reputation as some of the world's very best land had a profound appeal to Europeans who knew farming, and wanted only to be left in peace to make a better living at it.

4. **Colonialism.** An additional pull factor was represented by the growth and expansion of the European colonial holdings themselves. The nineteenth century was the high tide of colonial activity in Africa and much of Asia, with some leftover pieces of the Americas as well; and the demand for people (or at least European people) to serve in the military, in colonial administration, or as settlers to "show the flag," was substantial. It was generally believed that there was greater scope for advancement and enrichment "out there" than in the more circumscribed societies of home; and these opportunities further reinforced many peoples' resolve to emigrate.

5. **The steamship.** With all of the elements for a massive outpouring of Europeans to nearly all corners of the world in place, a final question remained: How? Since most areas of immigration were literally overseas, ocean transport was a necessity, and here the development of the steamship represented a technological advance that tied the migratory process into a neat package.

Until the arrival of the steamship, ocean travel had two major drawbacks: uncertainty and a high price. By the early 1800s, the age of sail had culminated in the famous clipper ships, which were very fast, but bought speed at the price of internal space, and thus had a relatively small payload. The cost of passage was accordingly quite high, well beyond the price range of most of the average Europeans who were beginning seriously to contemplate emigration. Movement was generally restricted to the relatively wealthy few.

Additionally, there was a fair amount of doubt associated with ocean crossings under sail (especially at a level below the "top of the line" ships). Winds could blow ships far off course, or not blow at all; food and water supplies were marginal, and sanitary conditions could deteriorate rapidly, and there were few guarantees as to when or even where one might land. To veteran sailors, these were simply the minor inconveniences of the profession; but to an emigrating farm family, casting themselves and all their possessions upon an uncertain fate, they loomed very large indeed, and were in fact a major reason for second thoughts and decisions not to migrate after all.

The steamship was invented toward the end of the eighteenth century (Robert Fulton's first commercially successful model dated from 1807), and by the 1840s Samuel Cunard was offering transatlantic service in four ships that had been especially constructed for the trade. As the ships grew larger, the price of cheap fares fell dramatically, while the publication of sailing schedules and arrival dates made the whole enterprise of migration begin to seem feasible, if not actually pleasant. Steamship companies began opening booking offices in many cities and towns (and even villages) across Europe, the final inhibitions were overcome, and the rush was on.

Most European countries were strongly marked by the great waves of emigration. The time frame was not always the same, however, and it is conventional to recognize three major periods of migration: an early period (1820–70) dominated by large numbers from Great Britain (including Ireland) and Northern Europe; an intermediate period (1840–1890) when migration from the German lands was most important; and a later period (1890–1930) when the waves of migrants from Southern and Eastern Europe left—often literally—for greener pastures (see Figure 4.1).

It should be noted briefly that not all of the European countries joined in this mass exodus to the same degree. France, for example, from which the protestant Huguenots had fled two centuries earlier, counted only modest numbers of emigrants in the 1820–1930 period. The modern French remark with a smile that this was because anyone once familiar with France could never bear to leave; a more realistic explanation is that France has always been a country of generally low population density, enough land being available to meet the demands of an expanding agricultural population. Switzerland also contributed few migrants during this period, as did Spain and Portugal, both countries from which millions had left in earlier centuries.

PERCEPTION: IRELAND: A CLASSIC CASE-STUDY OF EMIGRATION

No European country contributed a greater proportion of its population to the great emigration of the nineteenth and early twentieth centuries than did Ireland. In fact, the Irish case has come to typify the migrations of this period, with their undertones of human suffering, desperate hope, profound social and economic change, and dramatic reorganization of the world's peoples.

Ireland in the 1840s was in reality a colony of Great Britain; the independent state of modern Ireland was not created until 1922. Several hundred prominent English families owned enormous estates in Ireland, and as absentee landlords were concerned primarily with the collection of rents rather than with Irish economic development. There was little industry and few cities or towns, and the almost entirely rural population had few prospects of advancement. Nonetheless, introduction of the white potato into Ireland (see agricultural discussion, Chapter 9) had provided a stable food base. Tenants grew grain to pay their rents and lived on potatoes combined with vegetables and a rare piece of meat. While hardly prosperous, a rapidly growing population was easily supported by available food resources, until disaster struck at the heart of this precarious economy in the 1840s.

The potato crop, basic food supply for virtually the entire Irish population, was attacked by a mysterious disease (it was eventually diagnosed—far too late to be of any help—as an airborne fungus) in 1844, when virtually the entire crop was lost. Following a meagre harvest in 1845, the crop failed completely again in 1846, and the entire economy collapsed. With no food income, no reserves, and no savings, the Irish tenants were forced to eat their grain crop to survive. Without grain, they were unable to pay their rents, and most were promptly evicted from their lands by the landlords. Thus, within a matter of a very few years, a modestly successful economy had been ruined, the resource base was no longer adequate to support existing populations, the Irish lost most of their few material possessions, and starvation became a grim fact of daily life. Literally thousands perished.

Faced with this situation, Irish options were few. Rural-to-urban migration was not realistic, since there was such a poor urban structure. English relief efforts were halfhearted at best (imports of hard, dried maize from America were little appreciated) and the prospect of moving to England appealed to few among the Irish. Thus, as the new steamship services began to link such small Irish ports as Cork to the New World, millions of Irish scraped together their last remaining money, spent it on a steerage-class passage ticket, and regretfully turned their backs on a beloved but ungrateful homeland. It is certainly ironic that the Irish, who had been skillful and successful farmers, arrived in North America with no money and no possessions, and were thus largely unable to take advantage of the "free" lands to the west that were such a powerful attraction to all Europeans. The Irish literally had to make their living where they landed; thus developed the large and powerful Irish communities of New York, Boston, Philadelphia, Montreal, and other North American seaport towns.

At the time of the British census of 1841, the population of Ireland was estimated to be some 8.1 million. In 1996, about 4.2 million were living in the same space (3.6 million in the Republic of Ireland and an additional 600,000 in Northern Ireland or Ulster—still a part of Great Britain). Ireland is thus in the extraordinary position of having scarcely more than half the population today than was the case 150 years ago. No other nation in the world can make such a statement.

INTRA-CONTINENTAL MIGRATION

In addition to modern emigration, by which means European ideas and influence have expanded in many parts of the world, the population of the continent has also been substantially rearranged by various processes of migration involving movement from one European country to another. These movements may be considered basically as two types: resettlement and refugee movements, and labor migration.

1. Resettlement and refugee movements. During such periods of international warfare, civil wars, and general political tension as have been a virtually continuous feature of the European scene in recent centuries, local populations, generally with very little to say in the matter, have often been forceably displaced as a matter of state policy, or have been forced to flee their homes and native regions as a result of events entirely beyond their control. These movements are the sum of many often sad human dramas; collectively they serve to redraw the cultural map of Europe in various ways.

In the years during and after World War II, for example, the boundaries of Poland were changed dramatically and vast resettlement campaigns were carried out. When German military might was in control of this region, German settlers were established on what had been Polish lands, while Polish populations were forceably shifted farther to the east. After the war, when Soviet power was able to redraw the boundaries yet again, Germans fled to the west while Poles were also shifted westward (escaping those portions of their homeland that became Russian). It is estimated that some 8 million Germans and 2 million Poles were involved in these movements.

Earlier in the century, efforts to defuse a continually tense situation in southeastern Europe (where Balkan conflicts seemed to be continually breaking out, and where World War I had begun) led the major European powers to support in the 1920s a plan to resettle Greeks, Turks, and Bulgarians into their own national states, thus eliminating troublesome minorities in the other countries. Some 2 million people were affected by this rearrangment (1.2 million of them Greek), including families uprooted from homesteads that had been theirs for centuries. Countless other examples of resettlement could be cited.

Restless Europe: The Role of Migration

In addition to being moved at the whim of governments, many groups or individuals often find that the economic or political circumstances of their lives have altered for the worse to the point where a move, or even flight, may be indicated. These are refugees, and Europe's population contains a large number of modern communities whose origins are of this sort, much as in earlier centuries religious refugees did much to alter the European cultural map. Near the end of the Spanish civil war in 1938, for example, thousands of Spanish Republicans, fearing the reprisals of a fascist regime, fled across the Pyrenees to France, where many settled groups may still be found. More recently, refugees from tense political situations in Hungary (1956), Czechoslovakia (1962), and Poland (1983) have fled to Western Europe (and often beyond), bringing the variety of their skills and cultures to other lands. Even in 1989–90, harassment of ethnic Hungarians in Romania, and of ethnic Turks in Bulgaria continued the sad history of refugee migrations, while throughout the early 1990s the human disaster of Yugoslavia's violent disintegration forced many additional thousands of refugees to flee their traditional homelands. Moreover, as the rigid postwar frontiers of Eastern Europe have progressively crumbled, thousands of other Europeans, even though not politically threatened, have seized the opportunity to seek greater economic opportunity in the west.

2. Labor migration. Of all the forms of intracontinental migration, however, none has done more to alter the human landscapes of Europe in the late twentieth century than has labor migration, or the movement of millions of workers from southern Europe (and other countries of the Mediterranean fringe) to job opportunities

Figure 4.2 Opening of the frontiers in Eastern Europe (1989–90) led to large numbers of economic refugees.

in the north. This has been a migration of impressive numbers (perhaps as many as 10 million people have been involved), and has been for many regions and countless villages the "emigration" of modern times, much as movement to the opportunities of the New World was a century ago.

As the various European economies began to rebuild following World War II, it became clear that some were growing faster than others. West Germany, France, and Switzerland, for example, expanded very rapidly (even though Switzerland had not been a participant in the war, her economy stagnated nonetheless); while Italy, Greece, Yugoslavia, Turkey, Spain, and Portugal lagged behind. One immediate result of this uneven growth was the development of great disparities in employment: the northern countries, often with manpower depleted by the war, were unable to find enough hands to do the work, while unemployment remained very high in the south (nearly 4.5 million were unemployed or underemployed in Italy in 1952). Northern governments and individual industries began to actively recruit labor in the south, and another form of migration had begun.

The European labor migrations of the 1960s and 1970s have had both positive and negative effects. On the positive side, jobs were provided at good wages for millions of workers who would have had no such prospects in their own countries. These wages were the literal salvation of countless villages, where "remittance money" sent home by residents working in the north have not infrequently been the major source of economic support. Moreover, the budgets of many southern countries have been balanced, if at all, by the tax revenues and investment income earned by emigre workers abroad. Since most labor migrants intend eventually to return to their homeland, the money they earn is often invested at home; they may thus be building the future development of their country.

On the other hand, there is a certain dark side to this migration. Most of the migrants have been channeled into the types of low-paying, low-status jobs that citizens of the wealthier nations are increasingly unwilling to accept: street repair and cleaning, working on public transportation systems, unskilled jobs in construction and industry, and the like. This at once creates a second-class image for the migrant worker and also inhibits industry from modernizing into new labor-saving modes.

The influx of large numbers of linguistically and culturally different peoples into many cities of northern and central Europe has also raised levels of racial and cultural intolerance (many European nations have also welcomed immigrants from their former colonial holdings, which has compounded this problem). New ghettos of foreign workers have sprung up and questions of equal status and tolerance have been raised in many countries which once held the naive view that they were immune to such pressures. Also, as the economies of most European nations have stabilized in recent years, unemployment has become a nagging problem, and the question posed in 1960: "where can we get labor?" has become inverted by the1990s: "how can we induce these workers to go home?" As a prominent European statesman has remarked: "We sent for labor and received people;" and the complex, interrelated questions of assimilation, education, social security and other benefits, repatriation, and human rights have become vexing political problems in many countries.

Figure 4.3 Labor migrants in Europe have most often performed menial jobs: Street sweeper in Geneva, Switzerland.

In any event, one very visible result of this process has been the greatly increased diversity that today characterizes nearly every major European city. Formerly fairly stereotyped as to national ethnic types, the cities of most European nations, as they have grown rapidly over the past fifty years, have become increasingly cosmopolitan in their populations, cultures, and styles.

Thus, in recent years, and particularly since 1989, when dramatic events in eastern Europe and the then Soviet Union began to tear down rigid barriers, permit the exchange of ideas, and augment awareness of opportunities, Europe has again become a continent of immigration, while migratory flows within the continent have increased to levels rarely seen before. Annual net migration to Germany (then West Germany) in the early 1980s, for example, was about 3,000; but in 1989 the figure rose to a remarkable 378,000. At the same time, the relatively open borders of the southern European nations formed by the Mediterranean Sea offer a tempting invitation to potential economic migrants from North African countries, where rapid population growth is putting heavy pressure on economic resources.

Since the period of labor migration in the 1960–80 period (see discussion above), Europeans have generally become accustomed to the presence in their midst of modest minorities of differing languages and ethnicities. Larger numbers of immigrants, however, especially "non-European" ones, combined with the persistent recession and high unemployment rates characteristic of Europe during the early 1990s, have created a more dangerous political situation. Demonstrations (sometimes violent) and electoral success on the part of extremist anti-foreigner or even neo-fascist parties have become an unsettling part of the political landscape in Germany, France, Italy, and other countries. Faced with these pressures, most governments are now attempting to make more rigid demarcations between legitimate asylum seekers (nearly always admitted for humanitarian reasons), and economically motivated migrants who seek only to better themselves. As Europe's population inexorably ages, however, it may well be necessary to consider admitting higher numbers of working-age migrants simply to fill the jobs and pay the social security bills.

INTERNAL MIGRATION

A final type of migration that is serving to redistribute Europe's population on a dramatic scale is internal migration, or movement of population within a given country. Where economic opportunities are present, this sort of migration is generally favored, since even if they are forced to leave their ancestral homes because of various pressures, migrants are able to remain within a familiar linguistic and cultural context; leaving a poor village in the south of Italy for an industrial city in the north is clearly a less wrenching experience than leaving the country altogether, at least for most people.

By far the most prevalent type of internal migration has been the movement from rural areas to the cities of Europe. Towns have been an important feature of the European landscape for many centuries (see discussion of urbanization, Chapter 6), but they generally remained quite small: a place of ten thousand people was very large by the standards of the Middle Ages. Only since the impact of the industrial revolution, and the consequent destruction of countless rural livelihoods, has the rush to the cities taken on real dimensions. Many of the capital cities of Europe (London and Paris are good examples) grew at astonishing rates throughout the nineteenth and early twentieth centuries, eventually becoming so large as to be nearly unmanageable. In more recent years, emphasis has been placed on smaller regional centers or secondary cities within the orbit of the capitals; but the drift from the countryside remains the same in most cases.

In a continent where the landscapes have always seemed well organized and well tended, and where the human hand is everywhere in evidence, this is a disturbing trend. Land abandonment is growing in many areas, and the neatly manicured countryside is beginning to look a bit dog-eared in many countries. There is of course a countervailing migratory trend, in which retirees, people whose livelihood is not tied to a specific place, and those who can no longer support the pressures of modern urban life return to a rural lifestyle; but thus far their numbers do not compensate for departures toward the glitter of the cities.

As a related phenomenon, migration away from mountainous areas is also continuing at a steady rate. It is literally true that the population of Europe is now living at a lower average elevation than at any time in recent history! This pattern is obviously related to the more general drift toward cities, as economic prospects in highland regions are increasingly unattractive to many residents, as they are to rural folk in general.

In some countries, land abandonment at high altitudes is not considered a serious problem, and may in fact make available more land for recreation or wilderness areas, seriously deficient in most parts of Europe. In Switzerland, however, where the dairy products of high mountain pastures are a critical part of the national food supply (see discussion of the Alpine economy, Chapter 8), the decline of mountain populations is viewed with alarm. To attempt to stem the tide and keep farmers on the land, the Swiss government offers low-cost farm improvement loans based on altitude. The higher your farm, the more attractive terms for a loan you are able to negotiate.

In some cases, internal migration may be the end result of a longer process of international movement. Residents of remote villages in Greece, Turkey, or southern Italy, for example, might have participated in the labor migrations to Switzerland or Germany, saving their money and investing it in an urban small business or apartment building at home. At the conclusion of their labor contract, they may, on returning to the homeland, choose to live in the city where their investment has been made, rather than in the village. They thus become in a sense rural–urban migrants after an intermediate period of profitable exile.

ADDITIONAL READING

BOHNING, W.R., *Studies in International Labour Migration.* New York: St. Martin's, 1984.

HANSEN, M.L., *The Atlantic Migration, 1607-1860.* Cambridge, MA: Harvard University, 1940.

KING, R. *A New Geography of European Migration.* New York: Wiley, 1994.

OGDEN, P.E., and P. WHITE, eds., *Migrants in Modern France.* London: Unwin-Hyman, 1989.

SALT, J. and H. CLOUT, eds., *Migration in Post-war Europe.* Oxford: Oxford University Press, 1976.

WOODHAM-SMITH, C., *The Great Hunger: Ireland 1845-1849.* New York: Harper and Row, 1962.

DISCUSSION QUESTIONS

1. What were the push and pull factors that triggered the heavy waves of European emigration in the nineteenth and early twentieth centuries? What particular role did the steamship play in this process?
2. As a result of emigration, Ireland is arguably the world's only nation with a smaller population in comparable space today than in 1840. How can you account for this? What became of the Irish who emigrated to America?
3. "We sent for labor and received people." What does this mean for European societies in the context of the 1950–80 migration of labor within Europe?
4. Europe's population is getting lower in elevation. What are the factors that cause people to leave the mountains? What implications does this type of movement have for the European landscape?
5. What is the difference between refugee and resettlement migrations? Give some examples of both types from recent European history.

5
European Language Patterns and Religious Influences

Of the various cultural elements which serve to distinguish one human group from another, language and religion are among the most basic and deeply rooted. This chapter examines these factors and describes how they have profoundly marked the modern cultural landscape of Europe.

LANGUAGES

In North America, where relatively rapid European settlement and later the "melting pot" philosophy led to an essentially unilingual English-speaking continent, with substantial minorities only of French and Spanish speakers and a largely vanished group of Native American tongues, language plays a relatively small role in distinguishing one people from another. In addition, Americans are generally poor linguists, there being little incentive and few real opportunities to learn any other language in a real-world context. Since our decisions about people are thus made on other bases, we tend to have a rather insular view of language, seeing differences in accent or dialect as merely amusing characteristics of New Englanders or Southerners.

Such linguistic uniformity is far from being the case in much of the rest of the world. In those continents where human occupance stretches back for many thousands of years (that is Europe, Asia, and Africa), human groups often evolved complex and quite different cultures from other groups located only a few miles away. Even though starting from some common linguistic root, centuries of isolation and independent development meant that languages often became quite specific to a very small region, and were largely incomprehensible to groups outside. In modern times, although advances in transportation and communication technology have broken down many barriers and the development of a world language is at least a technical possibility, much of the world clings to language particularities as a sort of cultural badge: language becomes an expression of personal identity, and in many regions of many countries the language you speak says a great deal about you and your beliefs.

This pattern holds particularly true in Europe, where the language structure remains extraordinarily complicated even at the end of the twentieth century. Before reviewing the European language map, however, it might be of interest to consider why languages are more or less successful in the world; why some seem to become more widely spoken over time, while others seem to wither away, or to become only the preserve of academics, with no echo in the real world.

Factors of Language Competition

1. Usefulness. To be successful in the world, any language must have the quality of utility: that is, it must enable people to communicate to some useful purpose. The languages of everyday life, rather than those of a purely formal or literary nature, are those that survive and grow, whether or not they do much to contribute to the sum of knowledge. Latin, for example, is an "important" language because of its literature and the philosophies which it was used to express. It is also, however, "dead," in the sense that it is rarely spoken in modern day-to-day situations.

Particularly useful languages—those that convey ideas among large numbers of people for significant purposes—are often called marketplace languages. The term refers to one of the most basic human activities: coming together in a central location with a variety of different peoples to buy, sell, or exchange merchandise. In such a situation, some sort of common language will often be "invented" to get the job done, and many of the world's largest languages today (in terms of numbers of speakers) had their origins in precisely this way: Hindi and Bahasa Indonesia are two such tongues in Asia.

At a different level, we find that English—arguably the world's most important present language—owes at least some of its significance to being the marketplace language of science and technology. Since English is the natural language of many of the world's most technologically advanced societies, and since it is a language that copes relatively easily with new ideas, it has become the preferred language at the cutting edge of most fields of endeavor. Let us suppose, for example, that a Norwegian scientist develops a guaranteed cure for the common cold. If the findings are published in Norwegian, the total audience is limited to the 4 million men, women, and children who are natural Norwegian speakers, plus perhaps the same number of expatriates and foreigners who have learned the tongue, of which the number of scientists and others capable of interpreting the findings probably number only a few thousand. Result: obscurity. If our scientist chooses to publish in English, on the other hand, the audience will consist of not only the Norwegian scientific community (nearly all of whom can deal with English), but virtually the entire worldwide community of scientists and scholars. Result: possibly the Nobel prize!

2. **Administrative or "official" status.** In order to be successful—that is, to compete well with other tongues—it is important for languages to represent some authority: that is, to be the visible representation of power and influence. Languages that are the medium of decision, through which the rulers speak, have a powerful appeal not only because of their value in daily life, but also because they seem to offer one way of coming closer to, and perhaps even sharing, the authority they represent. As the official language of the Roman Empire, and later of the Christian church, Latin had enormous authority, and millions of Europeans struggled to learn it or modify it to their own needs more because of the power it represented than for any objective advantage of the language itself.

The most important way in which any language of authority is communicated to the people is through the process of education, and it is therefore scarcely surprising to find that one of the universal demands of linguistic minorities is to have at least some school instruction in their own tongue. Central authorities, on the other hand, see a universal language as important to the national educational process, in that it inevitably tends to bind all elements of a society more closely together. The reason for the importance placed on education is clear: children are hard to fool! The language of school instruction is quite clearly the language of their future: jobs, advancement, and even the glamour of some far-off capital are clearly linked to the language of power, while the language of the home may be seen as outdated and without influence, good only for a few stories from the grandparents.

In addition to demands for the instruction of schoolchildren (and eventually university students) in minority tongues, those concerned with the decline of local cultures also seek some control over the language of the media. Newspapers and other printed matter in countless languages abound, but the key is access to the more modern forms of popular communication: radio and especially television. Since these have until quite recently been close to state monopolies in many European countries, gaining a few hours per week of regional broadcast time in a minority language is a difficult process, and is thus considered very important by linguistic minorities where some arrangement has been worked out.

3. **Internal structure.** In addition to usefulness and authority, languages also compete with each other on the basis of their relative ability to cope with new situations or to express ideas more easily. To some extent, the success of one language and the decline or even disappearance of another can be traced to the structures of the languages themselves. While this is an entire field of knowledge in its own right, a few generalities may give some useful insights as to why some European languages have become international while others remain a mystery to all but a few speakers.

For example, English is certainly the most successful of European languages, and is by now the most widely spoken tongue in the world (although in terms of sheer numbers of speakers it may be exceeded by Han Chinese and Hindi). This stems in part, as we have noted, from the role of English as a marketplace language of science and technology; but it also relates to the fact that the structure of the language permits it to add new words relatively easily: whatever happens in the world, from a new musical style to a new development in space, English has a word (or more likely several words) for it. Many of the words thus coined ultimately find their way into the dictionary as permanent additions to the language. English thus has an enormous vocabulary (one characteristic that makes it a difficult language to learn) and can easily find simple, succinct ways of expressing almost any new idea.

For English speakers, it is difficult to realize what an advantage this gives their language over other European tongues, in which expression of the same new ideas is certainly possible, but with a result that is often much longer and more cumbersome. One result of this is that English words for a countless variety of twentieth century ideas and technologies have entered into nearly all other European languages, often to the dismay of linguistic purists who see the creeping advance of English as an attack on their traditions and cultures. To some extent, the perceived glamour of the English and/or American life style helps push the English language along; but it is often merely easier to use the simple English word or phrase than to struggle with the equivalent in a local tongue. The French, proud of their own language, are particularly upset by this phenomenon and have gone so far as to pass legislation mandating the use of proper French in all government documents and any other media over which control can be exercised. Language is a powerful force, however, and these efforts have met with little success: the average French speaker will still say "weekend" more often than "fin de semaine" or "hotdog" rather than "saucisson chaud."

French, however, far from losing ground, is another vital international language of long standing. In this case, the significance of the language stems from the complex but precise verb structure which is so characteristic of French (and so difficult to learn). The ability of a French speaker to subtly change a mood or slightly modify an idea by a slight alteration in verb usage has long made French the classic language of international diplomacy, where finely shaded meanings may have great importance.

4. **Rallying points for autonomous movements.** Finally, many minor European languages have survived and even prospered to the extent that they have become identified with a region or a group seeking greater autonomy within (or even independence from) a larger country. In these cases, language becomes the rallying point, the most visible expression of all the cultural differences that may set a certain people apart; the language spoken tells much about the person.

Small though they may be, most European countries over the centuries have seen the rise of strong, centralizing governments which tended to stamp out regional differences and minority aspirations (and languages) in an effort to promote national cohesion. France, Spain, and Great Britain have been notable examples of this trend, with the result that they now contain a number of restless minorities who, while adding a great deal of cultural diversity to these countries, have been increasingly strident in recent years in their calls for some greater degree of self-determination. Since many of these regions are around the edges, rather than at the heart, of their respective countries, they often tend to be economically disfavored with respect to better located regions. To the extent that nationalist minority groups are able to associate systematic economic shortchanging on the part of the national government with an attempt to deny freedom of cultural expression (as represented particularly by language), a truly explosive situation can be created. Thus in regions such as Scotland and Wales (Britain), Brittany and Corsica (France), and Catalonia and the Basque Provinces (Spain), political demonstrations and even widespread acts of violence have been seen throughout the late twentieth century, with the banners of minority language flying boldly in each case. It should be noted that these pressures have produced some positive results, as the nations struggle to accommodate regional aspirations without letting the entire fabric of their country come unravelled. Local parliaments have been established in several regions of Spain, greater regional autonomy is now a reality in both France and Britain, and minority languages are making at least modest headway in the schools and on the television screens.

The European Language Map

Reviewing the generalized map of European language types, we are again reminded of the complexity of the pattern. Most of the languages spoken in Europe today are derived from the ancient family of tongues known as Indo-European (the name indicating that the group also includes some of the many languages of India). As it moved into Europe with early arrivals from the East, cultural isolation gradually divided the Indo-European family into several recognizable groups, and ultimately into the dozens of individual modern and extinct languages.

  1. **Germanic languages.** The Germanic languages are essentially those of the northern and central parts of western Europe. Their extent is related to the movements of such tribes as the Goths and the Saxons, who brought a form of the tongue to Britain, where it slowly evolved into modern English. This group also includes the various Scandinavian languages, which are closely related, as well as Dutch and Flemish, the language of northern Belgium. It is interesting to note that, in Europe (excluding Russia), German is the language most widely spoken as a native tongue. This is because it is the dominant language of Germany, Austria, and much of Switzerland, and is also an important minority language in Belgium, Poland, Italy, the Czech Republic, and several other countries. If we consider second and third languages, however (that is, those that have been learned

Figure 5.1 Separatist movements often rally around distinctive languages: Ortigueira in the Galician region of Spain.

Figure 5.2 Europe: language groups and major languages
(Numbers of primary speakers, in millions, shown in parentheses)

- I. Indo-European Family
  - A. Germanic group
    1. German (100)
    2. English (60)
    3. Dutch (20) (Flemish, spoken in N. Belgium, is related).
    4. Scandinavian languages (Danish, Swedish, Norwegian). (Latvian and Lithuanian are also related.)
  - B. Romance group
    1. French (70)
    2. Italian (60)
    3. Spanish (38)
    4. Romanian (25) (also the language of much of Moldova)
    5. Portuguese (10)
    6. Romansch, Ladin, Provencal, Catalan, many other remnants.
  - C. Slavonic group
    (Russian and Ukrainian are also in this group.)
    1. Polish (40)
    2. Serbo-Croatian (18)
    3. Czech
    4. Slovene
    5. Bulgarian
    6. Several other minor languages with few speakers
  - D. Celtic group
    1. Gaelic (Scotch and Irish forms)
    2. Welsh
    3. Breton
    4. Several extinct forms, including Cornish and Manx
  - E. Greek
  - F. Albanian
- II. Ural-Altaic Family
  - A. Finno-Ugrian group
    1. Finnish (Estonian and Karelian are related.)
    2. Magyar (Hungarian)
- III. Semitic Family: Maltese (Arabic is widely spoken in France.)
- IV. Turkic Family: Turkish
- V. Basque (obscure language of the Spanish and French Pyrenees)

Immigrants to many countries from various parts of the world add great variety to the European language mix.

beyond the native tongue), English becomes by far the most widely spoken. Because of its usefulness and importance in the world (see discussion above), English is taught as the preferred second language in the school systems of most European nations.

2. Romance languages. The second major group of European languages is derived from Latin, and their modern extent closely reflects the spread of Roman influence and its persistence throughout several centuries. These are essentially the tongues of southern Europe, since although there are numerous Latin-derived words in all of the Germanic languages, the Romans did not subjugate Germany or England

European Language Patterns and Religious Influences

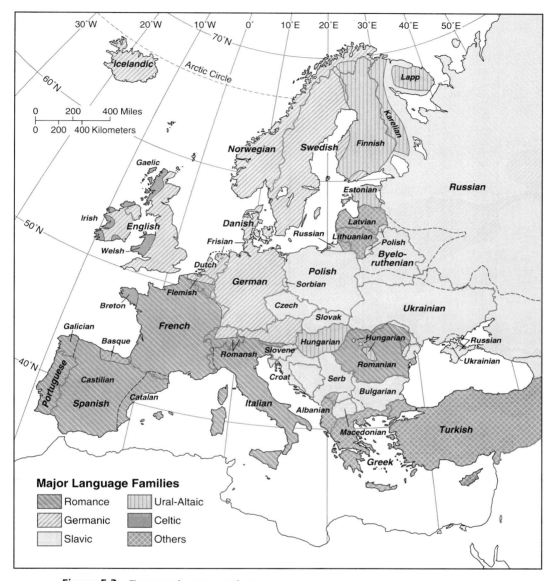

Figure 5.3 European language regions.

for a long enough period to ensure the supremacy of Latin. French, Italian, Spanish, Portuguese, and Romanian (reflecting the long Roman "watch on the Danube") are the principal Romance tongues, but the collapse of Roman authority some 1500 years ago left numerous isolated pockets of Latin speakers whose language subsequently evolved into quite unique patterns of speech. No other European language group has such a large number of small, remnant splinters of distinctive language scattered across the landscape. The Provençal, Langue d'Oc, Corsican, and Occitan of France, Catalonian, Galician, and Valencian of Spain, Romansch and Friulian of the Alps, and numerous others remain small pieces of the European language puzzle.

    3.   **Slavonic languages.** Throughout eastern Europe, the historic invasions, clashes of cultures, and fragmentation of territory that have made the region such a classic "shatter belt" have also resulted in a proliferation of languages. Predictably, most of these tongues have a strong Slavic component, some (such as Bulgar) even utilizing the Cyrillic alphabet more familiar in Russian. Apart from Russian and Ukrainian, Polish is the only language of eastern Europe with a substantial number of speakers. Serbo-Croatian, itself a composite, is only one of several languages spoken in the various Yugoslav states.

    4.   **Celtic languages.** The Celtic tongues derive from waves of migrants that arrived in Europe very early in history. They are thus substantially older than, and quite different from, the other major groups. Celts were once widespread across central Europe, but successive waves of subsequent invasion from the east pushed the Celtic speakers steadily westward, so that they and their languages survive today only on the western fringes of the continent. Irish Gaelic is an official language of the Irish Republic, while the Scottish form is widely spoken in the highlands of Scotland. In Wales, the language is a major rallying point focusing movements for greater autonomy, while such variations as Cornish and Manx have become virtually extinct in the last century, although they survive in the academic world. All of these varieties of Celtic speech compete more or less unsuccessfully with English in various parts of Great Britain and Ireland. The other form, Breton, struggles against the domination of French in Brittany, where it was brought back to the continent by Celtic groups fleeing from Cornwall in the fifth and seventh centuries.

    5.   **Other Indo-European languages.** This large linguistic group has other representatives in Europe, but the influence of these tongues rarely crosses national boundaries. Modern Greek, for example, is widely spoken only in Greece and among the large Greek expatriate community around the world. Similarly, Albanian is largely confined to its own small country along the Adriatic coast, although there is a substantial Albanian-speaking population in the Yugoslavian province of Kosovo where, as a minority language within the larger nation, it represents a focal point for dissent and demands for better economic treatment.

    6.   **Ural-Altaic languages in Europe.** Another large family of languages is known as the Ural-Altaic. Numerous tribes and groups of the northern and eastern parts of Russia speak various languages of this family, but there are only two important representatives in Europe. To the north, Finnish is a distinctive Ural-Altaic tongue that is totally unrelated to the speech of any of that country's Scandinavian neighbors. It is largely on the basis of this language difference, along with the distinctive cultural and literary heritage that accompany it, that Finland is distinguished as the "non-Scandinavian" part of northern Europe. The Russian region of Karelia, across the Finnish frontier, and the Baltic republic of Estonia are homes to related languages.

    Farther south, the arrival of the Magyars from the east in the ninth century brought a distinctive Ural-Altaic language to what is now modern Hungary. Despite the difficulties that history and geography have put in their path, the Magyars have been able to remain a cohesive cultural group and to retain occupance (although not always political control) of the middle valley of the Danube: a very different people than their Germanic or Slavonic-speaking neighbors. It should be noted that modern linguistic research has cast some doubt on the long-presumed close relationship between Finnish and Magyar. Nonetheless, they are still probably more closely related to each other than to the other languages of the continent.

   7. **Middle Eastern languages.** Europe is also the site of minor representation of some of the large language families that are much more important in the Middle East or North Africa. The Semitic tongues are represented by Maltese, spoken on that small island in the Eastern Mediterranean, while Turkic is spoken in that part of Turkey west of the Bosphorous (and therefore sometimes considered a geographical part of Europe); and is also an important minority language in Greece, Bulgaria, and on the divided island of Cyprus. Beyond these rather formal examples, the influence of Arabic and other Middle Eastern languages can be clearly seen in those areas which were occupied for many centuries, particularly Iberia and southeast Europe, where countless place names and everyday phrases betray this origin. In addition, it is worth noting that the recent years have seen heavy migration of Middle East language speakers to many countries of central and western Europe (see discussion of migration, Chapter 4). Thus Germany has a large Turkish minority population as "guest workers," while France has welcomed substantial numbers of Arab-speaking workers from its former North African colonies: Tunisia, Morocco, and especially Algeria.

   8. **Basque.** One of the world's most unusual and puzzling languages is Basque. On the list of European tongues, it is considered in its own "family"—fully equivalent to Indo-European—yet it is spoken on a daily basis by probably no more than a million people. Basque is the language of a population who inhabit the valleys of the western Pyrenees between Spain and France, and who have expanded to become an important part of the population in four provinces of northern Spain and three adjacent regions of France. It is a tongue whose structure and relationships to other languages defy analysis by even the most sophisticated means now available to students of linguistics, the most common judgement heard being that it must somehow predate even the Indo-European tongues in its arrival in Europe (a fairly unconvincing argument at best). The Basques (who have given the world, among other things, the beret and the game of jai-alai) made a traditional living largely by herding sheep in the Pyrenees, but many of them have today become industrial workers in such Spanish cities as San Sebastian and Bilbao. If the modern world has heard of the Basques, however, it is largely a result of their strong, language-based demands for greater autonomy within Spain or even, for a sizeable minority, total independence. Although most Basques seek to achieve these goals within the political framework of a democratic society, there is a violent fringe (known as E.N.A.) that pursues its ends through bombings and assassinations, which draw all the wrong kind of attention to the Basques and their unique language and culture.

PERCEPTION: ACCOMMODATION IN SWITZERLAND; STRESS IN BELGIUM

As has been noted, most European countries contain numerous linguistic minorities whose feelings of cultural cohesiveness beyond the national framework may pose substantial problems of a political and economic nature. To emphasize this complexity, let us consider two specific European nations with very complicated language patterns, one of which has generally been able to deal with the stresses without endangering the national structure, the other of which has been much less successful, and in fact periodically seems on the verge of partition along purely linguistic lines.

The first of these nations, and perhaps the world's outstanding example of a successful multilingual country, is Switzerland. This small Alpine nation is surrounded by larger countries with strong languages and distinctive cultures: Germany and Austria, France, and Italy. Reflecting this orientation, Switzerland has major groups speaking each of these tongues, and the genius of the Swiss system (as in so many other things) has been to allow local regions to make their own decisions about language; there is no single national language.

In the northern and eastern parts of the country, some 65 percent of the population speak German (although a form quite distinct from that found in Germany or Austria), and this is the official language in thirteen of Switzerland's twenty-two cantons, or states. About 18 percent of the population, mainly in the west, are French speakers, and French is now the official language of six cantons. To the south, 12 percent of the Swiss speak Italian, and the canton of Ticino has this as its official tongue.

The most interesting situation of all exists in the eastern canton of Grisons, where Romansch, a minor Latin-based language that was rapidly disappearing

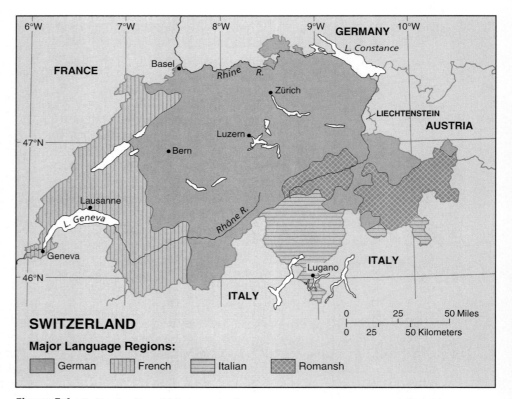

Figure 5.4 Switzerland's major language regions.

in competition with German, was saved from extinction in 1938 by being given official status (note the importance of administrative authority) within its canton. This is probably the most outstanding example of the deep Swiss concern for the rights of individual cantons . It is doubtful that more than a few hundred people speak Romansch to the exclusion of other languages (most would also speak German), but the people of the Grisons are proud to identify with this ancient culture and happy to have their language recognized as important and well-suited to their needs.

Far from being a static situation, the language patterns of Switzerland are constantly changing, and new sensitivities are always arising. As recently as 1978 an entirely new canton, Jura, was created to satisfy the demands of a French-speaking group who felt themselves economically and culturally submerged in the large German-speaking canton of Bern. In another instance, large-scale Italian labor immigration to jobs in the German- and French-speaking regions was severely limited in the 1970s in an effort to reduce linguistic and cultural tensions, even though the economy was still growing rapidly and could have absorbed many additional workers. Another way in which the Swiss ensure relative linguistic harmony is by insisting that school children in all parts of the country learn at least one of the other major languages starting with the earliest years of their education. Finally, all of the documents issued by the national government must be published in German, French, and Italian (publication in Romansch is only required in the canton of Grisons), thus ensuring that speakers of each language feel equally important.

The second example of a European country with deep linguistic divisions, and one which has been much less successful in dealing with the problems posed by this, is Belgium. Belgium as a nation is a relatively recent creation, having been pieced together as an independent state in 1830. As a result, two quite different groups of people suddenly found themselves coexisting in the same country.

South and east of a line generally dividing the nation in half, the people are French speakers known as Walloons. To the north and west, the language and culture are Flemish, a Germanic tongue related to Dutch. Perhaps the most interesting lesson of the language conflict in Belgium is the way in which language, serving as the "flag" or focal point for a people, masks deep differences which are really at the heart of the problem. While generalities are risky, of course, it is possible to characterize the Walloons as having a low birth rate, being largely Protestant, working mainly in industry, and voting socialist. The Flemish, on the other hand, have exactly opposite characteristics: their birth rate is higher, and they have traditionally been Catholic in their religion, agricultural in their employment, and conservative in their politics. Language differences are thus in reality merely the tip of an iceberg of cultural and economic disparities.

For most of modern Belgian history, the balance of economic power in the nation lay in the south, where its coal mines and steel mills made Belgium a surprisingly important industrial power. Thus the Walloons, who during the 1800s had a slight majority of the country's population, controlled the economic and therefore the political levers of power within the nation. In recent years, however, the balance has shifted significantly in favor of the Flemish region. The decline of the coal mining and steel industries in Wallonia, combined with the rapid

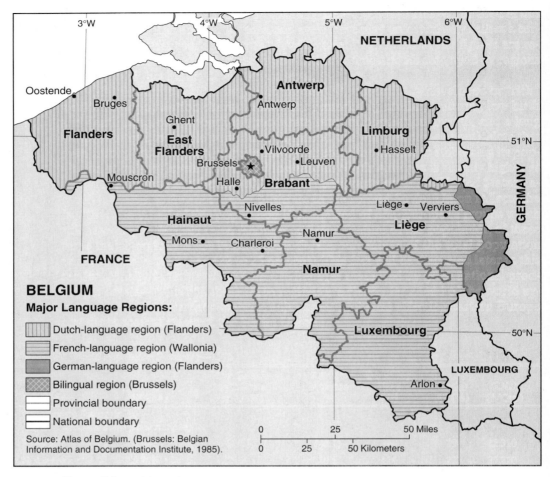

Figure 5.5 Belgium's language structure.

development of the Flemish port of Antwerp and the installation of more modern industries in the north, has changed the economic structure of Belgium significantly. At the same time, somewhat higher birth rates have now made the Flemish a slight majority of the nation's population. Based on these developments, the Flemish are demanding a greater share in government and national decision-making, while the Walloons are struggling to hold on to their traditional power.

Highly visible, language is pushed to the forefront of the struggle, and the efforts of each group to preserve their rights results in some bizarre patterns in the Belgian cultural landscape. Children living on either side of the language line (a very clearly defined division) must be educated in the language of that side. National political parties are divided into two sections: one for each language region. Even the ancient and prestigious University of Louvain (Leuven), which is divided by the line, is forced to duplicate all its colleges and programs so that both languages are equally accommodated. Protests and rioting over the language issue (or rather over language as a symptom of underlying differences) have been commonplace, and more than one Belgian government has been forced to resign as a result of inability to deal with language tensions or to forge coherent programs that would be acceptable to both sides. Moreover, a small German-speaking minority in eastern Belgium has also been very vocal, making it clear that their rights must also be considered.

One bright spot in a country whose language divisions seem at times serious enough to threaten the nation's integrity has been the rapid growth of Brussels, and the capital's emergence as one of the most international of European cities. Headquarters of the North Atlantic Treaty Organization (NATO) and many of the important institutions of the European Union, as well as countless multinational corporations, Brussels has little time to engage in petty squabbling about language. In fact, multilinguism is quite common in Brussels, and it is possible that some means of giving both groups a sense of meaningful participation in national affairs may be emerging. It has even been suggested that a totally new sort of Belgian language may very slowly be evolving in the Brussels region. As of 1994, the country has been divided into three semi-autonomous regions (the two major language areas plus Brussels). Whether this will produce sufficient independence of action to satisfy Flemish and Walloon aspirations, or whether the nation will be entirely fragmented, remains to be seen.

ADDITONAL READING

COLLINS, R., *The Basques.* Oxford: Basil Blackwell, 1986.

DUGDALE, J.S., *The Linguistic Map of Europe.* London: Hutchinson, 1969.

HELIAS, P-J., *The Horse of Pride: Life in a Breton Village.* Tr. June Guicharnaud. New Haven, CT: Yale University, 1978.

MURPHY, A.B., *The Regional Dynamics of Language Differentiation in Belgium.* Chicago: University of Chicago, 1988.

DISCUSSION QUESTIONS

1. Despite being the native tongue of only about 60 million Europeans, English has become the continent's most widely spoken language. Why? What advantages does English have to make it so popular?
2. Switzerland is able to manage a complex language pattern in relative harmony, while in Belgium linguistic stress threatens the stability of the nation. What are the differences between the two societies that lead to this situation?
3. Language is often the "flag" for nationalist and separatist movements. Why is this true? Give some examples of linguistic minorities in Europe that have been agitating for greater recognition.
4. Although Finland and Hungary are widely separated, their languages are closer to each other than to other European tongues. What is the reason for this?
5. Who are the Basques? What theory can you suggest that might account for the unique character of their language?

RELIGIOUS INFLUENCES

Among the cultural institutions which have strongly marked the character of the European region, religion certainly occupies a prominent place. Although in our time, religion plays only a modest role in the day-to-day lives of most citizens, almost no critical national decisions are made on the basis of religion, and religious tolerance (if not actual indifference) is the rule in all but a few small areas. It is well worth remembering that religion was for many centuries the central focus of both personal and national life. Some of the most brutal wars in European history were fought largely over religious issues, most of which seem trivial to us, or at least matters of theological debate rather than causes for outright warfare.

To students of geography in the late twentieth century, religion in Europe is perhaps most impressive in terms of the vast amount of monumental architecture, the persistent patterns of culture, and the peculiarities of political divisions that reflect the values of earlier ages, yet which have been maintained into modern times. The soaring Gothic cathedrals of the thirteenth and later centuries—certainly among Europe's best known buildings—the spectacular reminders of Muslim occupation in Spain and southeastern Europe, the puzzling division of certain Swiss cantons into two smaller sections: all reflect the importance of religion in the lives of earlier generations of Europeans. Therefore, although today religion may fairly be said to play no greater role in the lives of most Europeans than it does in the lives of most Americans, the European landscape reflects to a remarkable extent the fact that this was not always the case.

At the same time, one area of modern Europe—Northern Ireland or Ulster—is one of the few parts of the world where critical personal and national decisions are still made to a startling degree on the basis solely of religion (see following "perception" section). The vexing question of Ulster is one of the most persistent and dangerous in all of Europe, and serves perhaps as a window looking backward to earlier centuries, when almost everyone's entire life was narrowly circumscribed by the spiritual and temporal power of organized religion. A fair amount of the physical and cultural devastation currently afflicting the Yugoslav states is also based on religious intolerance. It is clearly worth examining some of these themes in more detail.

As a very basic generality, it is reasonable to say that Europe is a Christianized continent. Most of the triumphs and tragedies of the Christian faith have been played on the European stage, and it was from Europe that Christianity spread to the Americas and most other parts of the world; missionary zeal was always a major component of the colonial urge.

From its origins in the Middle East, the faith spread rapidly throughout nearly all of Europe, particularly when, through conversion of the Emperor, the Roman Empire became Christianized. The combination of the powerful appeal of a personal, optimistic religion and the organized might of Rome accounts for the great success of Christianity in rapidly displacing the animism and mythology that had been Europe's earlier religions. Only in a few folk traditions (such as the Christmas tree) are there any reminders today of the pre-Christian faiths that were largely swept away two thousand years ago.

This is clearly not the place for a detailed survey of the religious history of Europe, nor of a discussion of comparative religious philosophy. What interests us as

Figure 5.6 Religious affiliation in Europe*

Faith	Est. Numbers (thousands)
Christian	521,271
Roman Catholic	265,745
Protestant	83,277
Orthodox	131,510
Anglican	32,630
Other sects	8,109
Muslim	56,394
Jewish	3,448
All other religions	4,000
Atheist and Nonreligious	69,071

Source: Adapted from *Encyclopedia Britannica Book of the Year,* 1994.
*Includes the former Soviet Union.

geographers are the tangible pieces of evidence in the modern landscape that reflect twenty centuries of religion as a central focus of human activity. For example, in thousands of European cities and towns, some sort of religious building—church, cathedral, abbey—is at the very center of the community, and has clearly been the focal point around which the place has developed over the centuries (see discussion of urban growth, Chapter 6).

Upon the collapse of Roman authority, the Church remained as the only well-organized, disciplined institution over much of the continent, and thus played a major role in regulating the temporal, as well as spiritual, affairs of local populations. This meant that the church building itself became a major focus of activity not merely on Sundays and other religious dates, but on a daily basis.

The turbulent history of Christianity in Europe has also left abundant physical and human evidence. The twelfth century schism between East and West, for example, with both Rome and Constantinople claiming to represent the Church, led to a separation of ideas, ceremonies, and religious cultures that are still an important aspect of east–west differences in Europe. More significant geographically, and certainly one of the most critical events of European history, was the division of the western church into Catholic and Protestant groups. Martin Luther nailed his theses to the church door in 1517, launching Europe on nearly three centuries of reformation and counter-reformation, inquisition and persecution, warfare and destruction, all in the name of religion, which was still very much at the center of personal and national life.

There is no need to trace the course of these clashes in any detail, but several aspects of the modern European scene still reflect the bitterness and turbulence of the times. Economic historians have estimated, for example, that the energies and resources expended by Europeans fighting against each other in the name of religion retarded the economic development of the continent by at

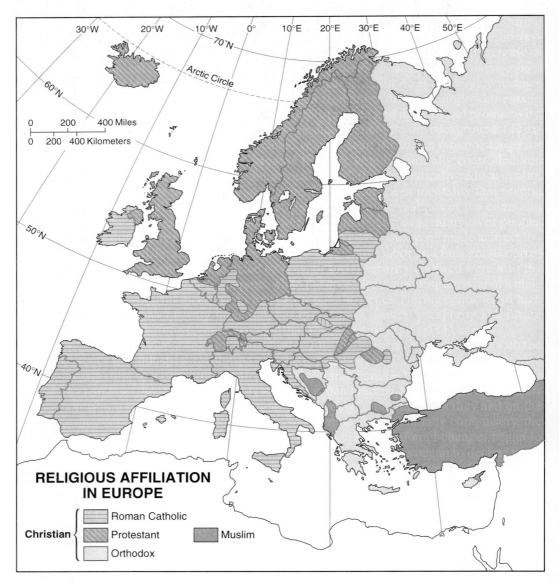

Figure 5.7 Europe: Religious strength.

least a century. At the same time, the mass migrations of various peoples to religiously "friendly" areas as Europe's nations became either Protestant or Catholic, enriched some countries while impoverishing others. The rise of Prussia as the most important state of North Germany was helped considerably by the influx of capital and skills represented by refugee French Protestants (Huguenots). Similarly, Amsterdam owed its status as one of Europe's most wealthy and successful cities at least partly to a policy of religious toleration.

European Language Patterns and Religious Influences

Figure 5.8 Religious faith often translated into dramatic architecture: Notre Dame cathedral, Paris, France.

The amount of senseless damage done to religious art and architecture during this time was also enormous, to our great loss today. The dissolution of religious orders in England by Henry VIII in the 1530s, for example, means that today one can only speculate among the ruins of most of the great abbeys and monasteries of that country. The modern political geography of Europe also reflects ancient religious stress in several instances. Among the cantons of Switzerland, which are already quite small, three are further subdivided into half-cantons, with distinctive flags and identities. This puzzling pattern is explained by old religious struggles: unable to agree on a single religion for the canton, its inhabitants chose to divide into Protestant and Catholic halves, rather than go to war about it!

Other religious influences have also played a major role in the development of Europe and have left visible traces on the modern landscape. Judaism spread into Europe along with Christianity and Jewish minorities soon began playing a major role in the commercial and cultural life of most countries. Because of differences in religious dogma (although Christianity and Judaism clearly spring from the same roots) and cultural patterns, Jews were generally viewed with deep suspicion by Christian Europe. Normally unable to acquire land, they became of necessity urban dwellers, and even in the more liberal atmosphere of cities their distinctiveness caused them to be set apart: the term "ghetto" probably comes from the Italian term for the districts of cities where Jews were required to live. For centuries, Europe fluctuated between grudging toleration (especially since successful Jewish merchants and bankers financed a significant share of Europe's commerce and even military campaigns) and outright persecution.

Culturally and philosophically different, often wealthy (or presumed to be so), and without a European homeland of their own, the Jews have unfortunately often been cast in the role of scapegoat when political or economic difficulties have beset one or several European countries.

Another major religion that has left its mark on the modern European scene is Islam, the most recent of the great modern religions and, like Christianity and Judaism, a product of the Middle East. Mohammed died in 632 by the Christian calendar (the Moslem calendar is lunar and is reckoned from the hegira or flight of the prophet to Medina in the Christian year 622), and less than two years later the first Islamic armies were sweeping out of Arabia to both east and west, beginning a campaign of religious and cultural conquest that for sheer speed is probably unequaled in world history. Only a century later, in 732, Islamic forces had swept across North Africa, conquered nearly all of Iberia, crossed the Pyrenees, and been narrowly defeated by the Franks at Poitiers, in Central France. Islamic authority remained in parts of Iberia for more than seven centuries.

In the turbulent history of southeastern Europe, Islamic influence arrived somewhat later. It was in the fifteenth century that the Muslim Ottoman Empire, which had gradually consolidated its hold over much of the Middle East, took control over most of what is today Greece, Serbia, Bosnia, Macedonia, Montenegro, Albania, Bulgaria, Romania, and southern Hungary, retaining authority in some of these regions until the collapse of the Empire in 1918.

The methods and objectives of Muslim control varied considerably between east and west. In Spain, the Moorish governors acted relatively independently, and were much concerned with progress and development. Universities, irrigated agriculture, and numerous public works projects made the Muslim period—despite tenacious Spanish resistance that eventually drove out the Moors in 1492—one of relative prosperity. In the east, on the other hand, Ottoman concern was mainly with keeping order and collecting taxes. Local development, notably of industry, was never favored, and some authorities ascribe the relative underdevelopment of southeastern Europe today to the long centuries spent under unenlightened Turkish rule.

From the standpoint of modern geography, the Muslim period in Europe has left a wealth of tangible reminders. Many mosques and minarets, often converted to other uses, are still seen, while the elaborate palaces and gardens of Grenada, Cordoba, and other Spanish cities are among the architectural wonders of Europe. Countless place names and local customs reflect Muslim origins, while at least one country, Albania, is still nominally Muslim (despite the efforts of a harsh communist government between World War II and 1990 to reduce the significance of religion). The Kosovo region of former Yugoslavia is peopled mainly by Muslim Albanians and is a source of great concern to the controlling Serbian government; while the large Muslim population of Bosnia has been among the most victimized groups of the ethnically-based civil wars that have devastated this region during the 1990s (see discussion of the Yugoslav states, Chapter 23). Additionally, the arrival of Islamic migrant workers (such as the Turks in Germany) and the movement of ex-colonials back to the former mother countries (such as the North Africans in France) have expanded the Muslim presence in many west European cities.

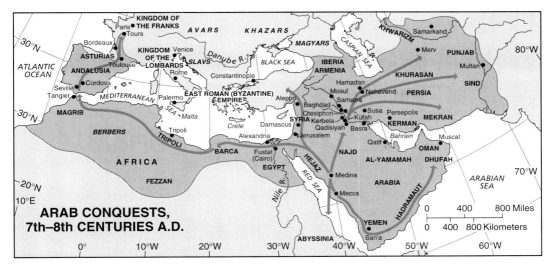
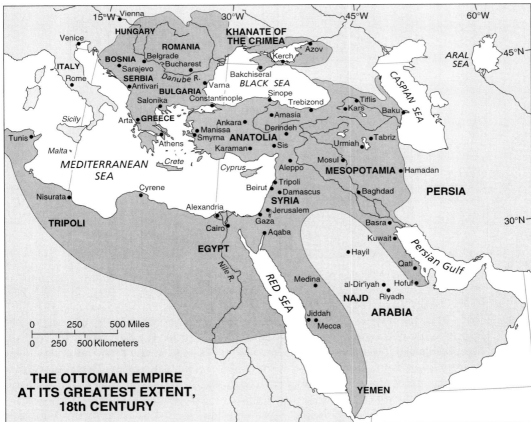

Figure 5.9 The spread of Islamic influence into Europe. The top map illustrates the dramatic expansion of Islam into North Africa and Iberia during the period 632–732 AD. The bottom map shows the later growth of the Turkish empire, which continued to control much of southeastern Europe into the 19th and even 20th centuries.

Figure 5.10 Reminders of Islam: Minaret in Pecs, Hungary.

Today, nearly every religion known to man is represented by at least a small group somewhere in Europe. Religious toleration is virtually universal, and the deeply felt religious passions that controlled the lives of nearly everyone and plunged Europe into centuries of irrational conflict have waned substantially. Religion—like politics, employment, family life, and even amusements—is simply another factor in the lives of most people. Even the Yugoslav conflict is essentially ethnically rather than narrowly religiously based. The exceptions to this rule are in Bosnia and in Northern Ireland, where the religious question is nearly always uppermost in the mind of everyone, where most of the basic decisions of everyday life are made with some reference to religion, where people still die as a result solely of religious beliefs, where the clock seems in effect to have been turned back several centuries.

PERCEPTION: THE TANGLED WEB OF NORTHERN IRELAND

Few European nations have as dark and troubled a history as Ireland. Following a brilliant period of flourishing art and literature based on the Christian church (Ireland was Christianized from the fifth century AD., and became a veritable mecca of learning and enlightenment during the early Middle Ages), the island was politically conquered by the Anglo-Normans at the time of Henry II of England (twelfth century). Down the centuries, punctuated by periods of doomed and despairing revolt, Ireland remained essentially a colony of a richer and more powerful England.

In the seventeenth century, however, a crucial change occurred in England which was later to lead directly to the modern Irish "troubles." Following the death of Elizabeth I (a Protestant), the throne reverted to James I of Scotland (a Catholic). Among the numerous ramifications of Catholic rule in England (four kings, divided by Cromwell's "glorious revolution"), was the emigration to Ireland of large numbers of Protestants, notably from Scotland. Most of these settled in the northern province of Ulster, and when England returned to Protestant rule under William and Mary of Orange, took part in a successful campaign to defeat the armies of the ousted Catholic King James II. Thus was created "orange" or Protestant Ulster, a community often seen by itself as an embattled culture in the midst of a Catholic-dominated island.

When England, finally weary of the "Irish question," at length decided on independence for Ireland in 1922, the question of the Protestant north became acute. Under terms of a referendum, six counties of the north were allowed to remain part of the United Kingdom, while the rest of the island became the independent Republic of Ireland (Eire). Bitterness over the resolution was expressed almost immediately, and has continued to be a central fact of Irish political life down to the present day.

Basically, the Ulster Protestants wish to remain part of a Protestant Britain, seeing in that association their best defense against potential Catholic cultural pressures should they be submerged in the larger nation. At the same time, most Irish Catholics have as their basic goal the reunification of the island under one government. The issue, however, is more complicated. In Protestant-dominated Ulster, the large Catholic minority has a long history of being systematically deprived of jobs, votes, and political influence in general. This leads to despair and frustration, and eventually to acts of defiance and terrorism, which are in turn countered with equal atrocities by Protestant extremists.

Of all of the countless problems facing modern Europe, the bitter question of Northern Ireland is (except for the disaster of Yugoslavia) one of the very few that routinely kills people, and is also one of the most intractable. The presence of British soldiers in a peacekeeping role pleases neither of the factions, while countless well-meant efforts to arrive at some agreeable political compromise have ended in failure. In a post-cold war Europe, where armed force is diminishing as a political option, and where human rights are gaining a new dignity, the attitudes and bitterness of the Irish question become increasingly anachronistic. This fact, however, so apparent to the outside observer, does not mean that the problems are any closer to resolution.

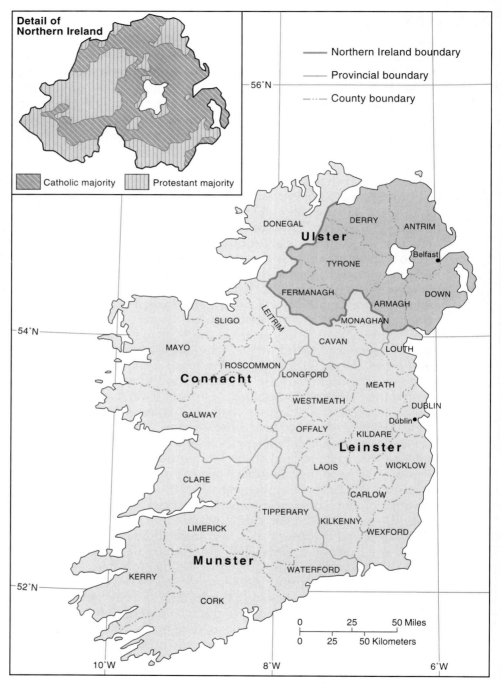

Figure 5.11 Ireland and its politico–religious divisions.

ADDITIONAL READING

ARCHER, J., ed., *Reader on Islam*. The Hague: Mouton, 1962.

DICKENS, A.G., *Reformation and Society in Sixteenth Century Europe*. London: Thames and Hudson, 1966.

DURANT, W. *The Story of Civilization: The Age of Faith*. vol. 4. New York: Simon and Schuster, 1950.

PRINGLE, D.G. *One Island, Two Nations? A Political Geographical Analysis of the National Conflict in Ireland*. Letchworth, Herts.: Research Studies Press, 1985.

SOPHER, D. *The Geography of Religions*. Englewood Cliffs NJ: Prentice-Hall, 1967.

DISCUSSION QUESTIONS

1. The period of Islamic occupation of Spain was relatively enlightened, while that of southeastern Europe offered little scope for development to local residents. Why was the pattern so different? What parts of Europe are largely Muslim today?
2. Why did Jewish populations in Europe become generally urban dwellers? What is a "ghetto"?
3. Northern Ireland (Ulster) is by far the most violent case of purely religious strife in Europe today. What are the origins of this conflict? What possible solutions can you suggest?
4. Europe essentially became a Christianized continent in the later centuries of Roman domination. Why did Christianity become the preferred religion?
5. Through the centuries, religious conflict in Europe took a dreadful toll of life, property, and economic advance. Who were some of the groups that became refugees during this period?

6
Settlement and Urbanization of Europe

Europe is one of the "old" continents, in the sense that human beings have been in the region in substantial numbers for a very long time (compared, for example, with the Americas). As a result, humanity has been leaving its mark on the European scene over an exceptional time frame, and the modern landscape of the continent clearly reflects in various ways the many centuries of human attention.

For example, as farming spread across Europe to replace a hunting/gathering economy some four thousand to five thousand years ago, various solutions to the problem of how the land should be divided and farmed occurred to people as knowledge lengthened and technology expanded. Originally, much of the land was farmed according to a simple "one-field" system, in which the settlement was surrounded by a zone of constantly cultivated space (the "infield") in which community members shared land of varying quality; and then by the pasture zone (the "outfield"), where animals were grazed.

By the Roman period, this primitive system had largely given way to the "three field" arrangement, which reflected increased understanding of the need to conserve soil quality. In this arrangement, the cropped land surrounding the village was divided into permanent holdings, which could ultimately be bought, sold, inherited, divided, taxed, and otherwise manipulated. Each farm holding was then subdivided into three parallel fields which ran from the village edge (for ease of access) to the end of the cropped surface. One of the fields was sown to the principal food grain (e.g., wheat), the second to an alternative food crop, while the third was left fallow in recognition of the fact that the land yields better when it is occasionally rested.

The result of this practice was a network of long, thin lots which spread across agricultural Europe and became a permanent feature of the landscape. Today, agriculture is vastly more scientific and productive, and the archaic three-field system has long vanished (although it persisted in parts of Russia into the twentieth century). The long, thin lots remain in many areas, however: a classic example of human impact on the landscape and its persistence through time.

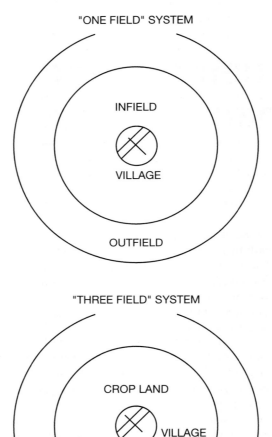

Figure 6.1 Evolution of field patterns in Europe. The communal "one field" system was the original pattern as agriculture spread into Europe some 6,000–8,000 years ago. Crop land (the "infield") was continuously devoted to food production, while the "outfield" was generally pasture land. By classical times (about 2,000 years ago), the "three-field" system had become general; this persisted in some parts of Europe until the early 20th century. Here, the crop land was individually owned, and farmers practiced a crude triennial rotation in a series of long, narrow fields extending away from the village. The "Common" was still pasturage for the community. Although modern agriculture has become much more sophisticated, the trace of "three field" farming is still visible in the landscape (see Figure 6.2).

DISPERSION VERSUS AGGLOMERATION

Apart from the way the land itself was divided, rural settlement in Europe took one of two forms: agglomerated or dispersed. Dispersed (or isolated) settlement, in which each farm or small cluster of farms is located at some distance from its neighbors, is uncommon in Europe. It has been linked to Celtic traditions, to some sparsely inhabited regions, and perhaps most convincingly to rare periods of history when peace and tranquility reigned in Europe and the cost of living in towns and villages (including especially the collection of taxes) was relatively high.

Much more widespread is the pattern of agglomerated or nucleated villages, with populations ranging from one hundred or fewer to several thousand, that spreads across most parts of Europe even today. These farm villages are a characteristic feature of the rural scene in nearly all European countries, and may of course group virtually the entire populations of large areas at only a few sites. The visitor to Europe, wandering through productive and well-tended acres of farmland, is apt to assume mistakenly that population density is very low, whereas it is in reality merely concentrated in the large village that suddenly appears around a bend in the road. Rather than living on their land, the farmers live in the village, "commuting" to their fields daily. Even livestock are driven from the village to their grazing lands and back each day, and it is by no means unusual for the motorist on back roads to become tangled in herds of animals both morning and afternoon!

Human groups tended to form agglomerated settlements for a variety of reasons, but the most significant was clearly defense. During the long centuries of its history, Europe has most often been unsettled, lawless, and downright dangerous, especially for the weak and isolated. Grouping together provided an obvious margin of safety, since many hands and many eyes gave even a fairly small village a certain sense of security. Farm villages, no matter how large and well protected, had no defense against large, well-equipped armies, but could put up a good front against the marauding bands and casual banditry that were common in Europe down to relatively recent times.

Defense could also be augmented by an appropriate choice of site. High ground, such as a hilltop, was favored, as were islands or any other physical location that would force invaders to fight at a disadvantage. If a manmade structure—a castle, manor house, fortified church, etc.—were constructed so as to supplement the natural advantage of site, a significant degree of protection could be assured. Thus in Europe today, literally thousands of places, including villages and cities that have grown quite large, are laid out around a hill upon which may be found the ruins or restoration of the fortress on which the local citizens pinned their hopes of survival for many centuries. Having a fortified hill right in the middle of town, of course, does very little to resolve modern problems of access, transport, and communications; and no town would be founded on the criteria of defense today. Once again, however, man's judgements and decisions of earlier centuries are still very much a part of the currently visible European landscape.

Other place locations in Europe also generally relate to site advantages provided by the natural environment. Waterways are a striking example. Since they were at the same time the basic source of water supply, virtually the only means of transport access, and the leading source of energy throughout long centuries of European history, waterways tended to focus settlement in most countries. Where there was some

Figure 6.2 Relicts of the ancient three-field system, long, narrow fields are still part of the European landscape: Zillertal, Austria.

extra advantage, for example where a river could be crossed comparatively easily or where falls or rapids augmented the power potential, places were founded which have in many cases become large cities: Paris owes its location exactly to the point where the Seine river could be most easily crossed by means of some small islands in the channel. Pass routes through major mountain ranges, or any other physical feature that tended to channel or concentrate the movement of goods or people, were also favored locations for early settlements. In short, wherever the natural environment provided some advantage of location, whether defensively or economically, early Europeans saw it in their best interests to found a settlement.

URBANIZATION

Europe today is a continent dominated by its cities. They concentrate its populations and its energies, define its cultures and its problems, and generally reflect Europe to the world. Great Britain, by the mid-1800s, was the world's first major country to have over half of its people living in cities, and if we travel in Europe, study the European scene, or even think casually about Europe, cities are what come to mind: London and Paris, Madrid and Rome, Stockholm and Berlin, Amsterdam and Vienna; they evoke the images of their respective countries to a remarkable extent.

This being so, it is often difficult to realize what a recent phenomenon large-scale urbanization actually is, in Europe as elsewhere in the world. About two thousand years ago, ancient Rome may have reached a population of over one million (estimates vary widely); but with the collapse of the Roman empire, a "millionaire" city was probably not

Settlement and Urbanization of Europe

Figure 6.3 A fortified structure on high ground is still a feature of countless European cities and towns: Sion, Switzerland.

seen again in Europe until London reached that magic figure early in the nineteenth century. During the intervening centuries, cities were the exception, rather than the rule, and some very famous places had populations that seem to us today ridiculously small. Throughout the Middle Ages and the Renaissance, only a handful of important capital cities (Paris, London, Constantinople, Cordoba) ever exceeded one hundred thousand inhabitants, and towns of ten thousand or even five thousand were often very important.

Although Greek and Roman urban models are certainly well known, if often only through archaeological excavation, the modern European city began to take shape during the Middle Ages and was based on certain elements that were very consistent and which have done much to give character to the cities of our own time. Once again, the decisions taken by human groups at a remote point in the past have substantial implications for the urban landscapes of today. In particular, most early cities were based on the interaction of three elements: the church, the market, and the wall.

With the fall of Rome in the fifth to seventh centuries AD, the Christian church became the only institution with any organization, stability, and authority over large parts of Europe. As such, it found itself cast in the role of temporal as well as spiritual focal point for the community, sorting out day-to-day problems for its parishioners as well as conducting Sunday services. This of course meant that not only the church as an institution, but also the church as a building, became very important in the affairs of countless European communities. Church architecture inevitably focused both the talents and the aspirations of local residents, thus adding to the physical importance of the church as the center of local life. Cities and towns therefore often began to grow around the church, which still remains at the heart of many European cities today.

Figure 6.4 European urban evolution

Date	\multicolumn{4}{c	}{City Size}	Largest City
	50,000	100,000	500,000	1,000,000	
	\multicolumn{4}{c	}{NUMBER OF CITIES}	
800	1	2	—	—	Constantinople[a] (300,000)
1000	2	2	—	—	Cordoba[b] (450,000)
1200	5	4	—	—	Constantinople (350,000)
1300	6	7	—	—	Paris (228,000)
1400	10	6	—	—	Paris (275,000)
1500	20	6	—	—	Paris (225,000)
1600	22	12	1	—	Constantinople (700,000)
1700	19	10	3	—	Constantinople (700,000)
1750	23	14	3	—	London (676,000)
1800	37	13	3	—	London (861,000)
1850	59	37	2	2	London (2,320,000)
1900	c	c	12	9	London (6,840,000)
1950	c	c	c	23	London (8,860,000)

Source: Chandler and Fox, *3000 Years of Urban Growth*.

[a] Although in modern Turkey, Constantinople was generally considered "European." Estimates of its population vary considerably.

[b] Cordoba was the most important center of Moorish Spain.

[c] Numbers not calculated.

If the church represented the spiritual and administrative focus of many a European town, its counterpart was the market, where the vital (if more sordid) everyday commercial dealings among the citizenry took place. The market was of course also a physical place, often in a large square, and as such tended to rival the church as a natural focus of community life, and thus as the center around which the town expanded. Markets were of many kinds, from the small daily variety to the periodical or annual affairs which often attracted great crowds from far distances. Towns competed eagerly for the financially rewarding right to hold one of the major fairs or markets.

The final basic element in the design of medieval European cities, and one which has done at least as much as the other two to determine the character of modern cities, was the wall. Most European cities were walled, often several times as they repeatedly outgrew their "shell." Walls were built, of course, primarily for defense. As cities began to grow and to concentrate the profits of trade and manufacturing, they immediately became favored targets for attackers attracted by the cities' increasing wealth. The glittering prize of a rich city drew invaders much more surely than a rural landscape of poor farming villages. Much European history, in fact, can be understood as a technological struggle between the architecture of walls and the evolution of military hardware.

In addition to their vital role in urban defense, walls also served to control movement into and out of the cities. This control was important in enabling cities to turn away craftsmen who might challenge the authority of local guilds, and it was particularly vital in ensuring collection of the taxes on food, animals, and merchandise on their way to market—virtually the only source of income to the cities of those days.

Settlement and Urbanization of Europe

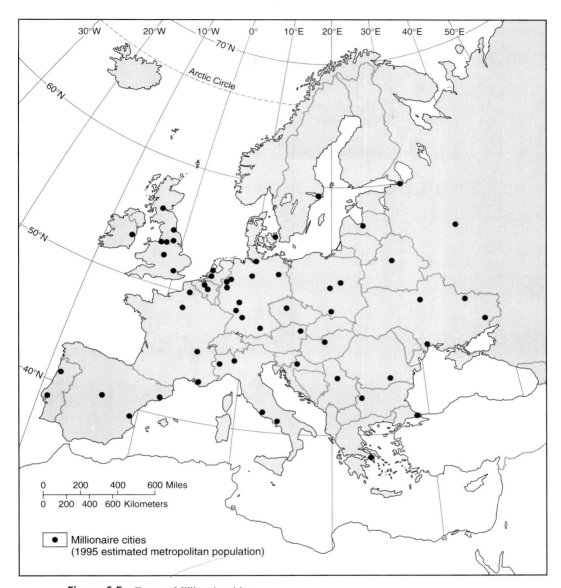

Figure 6.5 Europe: Millionaire cities.

Although the concept of walled cities in Europe finally succumbed to the realities of modern warfare in the nineteenth century, their legacies live on. Since, except where some physical barrier constrained them, cities expanded in an unplanned "amoeba-like" shape, circular walls were the cheapest and most effective way to enclose them. As successive wall lines were abandoned, the stone was often used in a new wall, while the freed space was turned into roads or occasionally green space. Thus, a majority of European cities today have a distinctly circular shape, and many important networks of boulevards or circumferential roads are largely built on the traces of the old wall lines.

Figure 6.6 The marketplace remains a feature of most European towns: Valletta, Malta.

Rapid growth of European cities began with the onset of the Industrial Revolution, and in fact the urbanization of various nations can be timed fairly precisely to coincide with the arrival of modern industry, with its great emphasis on the concentration of market, labor, and capital. The most spectacular rates of urban growth generally occurred from the mid-nineteenth century until World War I, as the impoverished artisans of the countryside flocked to the cities, where all the jobs had gone. Today, both London and Paris are among the world's largest cities, but the overall growth rate of the larger cities is low, as the problems and inconveniences endemic to twentieth century urbanism have slowed their expansion. Suburban growth, and especially the expansion of smaller traditional centers and the development of "new towns" or planned communities, are taking a larger share of urban increase, while rural populations are close to stability in many countries and continue to decline in others.

PERCEPTION: THE EVOLUTION OF PARIS

Each European city has its own distinctive character and unique history, and each has brought the heritage of good times and hard times, politics and warfare, power and decline, to create the reality that can be observed at the end of the twentieth century. Nonetheless, there are common threads that link the development of most European cities, such that analysis of the evolution and

Figure 6.7 Well-preserved walls are among the historic attractions of many European cities today: Avila, Spain.

problems of one can provide insights into many others as well. Although one of Europe's largest cities, Paris remains in many ways a good example of how urban places have responded to change over the centuries.

Few cities show as clearly the advantages of site location as does Paris. The original settlement, in about the third century BC, was on an island (today Île de la Cité) in the middle of the Seine. The Parisii tribe thus gained a food supply, a water supply, and a measure of safety from attack. The Romans captured the site in 51 BC, seeing in it an ideal place to cross the river by means of two short bridges. The settlement thus quickly became an important administrative and commercial center on the main north–south Roman road.

The first wall was built on the north (or right) bank in the eleventh century (it enclosed some fifty-two acres), and continuous expansion of the city led to additional wall building in 1200 and 1370, as a result of which such modern landmarks as the Louvre (now a famous museum) and the Bastille were added to the urban scene. In the sixteenth and early seventeenth centuries, Paris had again largely outgrown its boundaries, and yet another wall line was constructed. At this point, material from the earlier wall was used to partially construct the new one, and the empty space became in time a ring of *grands boulevards* that formed a vital transportation link and confirmed the circular shape of the city.

At about the time of the Revolution, much of the existing wall was again demolished and turned into a second ring of wide avenues that are still a major part of the urban road net. Yet surprisingly (considering the decline of the wall as a defense against modern artillery), Paris was again encircled by walls and fortifications in the 1840s. During the disastrous (for the French) Franco–Prussian war of 1870–71, these defenses saved Paris from invasion, although not from starvation and defeat. In modern times, this ring of defenses has been transformed into an eight-lane freeway, which is nearly always completely jammed with automobile traffic. (Cynics point out that it is probably a more effective barricade to movement in and out of the city than were any of the previous wall lines!) Even at that, the notion of circular defenses died hard. In the late 1800s, one final ring of forts was built on points of high ground around the city. Expensive and never really functional, these positions were not totally abandoned until the advent of the airplane put to rest forever the idea of defending a city with a wall.

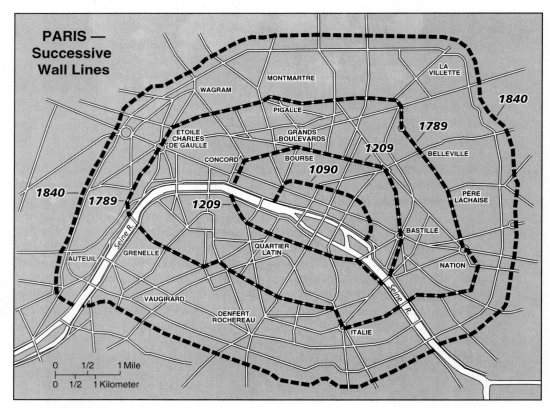

Figure 6.8 Paris: Successive wall lines.

Figure 6.9 From the Eiffel Tower, the modern complex of La Defense seems to be the center of Paris. Actually, it is in a suburb outside the city limits.

As Paris expanded beyond its successive lines of walls, other problems, most of them also endemic to other European cities, began to emerge. The medieval city center evolved in a random and haphazard fashion, with no thought of rational planning or the needs of future users. It was a pedestrian city, with no provision for even basic services, to say nothing of the demands of more technologically advanced civilizations. As cities grew, the center remained vital only so long as its economy could function and its inhabitants' requirements remained modest. Paris, however, like most European cities and in sharp contrast to most American cities, did not grow up with the automobile, and is thus very compact: the center could not really be abandoned to the suburbs.

The struggle of successive generations of French planners to "do something" about an ever expanding and demanding Paris need not be detailed here. As elsewhere, it is basically a long-running drama of need versus resources, vision versus pedantry, and a dim appreciation of the common good, with politics looming large at every turn. From the 1850s, when the famous Baron Haussmann established much of the grandeur of the central city with the approval of Emperor Napoleon III, to the arrival of the railways, to the building of the subway system around the turn of the twentieth century, to the contributions of the nation's (and the world's) most noted architects, Paris has attempted to live up to its reputation as one of the world's great cities, both contemporary and human.

From the mid-twentieth century onward, however, and especially since the advent of mass automobile ownership in the 1950s, Paris has been strangled by its own success as a magnet for immigrants from other regions, tourists, business, and international institutions. The question—again, one posed by many other European cities—has become: How is it possible to construct a competitive twenty-first century city, with all of the high-technology infrastructure required, without destroying the human scale, the historic architecture, the "quaint" old quarters, and the traditional values that make the central city of great significance to citizens and tourists alike?

For some cities, the choice is simple. Those devastated by warfare, especially the air raids of World War II (Rotterdam in the Netherlands or the Ruhr cities of Germany, for example), have had little option except to rebuild in the most modern style, recreating their cultural heritage as best they can. For Paris and many other relatively undamaged cities, however, the problem is in a sense more acute. The French have generally tried to cope by modernizing various districts while retaining as much of the architecture and local flavor as possible, often using a new museum or theatre as a focus, while improving transport connections and encouraging commercial redevelopment. Some of these projects have been successful, others much less so; most have had the result of changing the face of the city at local scale while retaining the overall image of the "city of light."

One specific project among many will serve as example. Paris has never had a "downtown" in the American sense of multiple high-rise buildings clustered in the midst of the central city. This was largely the result of a height limitation of 30 meters (102 feet) placed on buildings during the time of Louis XIV for reasons of aesthetics and public health. Paris to this day has one of the "flattest" profiles of any major city. This put the city at a great disadvantage, however, in attracting the sort of corporate headquarters and other large-scale business initiatives crucial to the vitality of any modern metropolis. Since the late 1950s, therefore, the French have been constructing one of Europe's largest, most modern, and most technologically efficient "central cities" at a site known as La Defense, which is actually outside the city limits of Paris to the northwest. With all modern facilities, connected to central Paris by a new express subway link, La Defense is a harsh, dehumanized world of glass and concrete; but its success is evident, and its critics are forced to admit that it is, at least, better than having done the same thing to the very heart of Paris.

ADDITIONAL READING

BURTENSHAW, D., M. BATEMAN, G. ASHWORTH, *The City in West Europe*. London and New York: Wiley, 1981.

EAST, G., *An Historical Geography of Europe*. London: Methuen, 1935.

EVENSON, N. *Paris: A Century of Change, 1878-1978*. New Haven: Yale University, 1979.

HOHENBERG, P. and L. LEES, *The Making of Urban Europe, 1000-1950*. Cambridge, MA: Harvard University, 1985.

HOUSTON, J.M., *A Social Geography of Europe*. London: Duckworth, 1953.

POUNDS, N.J.G. *Historical Geography of Europe*. London: Cambridge, 1990

SMITH, C.T., *An Historical Geography of Western Europe before 1800*. London: Longmans, Green, 1967.

SUTCLIFFE, A., *The Autumn of Central Paris: The Defeat of Town Planning, 1850-1970*. London: Edward Arnold, 1970.

VAN DEN BERG, L., et al., *Urban Europe: A Study of Growth and Decline*. Oxford: Pergamon, 1982.

DISCUSSION QUESTIONS

1. The church, the market, and the wall, were basic design components of most early European cities. What was the importance of each?
2. The "three-field" system of agricultural land division has long vanished from Europe, but its traces live on in the modern landscape. Describe the system.
3. Riverside location was a particularly important criterion for place location throughout most of European history. Why was location on a river so important for a settlement?
4. As many European cities, Paris in the late twentieth century has been faced with the problem of renovating its residential and commercial infrastructure without destroying the character and charm of its older districts. How has Paris sought to accomplish this?
5. A fortified site on a hill was a classic location for many European settlements. Why? Would you plan a new city around a similar site today?

PART THREE
Patterns and Implications of European Economic Activity

7

Pre-agricultural Europe; Plant and Animal Domestications and Their Diffusion into Europe

Europe's importance in the world stems largely from its long tradition of economic strength and its unparalleled success in harnessing the rest of the world to its economic system. Through emigration, colonial intervention, and control of financial institutions, European nations forged a worldwide network of wealth and influence which has persisted down to the present time. It is worth remembering, however, that Europe's exceptional impact abroad was solidly founded on a long, gradual accumulation of wealth at home. The evolution of the European economy is a lengthy history of invention and innovation, self-interest and national rivalries, exploitation and technological change.

PRIMITIVE ECONOMIES

Throughout nearly all of human history, the basis of every economy was simplicity itself: hunting and gathering—literally living off the land—was everyone's way of making a living. For those of us caught in the economic complications and stress of modern times, the idea of returning to simpler conditions and letting nature provide is often an appealing one; but the problem with hunting and gathering as a system is that it can support very few people. The standard estimate is about one person per square mile (more or less depending on the richness of the environment); thus even a small tribe or group required a substantial land base for its survival, wandering ceaselessly in search of basic sustenance. In fact, one of the basic reasons why total world population remained so low for so long was that even slight increases in numbers would soon begin to press hard on the small resource base. Effective overpopulation on a thinly peopled planet seems an unlikely problem, but it was very much a Stone Age reality.

With the passage of time and the evolution of human societies, more specialization began to appear in many parts of the world as people began to interact more closely with a limited number of resource animals and plants, thus ensuring a more stable food supply and permitting modest increases in population densities (from five to ten people per square mile, depending again on the quality of the ecosystem). Cultural geographers and anthropologists recognize such relationships as the Plains Indians and the buffalo of North America, the circumpolar tribes of the Arctic and the seal and polar bear, and others.

PERCEPTION: THE LAPPS AND THEIR VANISHING WAYS

In Europe, one specialized hunting and gathering society that still marginally survives is that of the Lapps (Sami) of northern Scandinavia, Finland, and the Kola Peninsula of Russia; and their traditional resource animal: the reindeer. For centuries, the Lapps have built an economy, and indeed an entire way of life, on following the reindeer migrations, as the animals move in a large circle, wintering along the relatively mild shores of the Atlantic, then moving inland to breeding and grazing grounds during the short Arctic summer. Food, clothing, shelter, tools, and cultural symbolism all came to the Lapps by way of the deer.

In our times, the lure of the outside world with its promise of more rewarding ways of life has reached the Lapps and largely destroyed their traditions. Reindeer are herded by snowmobile, the young people leave for jobs in mining or industry, and tourism brings in the money that makes the quaint costumes of yesterday marketable but rapidly changes human values. Nonetheless, the small number of Lapps that continue to practice the traditional herding lifestyle are viewed with a respect bordering on veneration by their "modernized" relatives.

Figure 7.1 Some Lapps still retain their ancient ways: Landscape in northern Scandinavia.

With the colorful exception of the Lapps, Europe's economies have passed from simplicity to complexity, from rural to urban, from marginality to affluence. The process has been a lengthy one, flowing through 9000 years of time and involving at one point or another nearly the entire world. It is to a large extent the story of Europe's emerging dominance of the global economy.

DOMESTICATIONS

The evolution of modern economies capable of supporting the dense populations that characterize the earth today had its origins in the beginning of agriculture. The history of farming is a complicated one, involving many gradual changes in crops, implements, land tenure and marketing systems, etc. Since our concerns are primarily with modern landscapes, we can keep much of this history in the background; however, the vital importance of agriculture in all regions of the modern world means that we should consider certain key processes in some detail. Foremost among these are the plant and animal domestications without which agriculture as we know it, and as it has formed the basis of nearly all other economic success, would never have been possible.

Plant Domestications

Of the hundreds of thousands of plant species known to modern science, it is intriguing to note that only a few hundred have been domesticated by humans to provide the basis of their food supply. It is even more interesting, and perhaps sobering, to remember that only three of these plants—wheat, rice, and maize (corn)—form in one way or another the food base for most of the world's people. Domestications, therefore, have involved only a small number of plants. They have taken place in many parts of the world, the domesticated plants diffusing away from these centers, along with farming ideas and technology, to other regions. Such areas as India, South China and Southeast Asia, and Mexico are among the more important centers of domestication. For Europe, the key region was the valley of the Tigris–Euphrates river system (also known as Mesopotamia) in the Middle Eastern country now called Iraq.

In nearly all cases of domestication, the key element seems to have been a combination of some random event in the plant community (such as a genetic mutation, for example) producing a unique opportunity, and the presence of humans in rather large numbers and of an intelligence capable of realizing the potential in what they were observing. Some 9000 years BP, Mesopotamia was a well-watered, rich grassland environment (glaciers lingering in the north of Europe brought cooler, wetter conditions to this region than is the case today) that supported a dense population of hunter-gatherers. Humans were thus on the scene, ready to take advantage of promising occurrences in their natural world.

All of the basic grain crops—wheat, rye, oats, barley—which so characterize European agriculture had their origins in genetic mutations taking place in the grasslands and of which humans were clever enough to appreciate the possibilities. One important such mutation, stemming from the natural cross-fertilization of grass species, caused the seeds or grains to remain clustered on the stalk of the plant even when ripe. This is a losing proposition for the plant, since its seeds are not diffused by the wind or other processes, and is thus a mutation that would be selected against and vanish relatively quickly. Humans, as prehistoric food gatherers rather than twentieth-century botanists, however, saw something different. If the grains remained clustered on the stalk, collecting was made much easier: more food could be gathered in less time. When woman or man had the wisdom to plant some of these seeds instead of eating them all, and when out of each year's growth only those plants with this particular mutation were selected for sowing, the laws of genetics gradually provided a higher and higher percentage of plants having this characteristic. Thus, over the centuries, humans gradually turned from gatherers into grain farmers, with astonishing consequences for the future of the world. Rice and maize, the great domestications of Asia and the Americas respectively, are also derived from grasslike plants, and also appear to have been domesticated by much the same combination of genetic mutation and human intelligence. It is interesting to remember that a modern wheat field, abandoned by the farmer, will revert to a variety of grasses and other plants within a few years: these are more successful than the human domestication in any sort of natural competition.

With some control over plant breeding, agriculture could begin, thus marking one of the great watersheds in human history. The key concept is that of surplus production. Instead of feeding only one person and a family as a gatherer, man (or woman) as a farmer was able to grow enough to support perhaps five, ten, or even more people. This in turn meant that these people, released from the everyday necessity of finding food, could turn to other occupations, becoming artisans and craftsmen, and eventually the priests, philosophers, administrators, soldiers, tax-gatherers, and other occupations associated with the rise of modern civilizations. Moreover, the growth of surplus production also led to the evolution of market economies and to trade in farm commodities, with the transport routes and financial institutions implied by these activities.

Additionally, the shift from wandering, hunting and gathering economies to sedentary, agricultural ones meant that people's ideas about land began to undergo profound and far-reaching changes. To a wandering tribe, the land itself has no value: only what it can provide is important. To a farmer, however, the land itself becomes a most precious resource. It can be bought, sold, inherited, taxed, mortgaged, and fought over. In short, all of our modern ideas about land and the almost mystic importance of "real estate" stem from the beginnings of agriculture.

Finally, the rise of agriculture was an obvious precondition to the growth of cities and thus the urban societies that dominate the world scene today. The hunter-gatherer requires space and lives by movement, scarcely the sort of lifestyle conducive to living in dense concentrations. Surplus agricultural production meant not only that not everyone had to be a food producer, but also that not everyone had to live on the land itself. The "city" thus became a feasible reality for the first time. It is certainly no coincidence that the world's first recognized cities began appearing in the rich, well-watered valleys of Asia and the Middle East at about the same time that agriculture was becoming widespread. Ancient cities found in Mesopotamia, such as Ur and Sumer, offer clear evidence of this relationship.

Between about 9000 and 6000 years ago (the dates vary widely from place to place), the domesticated crops, ideas, tools, and technologies of basic agriculture gradually filtered their way into Europe, carried along on the steady tide of east-to-west movement that characterized Eurasia for so many centuries. There were two principal routes for diffu-

Figure 7.2 Modern wheat: Its ancestor was domesticated in the Middle East.

sion of agricultural technologies into Europe. One was through the basin of the Mediterranean Sea, where the steady growth of sailing capabilities, along with the trade and commercial possibilities that began to open up, spread farming and the importance of surplus food production throughout southern Europe. The other was along the valley of the Danube River upstream from its mouth in the Black Sea into the heart of central Europe. As the Danube is the only major river in Europe flowing from west to east, its valley has always been a major corridor for invasion, commerce, and the diffusion of ideas.

The process of agricultural devlopment in Europe was a very lengthy one, complicated by the fact that the European environment was quite different from that of the Middle East, so tools and strategies that might work well in one setting had to be substantially modified in the other. In particular, the wetter, cooler climates and heavy, podzolic soils of most of northern and central Europe were vastly different from the warm grasslands and light soils of Mesopotamia. Archaeological exploration has revealed that the first areas of northern Europe to be farmed were not the heavy forest soils, but rather the light, sandy soils formed on glacial deposits along the North Sea. Although of inferior quality, these soils could apparently be dealt with more easily using the light, simple equipment then available to early farmers.

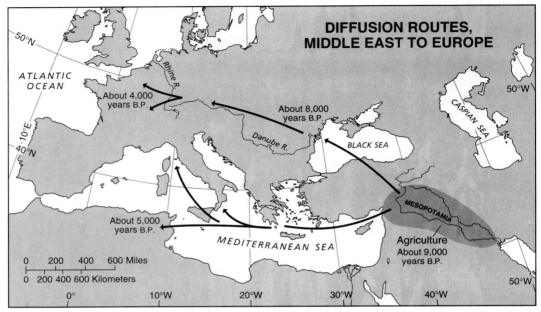

Figure 7.3 Diffusion routes of plants and ideas: Middle East to Europe.

Animal Domestication

In a time frame not inconsistent with that of plant domestication, and in many of the same parts of the world, humans began domesticating animals for their use, thus adding another new dimension to economic development and marking another crucial step in human history. In some respects, animal domestication is a logical step beyond the specialized hunting and gathering economies, which had already often focused on one or a few animals. Changing from a system involving following the resource animals on their seasonal movements to one involving control of the animals and their movements (even though the seasonal aspects may be exactly the same) is a fairly short step. Controlling the animals naturally implies providing food, water, protection, and other services, however, and the herding economies that are widely distributed over the modern world are vastly different in their complexity and organization from the preagricultural hunting-gathering societies.

Although a moderate number of animal species have been domesticated by humanity for one reason or another (it is certainly possible to argue, for example, that the dog actually domesticated people, rather than vice versa!), only a relative few have had a truly dramatic effect on the march of events and the development of human societies. In Europe, this short list would include cattle and horses, swine, poultry, sheep and (to a lesser degree) goats. The relationship of each of these animals to humankind represents a fascinating history (to say nothing, outside of Europe, of such species as camels, llamas, water buffalo, and many others), but this is perhaps not the place for such discussion. It is, however, important to remember that there are three major ways in which the domestication of animals has profited humanity, and which explain why the process is considered so vital.

First, and most obviously, animal domestication assured a more reliable supply of commodities crucial to the improvement of human food availability and vital to the expansion of other aspects of the economy. Meat, as well as fresh milk and such related products as cheese, has historically been a prized component of the European diet, adding needed protein to the various crops grown for human consumption. Beyond this, goats and especially sheep have provided the raw material for that most fundamental of all European industries: textiles. The sheep has, in fact, been a favored animal in most parts of Europe as a result of its all-purpose character: meat, milk, and wool.

Second, the domestication of animals provided an economic vehicle by which humans were enabled to penetrate into many of the earth's environments that had previously been inhospitable. Although, as an animal species, humanity is remarkably adaptable to most of the world's climatic regions, it is always faced with the necessity of providing food (or something to be exchanged for food) for the individual and the group. Since domesticated animals are inherently portable, and because of the variety of products they can provide, they have permitted humans to live in at least modest numbers in such inhospitable environments as the deserts and the mountains (of which such a surprising portion of the earth is composed). In Europe, the domestication of animals allowed people to develop distinctive and enduring economic systems in both the Alpine and the Mediterranean regions (see discussions of these classic economies, Chapter 8).

Finally, and most importantly, animal domestication made available a vastly enhanced source of power. For the first time, humans were able to break the energy confines of their own feeble muscle structure and, by harnessing animals of much greater weight and endurance to perform a variety of tasks, to expand dramatically their economic possibilities. It is worth reflecting on the fact that, from the era of animal domestication (some 9000 years ago) to the perfection of the steam engine in the 1700s, only modest additions were noted in human energy supply: the quantified importance of animal domestication is thus most impressive.

Predictably, it was in the expansion of agriculture that animal domestication as a power source played its most vital role. Even with the new crops, tools, and ideas, agricultural productivity was inevitably limited by a lack of power; a human alone (or even with a family) can only handle the farm chores on a relatively restricted amount of land. The addition of large, powerful animals to the equation (oxen, for example)

Figure 7.4 Wild pigs were among the animals commonly domesticated in Europe.

meant that more surface could be cultivated by the same work force and in the same time. Consequently, productivity and the volume of surplus production rose dramatically, and all of the socioeconomic spin-offs already discussed were speeded up or accentuated (see related discussion under energy resources, Chapter 10).

Thus the crops and the animals, the tools and the ideas, the problems and the profits of agriculture spread from the Middle East into Europe. Agriculture soon formed the basis of economic growth in all sections of the continent, and its steady productivity and reasonable prosperity provided a stable food base for an expanding population, which then began accumulating the capital that would be so vital for investment in the new industrial structures which were to transform Europe in future centuries. It is now possible to look in more detail at the basic model of agriculture that developed in Europe, where many familiar crops and ideas will be noted (American agriculture is, after all, largely an extension of the European model), as well as some more recent inputs based on Europe's colonial expansion in the 1500s.

ADDITIONAL READING

COLLINDER, B., *The Lapps.* Princeton, NJ: 1949.

GOOD, R., *The Geography of the Flowering Plants,* 4th ed. London: Longman, 1974.

HODDER, I., *The Domestication of Europe.* London: Blackwell, 1990.

ISAAC, E., *Geography of Domestication.* Englewood Cliffs, NJ: Prentice-Hall, 1970.

RICHARDSON, W., and T. STUBBS, *Plants, Agriculture, and Human Society.* Menlo Park CA: W.A. Benjamin, 1978.

UCKO, P., and G. DIMBLEBY, eds., *The Domestication and Exploitation of Plants and Animals.* London: Duckworth, 1969.

DISCUSSION QUESTIONS

1. Who are the Lapps? Describe their traditional mode of life. Why do you imagine the few Lapps who still practice the ancient form of economy are so respected by the many who do not?

2. Most of the crops and agricultural ideas that became common in Europe originated in the Mesopotamian region of the Middle East. How did these diffuse to Europe? What differences in climate and soil had to be overcome?

3. In addition to a more reliable food supply, animal domestication brought two other substantial benefits to humanity. What were these?

4. Plant domestication has been called the greatest single agent in the advance of human civilization. Why is there such extravagant praise for this process?

5. Conventional wisdom considers the average maximum population density of a simple hunting-gathering economy to have been about one person per square mile. Why is this number so low?

8
Classic European Economies: Alpine and Mediterranean

In considering the economic complexities of Europe, it is helpful to recognize that large sections of the continent display strikingly similar patterns of human activity. While respecting the great variety of local cultural conditions found in Europe, it is nonetheless possible to generalize two specialized economies that are broadly similar in whatever nation they are found. Based essentially on patterned human responses to similar environmental challenges, these are the classic Alpine economy and the classic Mediterranean economy.

The term "Alpine" implies regions of relatively high elevation and certain other related characteristics; it does not refer only to the Alps themselves. Thus the Rocky Mountains of North America or the Himalaya of Asia, for example, may be said to be "Alpine." The generalized use of this term relates to the fact that most characteristics and conditions of the mountainous environment were first identified and studied by European scientists working in the Alps; the specific examples then became universal. There is no minimum elevation above which "Alpine" conditions will automatically occur; but highland environments are widespread in Europe, and while the Alps, Pyrenees, Carpathians, Scandinavian mountains, Jura, Apennines, Scottish Highlands, and even the Urals of Russia are obvious examples, there are numerous other lesser ranges in nearly all parts of the continent. The similarity in economic patterns found in all Alpine regions is related to certain limitations imposed and opportunities offered by the environment.

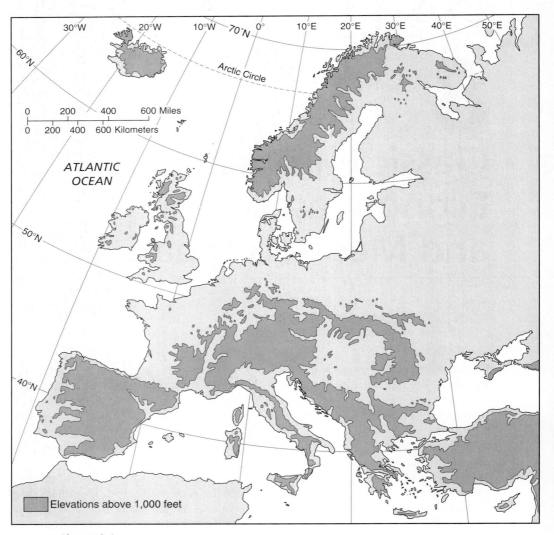

Figure 8.1 Europe: Areas above 1,000 feet in altitude.

Characteristics of the Alpine Environment

The problems posed to human economic adjustment by the Alpine environment are of several kinds:

1. **Slope.** Rather than elevation alone, percentage of land surface in moderate-to-steep slope is what more correctly defines "mountainous" landscapes. Areas of high elevation with flat rather than sloping relief constitute "plateaus" rather than

mountains, as in the case of Iberia. Excessive slope angles inhibit conventional agriculture in several ways. The use of machinery and even larger animals is difficult, if not impossible. More significantly, geomorphic processes such as soil creep tend to remove soil downslope by gravity as rapidly as it forms. Mountain soils thus tend to be thin and stony. Moreover, the action of glaciation, active even today in the Alps, Scandinavian mountains, and parts of the Pyrenees, erodes away whatever unconsolidated material might be in the path of the ice.

2. **Climate.** Mountainous regions, although not necessarily of great elevation, are invariably higher than the neighboring plain. This extra height modifies local climates, again generally to the detriment of agriculture, and produces more extreme and severe conditions. In particular, added elevation generally means longer, colder winters, later springs, and earlier fall frosts. This is less of a problem in southern Europe, where cooler temperatures may actually be economically advantageous, than in the central and northern regions, where colder mountain temperatures reduce growing season length below the minimum for most key crops. Beyond this, increased elevation normally results in increased precipitation (see climate discussion, Chapter 2), and while this might occasionally be useful (as in dryer parts of southern Europe), throughout much of the continent it leads to excessive snowfall totals at higher elevations, and to cool, wet conditions in summer and fall, which inhibit crop growth and timely ripening. However, it should be remembered that Highland climates are extremely complex, and that certain microclimatic enclaves (such as sheltered valleys or south-facing slopes) may present unusual opportunities for agriculture.

3. **Isolation.** With the increasing sophistication and urbanization of European economies since the Industrial Revolution, mountainous regions have become steadily more disadvantaged because of their physical isolation from centers of development. "Isolation" in this sense does not necessarily mean remoteness (the Alps, after all, are geographically in the very heart of Europe), but rather separation by time and expense to the detriment of the Alpine areas. In terms of accessibility and transport costs, these areas have fallen increasingly farther behind. Railways and modern highways are rare and very expensive to construct here compared to the regions of lowland Europe. It thus costs more and takes longer to provide industries with their raw materials, and to move products to market. The lack, for obvious geographic reasons, of water transport in mountainous regions is deeply felt. There are very few traditional industries able to overcome these drawbacks of isolation sufficiently to cope in a highly competitive Europe in which optimal location is usually the key to success.

Conventional Economic Supports: The Alpine Traditions

Although it is by no means true that geographic factors (the specifics of the Alpine environment, for example) determine absolutely the range of human activities, they certainly provide a framework of rewards and punishments for human economic decisions. Any study of comparative cultures reveals that, even given the spectacular diversity of the human condition, people in similar environmental circumstances have independently evolved similar strategies for optimizing their economic prospects. As enumerated above, the severe challenges posed by the Alpine lands have elicited related economic responses from human groups in most parts of the world, given their varied cultural and historical perspectives. In a traditional society, what can one do in the European mountains?

   1. Pastoralism. By far the most widespread economy of the world's mountain regions, Europe certainly included, is pastoralism, or the management of animals and the commercialization of their products. Animal husbandry has unquestionably played the major role in the settlement and humanization of the European mountains and, despite setbacks, remains today the principal economic support in many districts. Animals—generally cattle in the north, sheep and goats in the south—are raised precisely because they permit all of the Alpine problems stated above to be at least partially overcome. Animals first of all are portable. That is, they are an industry that can easily move to where its necessary resources are available. In the Alpine world, crops may not ripen, but a hay crop can usually be cut, and even in the highest and steepest pastures grazing is available for a few weeks of the year. By moving animals to where grazing is temporarily abundant, then moving them again, human groups maximize the limited food-energy resources of the Alpine environment. Refinements of this process have given rise to the characteristic human/animal movements known as *transhumance,* familiar to all the world's mountainous regions, but particularly well defined in Europe.

   The classic transhumant pattern reveals an extremely fine-tuned response to environmental possibilities. From a farm, or more often a village, at low elevation, animal herds are moved to higher pastures as snow cover leaves these areas in late spring and early summer. Villages may actually own these remote pastures, which are inhabited only during the short growing season. Traditionally, the animals move under their own power, although in modern times transport by truck is common, especially where fairly long distances are involved. Members of the family—older children or grandparents—may stay with the animals, living in cabins for the entire summer. Although hired herdsmen (not uncommonly students) now often perform this seasonal job, the extended family concept implied in such a division of labor has marked many social patterns in the mountains.

   While the animals are, so to speak, out of town, the lands around the home farm or village are used to produce two or even three cuttings of hay, as well as such other crops permitted by climatic conditions, thus assuring a winter food supply for the animals. All of the land and pasture resources of the mountainous region are thus fully utilized—a classic human adaptation to difficult circumstances. Transhumance movements are, of course, not necessarily restricted to the mountains and may involve horizontal as well as vertical movements. In earlier centuries, vast seasonal movements of animals were a common phenomenon in many countries (Spain as a classic case), drovers moving the herds in a constant search for food and in the process often conflicting with more settled populations.

   An additional economic advantage of animals for mountainous regions is that they are able to a considerable extent to overcome the problems of isolation and remoteness. Thus, the classic animal of European mountain regions from the Alps northward is the milk cow, producing a commodity which can be processed simply and

Figure 8.2 Transhumance persists in the higher Alpine pastures: Abtenau, Austria.

locally, and transformed into products which are relatively nonperishable and not time-specific. As an example, a wheel of "Swiss" cheese (Emmentaler is one common type) actually improves in quality with several months of aging, so that transport time to distant markets is not a factor. Transport costs remain a problem, of course, so products of this sort rely on lower production costs and a reputation for quality to remain competitive. Other nonperishable, milk-based products, such as milk chocolate and other candies, are also important, and many small mountain towns and villages are supported largely by modest industries processing milk from the high pastures. With improved road access and increased availability of refrigerated transport, provision of fresh milk to lowland urban markets has become a technical possibility, although competition from dairies nearer the cities is inhibiting.

In southern and southeastern Europe, where climatic patterns create sparse grazing, or in highland Britain (largely for traditional reasons), sheep or even goats are the dominant animals. In addition to being better able to extract a bare living from very unpromising landscapes (see later discussion of the Mediterranean economy), sheep are the classic "all-purpose" animal, valued alike for meat, milk, and wool. They have played a major role in Europe's economic, environmental, and even cultural history.

In the twentieth century, the animal economies of the Alpine regions have diminished steadily in importance. This is related to the overall decline in the population of mountainous areas, and particularly to the fact that it is extremely difficult to earn what might be termed a "twenty-first century living," with all the material comforts implied, by raising animals in the mountains. Thus, while there are still substantial transhumant societies in many regions, their decline is unmistakable. The loss of these economies, and indeed of much of Europe's highland population, is cause for concern in many European capitals. If no one lives in the mountains, no one is there to manage the landscape, deal with the problems, and generally create the orderly, manicured environment that reflects national pride to the Europeans and an enviable way of life to the tourists. Who could visualize an overgrown Alpine valley without its farms, flowers, and tastefully arranged cattle?

For some countries, the problem runs considerably deeper. In Switzerland, for example, loss of highland population poses a serious threat to basic national security. The Swiss would rely on Alpine products (notably cheese) as a fundamental component of their food supply in the event of any new European conflict in which their neutrality might again be tested and their importing possibilities restricted. If transhumance vanishes altogether, where will the cheese come from? As a result, Switzerland has on its books some of the world's most unusual legislation. Government agricultural loans, for example, are related to elevation: the higher a farmer's holding, the more abundant and advantageous the loans for which he will be eligible!

PERCEPTION: "ALPAGE" IN MODERN SWITZERLAND

The Swiss have also been active in promoting modern refinements on the transhumance theme, such as what is called in the French-speaking cantons *Alpage*. This system is employed in many cases where a village's higher lands have been abandoned by local farmers. The village leases its land for a modest fee to qualified applicants (often students or former farmers who have become city dwellers with no particular summer employment), who then contract with farms throughout the area to "board" cattle for the season. The milk may be processed into cheese at the site; this is then sold to cooperatives or markets in the area. As an extra source of income, some alpagists may also fatten hogs on the skim milk left from the cheesemaking process. At the end of summer, profits are divided, the cattle are returned to their owners, and the part-time farmers return to their homes and lives in the city. Thus the system is maintained, at least marginally, in a prosperous, thoroughly modern Alpine nation.

2. **Forestry.** Another widespread economic support of the Alpine lands is the cultivation, harvesting, and processing of forest products. As previously noted, Europe is a continent whose forests began to come under intense pressure very early in history. Since the consumption of wood and wood products has rarely slackened its rate of expansion, demand has largely outstripped production, and many European regions must now import their wood needs. In this context (discussed more fully in the forestry section, Chapter 9), the mountainous areas have profited from the isolation that sets them apart from Europe's mainstream. They tend to be the places, if any, where forests have survived, relatively safe from the land use pressures of the heavily populated lowlands. Moreover, where mountain districts were deforested (a process that lasted until the end of the nineteenth century), the resultant soil erosion and downslope

flooding were so catastrophic that even governments that had never thought of their environmental resources as anything but inexhaustible were moved to undertake reclamation projects. Reforested slopes and scientific attitudes toward forest management reflect some of Europe's earliest ecological concerns. Today, controlled timbering, based principally on the conifer vegetation of the Alpine environment, is widespread in mountainous regions, where it provides modest but locally important employment, which may also include local wood-processing industries and transport requirements.

These basic activities—pastoralism and forestry—are associated with Alpine environments the world over. They are, however, essentially extensive in nature, supporting small populations over large areas. Lack of more intense economic forms is, of course, the reason the European mountains are sparsely inhabited and becoming more so; but there are a number of traditional activities that do provide more concentrated economies, and these often account for the location and survival of larger towns and more densely populated rural areas within the mountainous setting.

3. **Intensive farming.** Although as a general rule problems of slope and climate make conventional farming in the mountains an unrewarding if not impossible activity, there are some striking exceptions. Where pockets of good soil have formed, as, for example, along river flood plains or in broad glaciated valleys, farming can be as intensive and productive as in more favored regions. In addition, the climatic complexity of Alpine regions creates local microclimates featuring longer growing seasons (valleys sheltered from storms, south-facing slopes receiving greater insolation), and in these conditions high-value crops may generate considerable local prosperity. The vine, for example, is successfully grown on south-facing lower slopes of the Swiss Alps (on Lake Geneva, for instance), and thrives on the sunny sides of the Hercynian-age hill formations as far north as the Rhine–Mosel region of Germany. Where agriculture is possible, maize for fodder, small grains, orchards, and truck-farmed small fruits and vegetables are commonly found; these may be supplemented by greenhouses to provide extra warmth and extend the growing season. Because these favored areas are comparatively rare and are scattered among the European mountains, they have taken on a greater significance than their size and productivity would merit in lowland regions. Importing basic foodstuffs is very expensive for mountain dwellers, and these areas thus fill a critical role in Alpine food economies, while at the same time supporting service activities and light industry that may be important to a vast tract of highland.

4. **Mining.** Even more concentrated than areas of conventional farming is another traditional economic mainstay of the mountains: mining districts. Mining is an activity that must obviously take place where the valued substance is found, and since the geologic processes of mountain formation (notably intrusive vulcanism) are those also associated with the deposition of metallic mineral ores, the European mountains originally contained a great volume and variety of commodities, from gold and silver to copper, tin, lead, zinc, iron ore, and many other industrial metals.

Since most of these minerals have been prized and worked by humans for many centuries, the large European deposits are by now generally exhausted (see discussion of European resources, Chapter 10). Only in areas of extreme remoteness and difficult access (such as the iron ore mines of Austria or northern Sweden), or in cases where the mineral in question had little value until the twentieth century (such as the uranium ores of the Central Massif in France) is mining itself a major economic

Figure 8.3 Where climate and soil conditions are favorable, farming can be intensive in Alpine regions: Ruediswil, Switzerland.

support of Alpine regions. However, many small mountain cities and towns owe their location and importance to early exploitation of minerals. Where these communities have been able to diversify their economies, or to acquire significant service and/or administrative functions, they have survived and even prospered.

5. **Water power.** Another basic concentrating factor in the generally dispersed life and economy of the mountains has traditionally been water power. In this case, two of the difficult problems facing humans in Alpine environments—slope and heavy precipitation—have been turned to a positive advantage. Water flowing down a steep slope represents an enormous energy potential, and where evenly balanced precipitation and a gradually melting snow pack create streams with very steady regimes (the case nearly everywhere in Europe outside the Mediterranean zone), the possibilities for simply and effectively harnessing this potential are widespread.

The technology of the water mill and its associated industrial activities generally became known as the humanization of the European continent took place (see discussion of energy resources, Chapter 10). Water power thus offered an attractive possibility for penetrating and earning a livelihood in the difficult world of the mountains, the abundance and low cost of power serving to compensate for Alpine isolation and high transport costs. Many early industries established on this basis were able to compete successfully with similar activities in more favored regions until the twentieth century, and many of the largest mountain towns today can trace their origins directly to an early mill site (which may, in fact, still exist). Combining power with output of the local dispersed economy, these towns specialized in the production of such items as wood products, leather goods, textiles and clothing, and shoes.

Despite the scattered focal points offered to Alpine development and stability by these few concentrated economic types, traditional life in the mountains has never been easy. With the decline of mining and the eclipse of many outmoded industries (textiles most notably), depopulation and hard times have been the lot of Europe's mountain regions throughout most of the twentieth century. The disadvantages of this environment for many activities, however, become positive advantages for others, and the European highlands are now witnessing at least three kinds of new developments which, while they can never bring back the population numbers or the diversity of the eighteenth century, are nonetheless providing stability to many mountain districts and creating a positive renaissance in others.

Modern Changes in the Alpine Economy

1. Water power to hydroelectricity. The traditional Alpine industries based on the use of water power to operate mills, saws, hammers, and other simple mechanical devices may have declined sharply, but the power potential still remains, and the march of technology has permitted mountain regions to capitalize on this fact through the generation of hydroelectricity. Hydropower is by now more fully developed in Europe than in any other continent, some countries in fact (notably Switzerland) utilizing nearly 100% of their potential. To many mountain districts, this resource has meant renewed growth and prosperity. With the increased efficiency of long-distance power transmission, surplus electricity can be profitably provided to heavy-consumption lowland regions, or even exported to neighboring nations. At the same time, an abundance of low-cost electricity is attractive to many modern industries, including metal smelting, which are willing to locate plants in the highlands despite the persistent disadvantages of isolation and high transport costs. However, the replacement of many small water power sites with a few very large hydro-generating plants is another process of concentration operating on the Alpine economy: most small towns decline, a few become prosperous cities.

2. Tourism. Perhaps the most original changes in the economies and landscapes of mountainous Europe over the past century have been those resulting from the spectacular growth of tourism (see discussion of European tourism, Chapter 12). Traveling to the mountains was virtually unknown before 1850, developed slowly until the rise of the paid vacation in the 1930s, and expanded dramatically with the growth of the winter sports sector beginning in the late 1950s. Here again, the disadvantages of the highlands for traditional activities—isolation, rugged relief, cool summers, winter snowfall—become positive advantages, luring the tourist from the hectic congestion of urban lowland Europe. Once again, however, the economic impact of this activity on mountain regions is uneven. Because of the need to provide an elaborate and expensive infrastructure (ski lifts, swimming pools, hotels, access roads, etc.) to support the tourist business where it functions on a substantial scale, tourism tends to be concentrated at a few often very well-known places. Such resorts as Zermatt, Switzerland and Chamonix, France are rich and world-famous, but a few miles from them can be found abandoned farms and decaying Alpine villages.

3. Return migrations. Most recently, the increasing acceptability of nontraditional lifestyles in traditional European societies, as well as a certain disillusion with the frantic pace of city life, has created a substantial drift of population back

Figure 8.4 Hydroelectric power is a major energy source in the Alpine regions: in the Austrian Alps.

to rural areas, and especially to the isolation and tranquility of the mountains. Retirements account for part of this trend, but increasingly involved are economically active families who seek to build or recreate a more personally satisfying way of life. While this trend is modest, by no means counterbalancing the out-migration from similar regions, it does create a certain animation in many districts, reflecting attitudes of those drawn to the challenges of Alpine life, rather than repelled by them. The growth of artisan and craft activities, the provision of skilled services learned in the city, the increasingly portable lifestyle of the computer age, and the revitalization of derelict farms are among the economic impacts of this small but growing group. In addition, a certain number of small industries, particularly those involved with advanced technology and thus less dependent on traditional location constraints such as transport costs (see discussion of industrial location, Chapter 11), have found the mountains a congenial atmosphere.

In Europe, as elsewhere, most mountain or Alpine environments display similar physical characteristics, pose similar economic problems for humans, and have given rise to recognizably similar patterns of activity, regardless of national frontiers. Extensive activities such as pastoralism and forestry remain universal in the European mountains, although they support a steadily decreasing population. Intensive activities, including mining, water-power-based industry, or farming on rare patches of good agricultural land, located the most dense populations and a majority of the towns of Alpine regions. In recent years, hydroelectricity and the growth of tourism have brought prosperity to a relatively few sites, while the mountains as a whole continue to lose population and vitality.

THE CLASSIC MEDITERRANEAN ECONOMY

The second traditional pattern of economy prevailing over a large section of Europe is that associated with the Mediterranean basin. This region comprises one of the world's most distinctive natural environments, which has historically posed substantial problems to human economic adjustment. Although not basically as inhibiting as the Alpine zones, and thus more heavily populated and more intensively utilized, the Mediterranean region still calls for some very specific compensations and, perhaps not surprisingly, human groups in nearly all parts of the basin have come to the same practical conclusions about what works and what doesn't. In Spain and Greece, all the intervening parts of the European Mediterranean coast, and even to a substantial degree on the African and Middle Eastern shores as well, the same traditional economic solutions to the same problems are found. Culture and languages vary widely, as do history and politics, but economically the Mediterranean lands are another of Europe's most consistent regions.

Characteristics of the Mediterranean Environment

As in the Alpine lands, both landforms and climate create a characteristic, often difficult, environment around the Mediterranean. The categories of problems suggested for the Alpine world (see previous discussion of Alpine economy) are also found here.

1. Slope. Most of the coast, especially on the European side, is rough and is bordered by hills or mountains. Only along the coast of Languedoc in southern France west of the Rhône delta, at the head of the Adriatic Sea in Italy, and at the mouths of the larger rivers (in Spain, for example) are there important areas of flat coastal lowlands, and even these are often fairly narrow. Slope limitations have thus been an important negative factor in economic development.

2. Climate. Perhaps even more significant has been the distinctive Mediterranean climate (see climate discussion, Chapter 2), which offers both daunting challenges and great potential advantages. With the Alps, Pyrenees, and other major ranges trending east–west, thus sheltering the Mediterranean from Atlantic storms and colder winter temperatures to the north, the region enjoys mild winters and warm summers, with abundant insolation. This would seem ideal for agriculture, but the persistent summer drought that reliably ranges up to four months in duration presents severe problems: plants are without rainfall during their natural growing season. This predictably dry summer has been the single most important factor in determining the evolution of the Mediterranean economy.

3. **Isolation.** Although the Mediterranean Sea itself has historically been a unifying influence, often at the very heart of European development, many areas near the coast suffer from the same isolation and costly access that poses such problems in Alpine regions. The hills and mountains of southern Spain, southern Italy, Croatia, Macedonia, or Greece shelter villages which, although no more than a few miles from the sea, are as isolated in time and space as if they had been in the most remote corner of the Austrian Alps. As the balance of economic power has shifted over the centuries from the Mediterranean basin to the North Sea plain, the problem of isolation (which is in economic terms a case of being in a noncompetitive location) has grown worse. Many districts that touched at least the fringes of wealth during the sixteenth or seventeenth centuries have by now become the true backwaters of Europe. Against this background, traditional economies that emerged in all sections of the Mediterranean devised similar strategies to overcome natural obstacles.

Dry-farmed Crops

In the Mediterranean dry-summer environment, great importance was obviously attached to those crops which could not only survive, but prosper without the addition of any moisture during the ripening season. These fell largely into three categories: grain crops, especially wheat; tree crops, especially olives; and the vine. All of these plants are native to the Mediterranean or the Middle East, where they were domesticated; and in combination—the bread, the oil, and the wine—they formed the indispensable food base for the expansion of Mediterranean cultures.

Wheat (see discussion of domestication, Chapter 7) spread from its Mesopotamian origins to become the basic food grain of Europe and, by extension, North America, Australia, Russia, northern China, Argentina, and other scattered regions. Its great popularity, productivity, and adaptability to a wide range of environmental conditions make wheat even today the most widely grown of all major food crops. Even with this broadscale success, however, wheat remains especially a Mediterranean plant, the mild temperatures, moist winters and springs, and dry summers exactly suiting its requirements. Wheat, and its hardier substitutes oats and barley, have been the staff of life for all Mediterranean peoples. The ancient city of Rome could not possibly have grown to its enormous size (probably at least a million inhabitants) were it not for the vast granaries of wheat cultivated in North Africa and linked to the capital by a continuous "bridge of ships."

The olive, a native Mediterranean plant, headed the list of dry-farmed tree crops which contributed largely as well to the food supply. A small, spreading tree, with thin grey–green leaves and the elaborate roots and thick bark characteristic of Mediterranean vegetation, the olive still today spreads over thousands of acres in most countries of the region. Although olives, suitably cured and processed, are widely consumed as fruit, it is their oil that has played the most vital role in the Mediterranean food economy, and that is still their most important product today. The pressing, refining, and transport of olive oil has been big business throughout the Mediterranean for thousands of years; jugs for the storage and shipment of oil are among the most widespread pottery remains found in archaeological excavations.

In addition to olives, many other tree crops—both fruits (such as apricots) and nuts (notably almonds)—will survive and produce crops despite the dry summers. These plants, again widely distributed throughout the region, have thus formed important nutritional and commercial supplements to the agricultural economy. Another tree "crop," although nonedible, has also played a significant economic role: the cork oak. Another typical native tree of the Mediterranean, its thick bark is designed to protect against excessive water loss through the trunk and limbs. This covering is periodically harvested as "cork" without damage to the plant, particularly in Portugal and parts of Spain.

The third vital component of the traditional dry-farmed Mediterranean agricultural economy is the vine. Another local domestication of the region, this plant has provided fresh fruit, raisins, and especially wine to all civilizations around the shores of what Homer termed the "wine-dark sea," and the Mediterranean has lately taken to providing much of the world with this highly prized beverage. The vine, as a native, is like wheat well adapted to the rhythms of the Mediterranean seasons. Moist, mild springs ensure proper growth and flowering, while the long, dry summers promote ripening and reduce water content, thus raising the sugar content of the grapes and eventually the alcoholic strength of the wine. Except in the North African and Middle Eastern nations where Muslim scriptures are rigidly enforced (and note that Algeria and Morocco, with their French colonial background, remain important wine producers despite the dominance of Islam), all Mediterranean nations remain today important wine producers, although much of the product of these sun-baked summers tends to be coarse and of modest quality (see discussion of European agriculture, Chapter 9). Not surprisingly, nearly all of the world's other major wine-producing districts also enjoy a Mediterranean dry-summer climate: California, South Africa, Chile, southeast Australia.

Figure 8.5 The olive, a typical Mediterranean plant, forms part of the classic economy: Near Delphi, Greece.

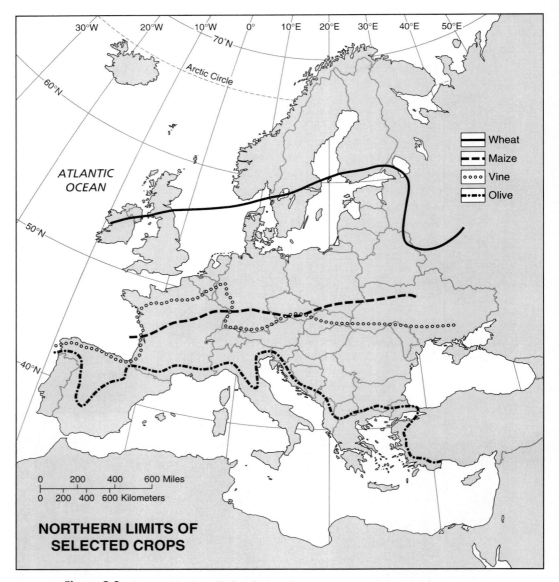

Figure 8.6 Europe: Northern limits of selected crops.

Pastoralism

Although these dry-farmed crops played a crucial role in producing the food supply that permitted the rise of Mediterranean civilizations, they were by no means the only component of the classic Mediterranean economy. As in the Alpine regions, topographic and climatic problems made pastoral economies a useful alternative to conventional agriculture. In contrast to the mountains, however, the significant differences

Classic European Economies: Alpine and Mediterranean

in the Mediterranean have less to do with the differences in seasonal temperatures (few frosts and only rare snowfall anywhere along the coastal zone), than with the unequal distribution of precipitation.

Rather than the lush summer pastures of the highlands of central and northern Europe, the Mediterranean offers only year-round grazing on more sparse and generally less nutritious scrub vegetation, which is especially rare during the summer. As a result, cattle, which do poorly on such fare, are replaced in the Mediterranean world by sheep and goats, which are better able to utilize the modest resources (although the entire Mediterranean has paid a heavy price for their foraging; see environmental discussion, Chapter 13).

In any event, the meat, milk, and wool of these hardy, multipurpose animals have been the principal economic support of countless communities in all Mediterranean nations. Transhumance in this case has generally meant a nearly continuous wandering—both vertically and horizontally—in search of food; and herds are still seen today combing the nearly barren hills of southern Italy, grazing and fertilizing the harvested grain fields of Spain, feeding in the open spaces between apartment blocks in Serbian towns, and wandering along the river valleys and roadsides of every Mediterranean nation.

The Sea as a Resource

The regions of the Mediterranean basin enjoy one vital traditional resource unknown to the Alpine world: the sea itself. This has given an entirely different dimension to the cultures and economies around the shores of the sea, nearly all of which have been at least periodically oriented toward the water.

Figure 8.7 Pastoralism is a way of life across the Mediterranean basin: Gattieres, France.

Initially, the Mediterranean has long been one of the great arteries of commerce and trade, linking Europe to Africa, the Middle East, and eventually to Asia. Thousands of ideas and innovations in all areas of human endeavor were diffused into Europe through the Mediterranean routeways; the evolution of water transport can be traced here down the centuries; and great civilizations from classical antiquity through the Renaissance owed their wealth and success in major part to controlling and exploiting the Mediterranean trade connections. Even today, the transportation significance of this water body is undiminished, and such great ports as Marseille, France, Genoa, Italy and Barcelona, Spain are among Europe's most important.

At a more local scale, the fishing industry of the Mediterranean has traditionally supported the economies of hundreds of small coastal villages in every nation around its shores (see discussion of European fisheries, Chapter 9). The rugged hills and mountains of most of the European side of the basin push down to the sea, dividing the coastline into numerous small embayments and tiny harbors. Fishing as an activity was thus always finely divided, "fleets" of no more than a few dozen small vessels being the rule.

Customarily, most fishing in the Mediterranean had only a daily rhythm, the boats prospecting local waters and returning to sell their catch fresh off the quay. This pattern was reinforced by the familiar factor of isolation: before the advent of refrigeration and quick-freezing, the little fishing villages, tucked in their respective coves, had no way of commercializing their catch in the populous north, which most often meant that fish and shellfish served only as a useful supplement to local diets that might otherwise be protein-deficient.

The Mediterranean has never been one of the world's great fishing grounds, and, with the decline of many species as a result of pollution and/or overfishing, this is more true than ever today. However, the color and animation of the ports, the classic cuisines utilizing fresh seafood, and the legends and mystique of the fisherman's trade are a fundamental element of the Mediterranean world from one end of the basin to the other.

Irrigation

A final element of the traditional Mediterranean economy, which again has spread throughout the basin, is the technology of irrigation. Without the availability of water during the summer dry season, agriculture is limited to the dry-farmed crops already discussed; with water, the Mediterranean environment, with its warmth, sunshine, and generally productive soils, becomes one of the world's richest farming regions.

The Romans were, of course, among the world's first great hydraulic engineers, bringing water by aqueduct over long distances to supply the enormous needs of the capital and many provincial centers throughout the empire. Some of these structures are still standing, or even in partial use; the famous Pont de Gard in southern France, which once brought water to the city of Nîmes, is a classic example. The Romans, however, made little use of water to irrigate crops, except on a lavish scale in private gardens. The large-scale, commercial use of irrigation was introduced to Europe by the Arabs, who brought this important technology with them from the Middle East as they swept across North Africa and into Iberia.

It was only where permanent sources of fresh water were found, of course, that irrigation could be practiced; thus it was in the valleys and flood plains of the major rivers (which also represented much of the available flat land in many cases) that irrigation spread rapidly. These areas thus became prosperous, concentrating the economic life of their regions to an extraordinary degree. In Spain, these fertile, rich, irrigated river valleys, which offer such striking contrasts to the semiarid slopes away from the water supply, are known as *huertas;* the area around the city of Valencia is a fine example.

The introduction of irrigation technology also made it possible to expand dramatically the range of crops grown around the Mediterranean. Perhaps the best known group of such plants are the citrus fruits. Native to southeast Asia, oranges, lemons, and their relatives require abundant supplies of water to survive and produce. Imported by the Arabs (who had long been voyagers to the East), they spread quickly into the irrigated Mediterranean lands, becoming such basic items of the local economies that it is difficult to imagine the Mediterranean landscape (including, of course, California!) without them. In Spain, Valencia even gave its name to a new variety of orange which became known throughout the world.

Dry farming, pastoralism, concentrated irrigation, and the sea itself: these then became the universal supports of Mediterranean economies, as familiar to one part of the basin as to any other. As in the Alpine economies, the traditional picture has altered somewhat in recent times, although many of the old ways still survive. What is significant about the modern changes, however, is that in general they still affect nearly all of the Mediterranean nations in a similar way. The entire region continues to show remarkable uniformity in its economic patterns.

Modern Changes in the Mediterranean Economy

Over recent centuries, the relative balance of economic power in Europe has generally shifted away from the Mediterranean to the lowlands around the North Sea. The comparative modern importance of Spain, Portugal, or many parts of Italy, for example, is only a fraction of what it was in the seventeenth century. As the Mediterranean has increasingly become equated with "poor" Europe, however (along with many of the eastern nations), several major developments have brought renewed prosperity, at least on a selective local basis. Once again, these changes are universal throughout the region.

Irrigation in the Modern Economy

The technology of irrigation, traditional key to agricultural prosperity in the huertas of Spain and at other points around the basin where permanent water supplies were available, has assumed an even greater significance in the region at the present time. As the nations of northern Europe achieved steadily greater and more widespread prosperity with the onset of the modern industrial age, their appetites for such "luxury" foods as fresh fruits and vegetables also began to increase. Unfortunately, however, the climates of these regions are only conducive to the production of such foods over a relatively short season: the bounty of summer and fall is followed by the cabbages, turnips, and potatoes of winter. This unfulfilled demand presented a substantial opportunity for the area nearest these thriving centers (to reduce transport costs) that could produce such crops over a significantly longer season: the Mediterranean.

Thus, during the twentieth century, with the steady advance of processing, handling, and transport technology and efficiency, the Mediterranean region has come to play a significant role in the provisioning of all Europe with many off-season fruits and vegetables—much the same role that Mediterranean California plays in North America. With the opening of new markets and the growth of consequent opportunities, many nations have taken major steps to enhance the productivity of their Mediterranean regions by better organizing or substantially augmenting the critical water supply. Utilizing surplus water from the Alps and Pyrenees, for example, has been a priority project, notably in France, where the Rhône and its tributaries have been embellished by dams, canals, and a generally complex system of water control and distribution facilities. The works at Donzère-Montdragon and on the Durance river are especially impressive.

Tourism

Around the Mediterranean, as in the Alpine regions of Europe, the most dramatic economic changes over the past century have been related to the spectacular growth of tourism (see discussion of tourism, Chapter 12). The modern summer travelers on the French Riviera or the Spanish Costa del Sol, who can be forgiven for imagining that half of humanity has joined them (more than 100 million visitors descend on the Mediterranean annually, the vast majority in July and August), find it difficult to believe that as late as 1820, virtually no stranger set foot on these malarial, pirate-threatened shores without some compelling motivation. It was not until both of these scourges were more or less controlled and the rise of a prosperous leisure class in industrialized northern Europe led to potential travelers with both time and money that tourism began to modify the landscapes and outlooks of the Mediterranean towns and villages, hitherto withdrawn deep into their traditional ways.

Beginning with the British aristocracy, who "discovered" the Riviera in the 1820s and 1830s, more and more affluent northerners came to shelter along the coast, especially in winter, when the cold, damp, polluted environment of the growing industrial cities became both physically and emotionally hard to bear. Later, as the twentieth century brought with it the paid vacation and a broadening of the distribution of wealth, the Mediterranean discovered its true modern vocation: a sort of promised land for summer tourists. As in the Alpine lands, those aspects of the environment which had so hampered traditional economic activities, notably here the dry summers, suddenly became extremely valuable assets. In an age in which the modern tourist frequently asks little more than a few weeks of "guaranteed" sun and sea during July and August, the Mediterranean provides an ideal destination, the more so since it is so convenient: no more than a day by superhighway, an hour or two by air, from most of the population centers of the north.

Whether the phenomenon of modern tourism in Europe is a blessing, a curse, or something in between is a question that can be argued from many perspectives; but it has unquestionably changed the face of the Mediterranean coast beyond recognition. A few miles into the rugged interiors, however, behind the glittering facade of the overdeveloped coast, the traditional elements of the Mediterranean may still be found, and declining villages still struggle to wrest a rewarding living from an inhospitable environment.

Industrial Development

Finally, the economies of the Mediterranean region have been selectively modified in recent years by new, often large-scale industrial implantations. Here again, the two major traditional attributes of the Mediterranean Sea itself: its routeway function linking Europe, Africa, and the Middle East; and its advantage as a low-cost transport facility, have been instrumental in promoting modern industrial growth. Thus, the transportation by sea of such products as oil from the Middle East make Mediterranean locations ideal for refining and petrochemical industries, as well as for transshipment operations. Iron and steel industries, also using imported raw materials brought by ship, are significant at several coastal sites.

Because of the general poverty of the Mediterranean regions as compared to more favored parts of several bordering countries, governments have encouraged industrial growth by a variety of supportive measures (see discussion of industrial location factors, Chapter 11). The complex French development at Fos, near Marseille, and many large-scale projects in southern Italy (steel at Taranto, petrochemicals at Bari) are examples of this influence.

SUMMARY

As in the Alpine case, the natural environment of the Mediterranean basin, notably its dry summers, has created particular challenges and opportunities for human economic systems. Once again, there is a remarkable similarity from one country to the next in both the problems and their solutions, so that it is indeed realistic to speak of a "universal" Mediterranean economy; only the pace of development, not its components, has distinguished "rich" from "poor" countries.

Dry-farmed crops, notably wheat, olive, and the vine, formed the basis of the Mediterranean food supply, while sheep and goats dominated the pastoral economy that spread over the rugged borderlands of the Mediterranean basin. The sea itself was a substantial resource for most bordering nations, providing a modest fishing livelihood for many communities and a vital transport route to the south and east. Irrigation, spreading westward with Muslim invasions, brought heightened prosperity to many regions of available water and led to the introduction of many new crops, such as citrus. In modern times, the spread of irrigation based on demands for off-season produce in northern Europe and the development of modern industries based on still-active trade connections have given new vitality to many parts of the Mediterranean; but it has been the dramatic growth of tourism that has perhaps done most to redefine the character of this most distinctive of all European regions.

Figure 8.8 Tourism has revived the economies of both the Mediterranean and Alpine regions: Nerja, Spain and Zermatt, Switzerland.

ADDITIONAL READING

HOUSTON, J.M., *The Western Mediterranean World.* London: Longmans, 1964.

LICHTENBERGER, E., *The Eastern Alps. Problem Regions of Europe series,* D.I. Scargill, ed., London: Oxford University, 1975.

LUCIANI, G., ed., *The Mediterranean Region.* New York: St. Martin's, 1984.

NETTING, R., *Balancing on an Alp: Change and Continuity in a Swiss Mountain Community.* Cambridge: Cambridge University, 1981.

WALKER, D.S., *The Mediterranean Lands.* New York: Wiley, 1960.

DISCUSSION QUESTIONS

1. What is "transhumance"? How does this practice enable humans to optimize their use of the Alpine resource base? How does it operate in a Mediterranean setting?
2. Most economic supports of the mountains are "extensive" rather than "intensive." What is the difference and why is this true?
3. Tourism has become a vital modern support of both Alpine and Mediteranean economies. Why is tourism particularly well suited as an economic base for expansion into these environments?
4. Wheat, the olive, and the vine formed the classic trilogy of the Mediterranean food base. What do these plants have in common?
5. Water power has long been, and remains today, a key Alpine resource. Why is this so? Why are the mountains more likely than other regions to rely on this form of power?

9
European Agriculture, Forestry, and Fisheries

AGRICULTURE

As the plants and animals domesticated in the Middle East (see discussion of domestications, Chapter 7), along with associated technology, farming methods, and ideas about land diffused into Europe, a distinctive pattern of agriculture began to emerge. If agricultural origins in Mesopotamia date to about 9000 BP, their slow infiltration into Europe (largely through the Mediterranean basin or along the Danube valley) meant that it was only about 4000 BP that agriculture was generally established over much of the continent. The process was a slow one, among other reasons, because of the climatic and soil differences between Europe and the Middle East. There is evidence, for example, that the earliest farming in northern Europe was undertaken on the light, sandy soils associated with certain glacial landscapes. Although these are relatively infertile, they could have been handled with the light plows familiar to more arid lands. Only much later, with teams of oxen and huge wheeled plows, could Europeans successfully tackle the heavy, podzolic forest soils that covered so much of the continent.

In the development and refinement of European agriculture, animal power, and its eventual extension into mechanical power, played a major role. Production often lagged behind demand, farm labor was a surprisingly scarce commodity in many countries at various times, and mechanization has thus been a constant theme. This has given rise to an extensive system of farming, stressing inputs of power, capital, and where possible, larger land units (although farm sizes in Europe remain relatively small), rather than labor. This contrasts markedly to the Asian, or intensive, system, in which large labor inputs have traditionally been used to achieve greater productivity. The European model of farming seems very familiar to most Americans, because it quite literally came off the boat with the early European colonists. With few exceptions, the crops, techniques, and ideas about farming that characterize American agriculture are European in their origin.

Grain Crops

There is a remarkable similarity among the peoples of the world in the kinds of things that make up a national diet, and in its traditional farming model, Europe is representative of many other regions. Some grain crop, for example, almost always forms the basic element of human food consumption: wheat, rice, and maize support, in one way or another, the vast majority of the world's population.

Figure 9.1 Europe: Basic agricultural data

	Production 1994 mil. metric tons	World Total(%)	World Rank	World Leaders	
CROP: WHEAT					
Europe	115	21		China	102 m.t.
France	30	6	5th	USA	63 m.t.
Germany	16	3	9th	Russia	37 m.t.
GB	13	2	11th	Ukraine	18 m.t.
Italy	8			Turkey	18 m.t.

Note: Fifteen European countries produce more than 1.5 million tons annually. European countries hold seven of the first nine places in yield/land unit (France's productivity is over three times that of the former USSR). In 1994, Europe accounted for more than 36% of world wheat exports.

CROP: MAIZE					
Europe	51	9		USA	254 m.t.
France	13	3	6th	(46%)	
Romania	10	2	9th	China	104 m.t.
CROP: RYE (1992)					
Europe	18	27		ex-USSR	20 m.t.
Poland	4	14	2nd		
Germany	3	10	3rd		
CROP: POTATOES					
Europe	66	24		ex-USSR	72 m.t.
Poland	23	8	3rd	(26%)	
Germany	12	4	7th	China	36 m.t.
CROP: WINE					
Europe (Mil. hectolitres)	170	66			
Italy	60	23	1st		
France	55	21	2nd		
Spain	18	7	4th		

Note: These three countries produce some 51% of the entire world total! Despite its famous production in California and elsewhere, the USA provides only some 7% of the world total.

*Russia and other ex-USSR nations not included in European totals.

Source: *Images Economiques du Monde,* 1995

European Agriculture, Forestry, and Fisheries

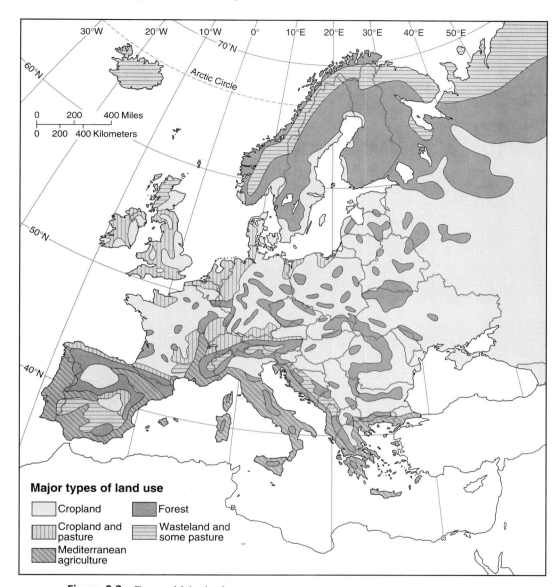

Figure 9.2 Europe: Major land-use types.

In Europe, the favored grain has always been wheat, and the pattern of its cultivation suggests the cultural importance its consumption has assumed. From the time of its Mesopotamian origins, wheat has been the classic crop of the world's semiarid grasslands; in fact one of its principal advantages is that it will make a productive and valuable crop in many regions which are climatically unsuited to any other significant type of cultivation. In Europe, wheat quickly became the classic crop of the Mediterranean basin (see discussion of Mediterranean economy, Chapter 8), yet it was in such demand throughout the continent that its cultivation gradually spread into all areas where it could possibly mature. Wheat's short growing season and modest moisture demands make it an ideal crop for many climates, but it does poorly in conditions of cool, wet ripening seasons, and thus cannot be grown successfully in many parts of northern and western Europe, where maritime and continental climates make its cultivation difficult.

The status of wheat, or light, bread is deeply engrained in European history and tradition. Thus, where wheat can possibly be grown, it will be, whether the setting is optimal or only marginal; taste creates value. The major wheat-producing country is France, where the rich soils and flat lands of the Paris Basin create one of the world's great agricultural regions. Wheat also does well, as has been seen, in all of the countries of the Mediterranean basin; but perhaps more telling is the fact that wheat is also grown by preference in such marginal locations as England and southern Sweden. Farther to the east, the semiarid climates and rich steppe grassland soils of Ukraine make this one of the world's most productive wheat growing nations.

Where wheat cannot be grown, other "lesser" grains are cultivated as the basic food supply. Chief among these is rye, the only grain other than wheat that will make a true rising bread. Rye is largely associated with the lands of northern and eastern Europe, extending into Russia, where it tolerates the cool, wet summers that make wheat cultivation impossible. The historic inferiority of rye creates an interesting inversion of modern values. For centuries, Europeans ranked themselves socially according to the type of bread that was being consumed. The rich, powerful, and successful demonstrated these virtues in part by consuming wheat or "white" bread; the poor or outcast had to be content with a crust of rye or "black" bread. In our times, many Europeans (and Americans) shun "white" bread and spend impressive sums at the delicatessen on various types of "black" bread! Such minor grains as oats (oatmeal is a classic food of Scotland largely because wheat will not grow there) and barley, all originally cultivated in the Middle East, fill in the European food gaps where wheat will not succeed.

Root Crops and Other Vegetables

In addition to a basic food grain, most world cultures rely on some sorts of root crops to supplement diet. This is particularly important because of the inground storage capabilities of many of these plants through periods of history before the advent of refrigeration, canning, or quick-freezing; it is still a vital concern in many areas where these preservation technologies are not well developed. In Europe, many types of beets filled this role, including particularly the large, orange fodder beets that were traditionally fed to livestock. For human consumption, such root crops as carrots, turnips, parsnips and rutabagas (definitely an acquired taste!) represent this aspect of diet, retaining a modest role in modern nutrition although the original reason for their cultivation has now passed.

Aboveground vegetables are also an important part of most diets, and in Europe these were also widely used as both human and animal sustenance. The hardy cabbage family of plants—including broccoli, kale, cauliflower, and brussels sprouts—thus found a major niche in the pattern of European cultivation. Today, fresh vegetables are as much a part of the year-round diet of most Europeans as they are of most Americans. Improved transportation and marketing techniques mean that vegetables that were once only seasonal (lettuce and tomatoes, for example) may now be

European Agriculture, Forestry, and Fisheries   **143**

Figure 9.3   Wheat and the vine in central France, Europe's richest agricultural region.

enjoyed throughout the year. In fact, a simple but reasonable guide to the relative economic success of European nations is to observe the variety of vegetables available in markets during the winter months.

Fats and Oils

All diets require some source of fats or oils. In Europe, this need was filled partially by such plants as the olive (notably in the Mediterranean region), and the sunflower, a traditional crop in many parts of eastern Europe and Russia. Otherwise, animals have been the source of this necessity, dairy products being for most of European history more important than meat as an animal product.

Beverage Crops

Finally, nearly all cultures seem to feel the need for a beverage crop; one of humankind's earliest skills appears to have been brewing! In Europe, where it was invented, beer has long been heavily consumed in the central and northern regions, where barley grows well. Various forms of cider have also been popular in England, France and elsewhere, leading to a proliferation of apple trees in fields and along roadways. Although a colorful part of the European scene, these trees (often very old and only marginally productive) inhibit the development of larger fields and modern agricultural practices in many regions; some governments have offered "bounties" on relict apple trees, sharing the cost of their removal with individual farmers.

In Europe, however, the principal beverage crop has long been the vine. This plant is another Middle Eastern domestication, and its cultivation around the Mediterranean extends back into the mists of history (see discussion of the Mediterranean

Figure 9.4 Because of its value, the vine has become a virtual monoculture in parts of southern Europe: Near Montpellier, France.

economy, Chapter 8). As did wheat, wine acquired a strong positive image in Europe, becoming the drink of the cultured and successful (and of anyone else who could afford a pot!). Thus its cultivation spread (often through the agency of the Romans) to the margins of northern Europe where the vine, again as wheat, cannot tolerate the cool, wet ripening season so characteristic of those latitudes.

Even to the north, however, any slight advantage of microclimate, such as a south-facing slope, was often given over to cultivation of the vine; the wines of the Rhine and Moselle river valleys are world famous, and even Switzerland produces a surprising volume of good wine on the relatively sunny slopes of Lakes Geneva and Neuchatel. Many northern wines were of strictly dubious character, however. One legend of Brittany, in northwestern France, notes that it requires three men and a wall to consume the local wine: one man to drink, two to hold him up, and the wall to keep him from falling backward!

In modern times, cultivation of the vine remains a major support of rural economies in nearly all districts where it will flourish or even survive. The reason for this is purely economic: more profit can be made on high-quality wine than on almost any other crop. Since there is (not surprisingly) a flood of cheap wine on the European market, most growers are working hard to upgrade their production and earn some sort of quality designation (strictly controlled by all governments) so that their product can fetch premium prices abroad as well as at home. It is not surprising that the two most classic of all European crops—wheat and the vine—are often found growing in close association with each other.

Arrivals from America

As European explorers moved out into the world in the fifteenth and sixteenth centuries, they discovered not only new nations and exotic cultures, but also a variety of food crops which had been unknown in Europe. While many of these remained only curiosities, or were unsuited for the growing conditions of Europe, at least two have had a major impact, and have established themselves as basic components of the European farming economy: the white potato and maize.

The potato is a domestication of the Andean highlands of South America, where it became the basic food crop of the Inca empire. Brought back to Europe by the Spanish following the American conquest, the potato was at first thought to be poisonous, and was viewed with deep suspicion. By the seventeenth century, however, its high nutrition values, storage capabilities over many months, and suitability for the cool, moist climate and sandy soils of the northern European nations made it a great success. We have seen how the Irish economy came to depend totally on the potato (see discussion under "migrations," Chapter 4), and a similar popularity prevailed all across Europe. In fact, the potato came to fill a vital food niche in the northern regions where wheat cultivation was marginal at best, and the area generally located along the shores of the Baltic Sea—Scandinavia, Finland, north Germany, Poland, the Baltic states and northwestern Russia—have come to be known as the "rye–potato belt," because of the importance of these two crops in local diets. Today, the potato is found on the menu of nearly all European countries: what Americans call "french fries" are called by the French merely "fries!"

The other major crop brought back to Europe from the Americas was maize (or, using the American term, "corn"). A domestication of the Valley of Mexico, maize cultivation had already spread widely across the Americas by the time of European impact. Whether or not the legend of Native Americans approaching Pilgrims in early New England with gifts of maize is true, the crop had certainly spread far from its origins by that time. Maize as a grain, however, is a crop not particularly well suited to most parts of Europe. A longer growing season than that of wheat, and a preference for long, hot, humid summer nights (as for example in the American "Corn Belt") meant that maize became a traditional food crop only in a few regions where climatic conditions were suitable, notably the Danube basin (especially Hungary, Romania and Moldova), southwestern France, the Galician region of Spain, and some parts of Italy.

Following World War II, however, part of the Marshall Plan, by which the United States helped the shattered nations of Europe rebuild their economies, included the introduction of the new varieties of hybrid maize which had been developed for American farming in the 1920s and 1930s. These high-yielding varieties proved quite successful, but even more significant was the realization that the entire maize plant could be converted to green silage and fed directly to animals, thus eliminating the need for the grain to ripen. Coinciding with a rapid increase in meat consumption in Europe and a consequently growing shortage of high-quality animal feed, this has meant that maize has been the most rapidly expanding crop in Europe over the past fifty years. It is now grown in nearly all environments, even to the shadows of the Alps, and has actually displaced wheat in some parts of the rich Paris Basin.

The Modern Role of Agriculture

The result of this variety of crops and the long centuries of farming as a way of life for most of Europe's population was the development of a strong sense of self-sufficiency: farmers ate and wore what they raised, and the more variety on the farm, the better life might be, or at least the more likely it was that life's disasters might be averted.

Thus the classic small European farm would raise a grain crop, root and vegetable crops, a beverage crop, and a variety of animals. This model came to the New World as the familiar (and vanishing) American family farm, with its strong emphasis on self-reliance.

At the present time, farming remains an important part of the European economy. Agriculture is productive and generally profitable, although employment in this sector has been declining. In general, the countries of southern and eastern Europe rely more heavily on agriculture for income and employment, while those of the north and west were quicker to turn more heavily to industrial and service sectors. Agricultural employment in the fifteen countries of the European Union (see discussion of

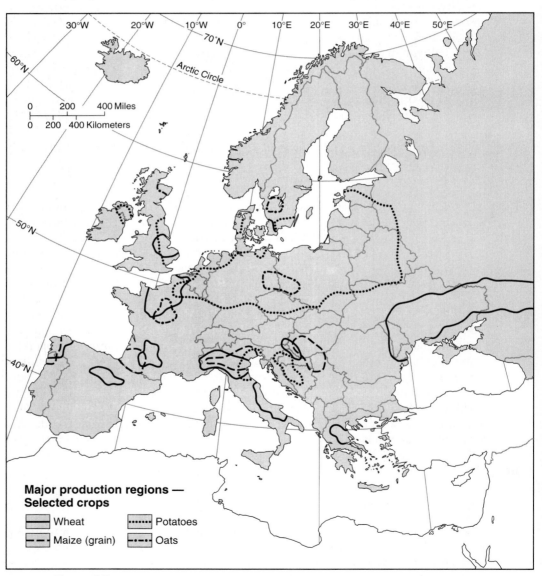

Figure 9.5 Europe: Major production regions of selected crops.

Figure 9.6 Maize is an American import that has become a common sight in the European landscape: Near Rakos, Hungary.

international organizations, Chapter 14) averages only 6.2%; but figures in the east are much higher: over 30% of Romanian labor is still employed in agriculture. Nonetheless, some of western Europe's most highly developed countries, such as Denmark, still feature a successful, export-oriented agricultural component as a mainstay of the economy.

Agriculture has played a vital role in the economic growth of Europe. It was, in fact, the long-term agricultural success of the continent through the centuries, based on the accumulating skills and developing technology of its farmers, that provided the surplus production, urban food supply, and investment capital that led to early industrialization and eventually to the modern image of Europe as the world's original rich and powerful continent. Today, while many countries remain importers of the basic foodstuffs that their crowded space (or, in the East, inefficient production) cannot provide, Europe has become a net exporting continent. With France and Italy leading the way, and with a strong emphasis on quality, Europe has contributed substantially to the current world food surplus. Moreover, the politics and economics of agriculture remain at the heart of many European debates and disagreements, and farm policy can arouse more passionate debates in national parliaments than virtually any other issue.

FORESTRY

Another primary economic activity widely dispersed in Europe is forestry, or the commercial exploitation of various woodland resources. Although statistically forestry plays a very modest role in the economies of most countries and employs only a small fraction of the labor force, nearly all nations have certain regions in which this activity is substantially more important. For a few nations, moreover, forestry is one of the most vital of all economic supports.

When humans arrived on the scene, Europe was largely a forested continent (see discussion of natural vegetation, Chapter 2), and the dense, mature woodlands presented both problems and resources for human economic development. In one sense, the forest was an enemy: dark, trackless, filled with wild beasts, the haunt of the supernatural. The European perception of "the forest," as it appears in literature and folklore, was until very recently a generally negative one: what is a good fairy tale without a witch's house hidden deep in a gloomy wood? On a more practical level, but still contributing to the sour European view of woodlands, was the fact that the forest had to be removed before any extensive agriculture could begin. Thus another classic confrontation: the long, tedious struggle to push back the trees (and they were always quick, during any periods of land abandonment, to invade the cleared lands and begin the complex process of succession, bringing the forest back again).

However, if the standing trees had little to recommend them to Europeans until the ecological vogue of the twentieth century (collecting nuts and other forest products and allowing animals to browse the vegetation within reach were common, but generally marginal, activities), the trimmed and felled trees offered an entirely different category of resources to an expanding European population. In addition to being the universal fuel, wood was used to make the vast quantities of charcoal that were the basic requirement of iron smelting for many centuries (see energy resources discussion, Chapter 10). Moreover, wood was used to build Europe's expanding cities and growing navies, and to provide the countless railway ties, pit props for mines, and many other material requirements of the rapidly emerging industrial age. Removing forests was thus viewed as a positive activity in both senses: clearing potential agricultural land while simultaneously providing a most valuable resource.

The result of this greatly accelerated assault on Europe's woodlands, however, was the general deforestation of much of the continent. Contemporary reports tell of complete loss of forest cover across much of the Mediterranean basin as early as classical times, and by the time the industrial revolution and its ravenous steam engines finally turned to coal as a fuel, Europe was one of the world's least forested regions. Nor has the demand for wood slackened in Europe with the passage of time; only the uses have changed. Wood as a building material remains important, but its roles as a fuel and in metallurgy, mining, and transportation have largely been replaced by an enormous demand for wood pulp, paper, and related products.

The great advantage of forests as a resource, of course, is that they will regrow, and while the ancient large trees that comprised the original forest are today very rare, twenty-five to forty years is sufficient to re-create a wooded landscape, and to produce a useful pulpwood tree. Thus, in nearly every European country, large tracts of land have been replanted to forests, and forest management practices have been improved and modernized. As well as providing a cash crop for many often isolated regions, reforestation also has numerous environmentally beneficial effects. Slower runoff of precipitation reduces soil erosion and permits more water to percolate into ground water reservoirs, wildlife habitat is extended and enhanced, and additional possibilities for human recreational activities are provided. Much early reforestation featured fast-growing conifer trees, which often replaced what had been broadleaf or mixed forests, thus creating a more sterile environment. Pressure by environmental groups has led to a change in philosophy in many countries, where reestablishment of something approximating the original forest is now more fashionable.

Figure 9.7 Forest resources are vital to the Scandinavian, Finnish, and Alpine economies: Grandvin, Norway.

While forestry on a sustained yield basis is now a goal of nearly every nation, consumption continues to increase, and pressure on available forest resources is strong. This means that those countries where large tracts of original forest have survived, through low population densities and historic isolation from major areas of consumption, are in an advantageous position today. This includes particularly the Alpine countries—notably Austria—and those of the far north—Sweden, Finland, Norway—which are among the world leaders in exporting both saw timber and wood pulp or its products. Karelia and the Kola Peninsula region of northwestern Russia also rely heavily on the forest resource.

Although forestry is obviously a minor component of the complex economy of most European countries, for some it plays an absolutely vital role. Finland, for example, has a population of only about 5 million, yet it ranks behind only Canada (29 million) and Sweden (9 million) as an exporter of timber and wood products. Over 40 percent of the country's exports by value consists of forest products, and when the additional value of wood processing industries, the manufacture of related equipment, the export of furniture and handicrafts based on wood, and the great volume of forest-related technical expertise available to the world are considered, the overwhelming value of its forests to Finland's economy becomes evident.

FISHERIES

Humans have fed themselves by fishing for substantially longer than they have practiced agriculture, and since Europe is a continent in which nearly all countries have a slice of coastline (or at least a variety of lake and river resources), fishing has been a vital as well as colorful component of the economy virtually as far back as human settlement itself can be traced. Since fish and other seafood tend to spoil quickly, consuming them fresh was long a possibility denied to all but the coastal dwellers themselves, while various methods of preservation created a variety of foods that Europeans still enjoy today. Smoked salmon and pickled herring are well-known delicacies, and along the quays of ancient fishing ports such as Bergen, Norway may still be seen the warehouses where millions of cod were salted, dried, and literally baled for shipment all over Europe.

The European fishery was stimulated by the location of many nations on the North Sea, one of the world's great fishing grounds. Later, European fishermen ventured farther afield, to the rich "banks" off the coast of Newfoundland, in North America. In both of these areas, shallow water, turbulence, and the mixing of warm and cold currents provide light and nutrient near the surface, thus stimulating production of plankton, the basis of the sea's food chain. Another useful biological characteristic of these northern seas is that although there tend to be relatively few species of fish, these occur in large concentrations, so it is easy to fish exclusively for a single species, e.g., cod, herring, sardine. Thus, the nations of the Atlantic and North Sea coasts, with easy access to large stocks of fish, and often with long seafaring traditions, became and remain among the leading fishing nations.

The southern countries, with the exceptions of Atlantic-facing Portugal and Spain, are less important in this industry. The Mediterranean and Black Seas, with their lack of water overturn and their increasingly polluted conditions, are comparatively poor fisheries, and a typical catch would contain only a few fish of several species. It is not surprising that some of the classical cuisine of the Mediterranean region consists of fish stews (*bouillabaisse*, for example), in which whatever comes to the net goes to the pot!

Since only a few European countries are landlocked, since there are only a few areas with abundant lakes (the north and the Alpine area), and since stream pollution has in many areas reached such dimensions as to endanger all fish stocks, freshwater fishing on a commercial basis is much less important than sea fishing. Many centuries of overfishing have seriously depleted fish populations in many cases, although sports fishing is still very popular throughout Europe. It was a great shock a few years ago to the Swiss and French residents on the shores of Lake Geneva to discover that the lake perch which are such a prized food delicacy in lakeside restaurants are now being largely supplied frozen from Canada! (See environmental discussion, Chapter 13.)

As might be imagined on a continent with so many political divisions, fishing is divided among many countries. Europe—except for Russia—catches roughly 15 percent of the world's fish, but only Norway, at just under 3 percent, ranks among the world's leading fishing nations (usually ninth in recent years). Denmark, Iceland, Spain, and the United Kingdom are the next leading producers, but nine other countries also make a significant contribution to the total catch. For most of these nations, as is the case for forestry, fishing is but a small footnote to the overall economy, but for a few it is a vital support. The importance of fishing to Norway, with only 4.3 million people, is obvious; but Norway has some other major resources, notably North Sea oil. A more interesting case is Iceland, Europe's fourth leading fishing nation (and the world's nineteenth) despite having a population of fewer than 300,000. Fish and fish products account for over two thirds of Iceland's export income, and it is scarcely surprising that the Icelanders are very touchy about fishing craft from other nations exploiting their territorial waters.

Figure 9.8 Fishing is an activity divided among many small European ports: Lastres, Spain.

Fishing is a colorful, animated, and timeless element of the scene in countless small European coastal cities and towns. It also provides a welcome component of diet and often a source of export income. Unfortunately, however, fishing in the traditional sense is a dying industry. There is really no living to be made on the small fishing vessel that sells its meagre catch off the dock at the end of each day, and although small fishermen cherish their independence, their children often seek better rewards in land-based industrial or service jobs. Most of the world's fish are now being caught in a highly automated fashion, by large vessels with abundant and expensive equipment, and few European countries except Norway have been able to afford the massive investment necessary to compete with the giants of the long-distance industry: Japan and Russia.

Mention should finally be made of the important shellfishing industry conducted along many European coasts. Oysters, clams, mussels, and other animals that thrive in the shallow waters of the coastal zone have historically been a featured item of diet in Europe, and their production has yielded a good living for many fisherfolk. Recently, however, increasing pollution of the coastal environment (an especially vulnerable zone where rich biological productivity and human development inevitably clash) has sharply curtailed the industry in many areas. Imaginative ideas, such as raising shellfish on artificial "islands" offshore, have restored production to some extent; but the era when a half dozen large oysters, black bread and butter, and a glass of white wine constituted an inexpensive snack at the local cafe is long past.

ADDITIONAL READING

ALEXANDER, L.M., *Offshore Geography of Northwestern Europe.* Chicago: Rand McNally for the Association of American Geographers, 1963.

BOWLER, I.R., *Agriculture Under the Common Agricultural Policy.* Manchester: Manchester University Press, 1985.

CLOUT, H., *Agriculture: Studies in Contemporary Europe.* London: Macmillan, 1971.

CLOUT, H., *A Rural Policy for the E.E.C.?* London and New York: Methuen, 1984.

JONES, J.,ed., *East-West Agricultural Trade.* Boulder CO: Westview, 1986.

MARSH, J.S., and P. SWANNEY, *Agriculture and the European Community.* London: George Allen and Unwin, 1980.

DISCUSSION QUESTIONS

1. Two food plants brought to Europe from the Americas in the sixteenth century have become important elements of the agricultural landscape. What are they, and why did they become important?
2. Rye, oats, and barley largely replace wheat as the most important cereal grains in northern and northeastern Europe. Why is this? Why are the lands along the Baltic Sea often called the "rye–potato belt"?
3. What role do livestock play in the modern European agricultural economy? Why has animal feed of various kinds become a major American export to Europe?
4. As a proportion of its total economy, Finland relies on wood and wood products more than does any other European country. Why is this industry more important here than closer to major markets?
5. Fishing is to the Icelandic economy what wood is to the Finnish. Explain. In terms of employment, fishing is a declining industry in many European countries. Why is this?

10

European Resources

The successful development and economic growth of any of the world's regions depend on intelligent utilization of available resources, or the acquisition of those that are in short supply. "Resource" is of course a very broad term, which may imply population, land, water, or many other items; in a more conventional sense, it applies to mineral deposits of various kinds, and it is in this sense—with some exceptions—that the word will be used here. Most European countries were at least modestly endowed with a variety of mineral resources, but a long history of economic growth and resource consumption has meant that in modern times the continent has become a heavy importer of most such commodities. Discussion of European resources and their utilization can usefully be divided into two sections: power and fuel sources, and metallic and nonmetallic minerals.

POWER AND FUEL RESOURCES

Traditional Energy Sources

1. Animal power. For most of human history, Europeans, as well as the rest of the world's peoples, were limited to three basic power sources beyond their own muscle power: animal energy, the wind, and running water. Although these sources all evolved in technology and complexity, it is remarkable that the enormous energy impact of animal domestication (see discussion of domestications, Chapter 7) was not duplicated again until the perfection of the steam engine nearly 9000 years later. Consider the fact, for example, that the speed at which a person could move did not exceed the steady gait of a horse (about 8 mph) until the rise of railways less than two centuries ago!

Although animal power was a vast improvement over the comparatively weak human muscle structure, its applications were necessarily limited, being restricted largely to agricultural uses. Animals could be harnessed to turn grindstones, pump water, or produce other simple motions; but it was on the farm, as plough teams or hauling "machines," that their impact was really felt. A particular importance of animals, however, is their portability. It has been noted that animals opened up many of the world's marginal lands to humans by providing some sort of economic base that could lend value to these difficult environments. In Europe, animal adaptation to a wide range of climatic conditions meant that they became an almost universal power source.

    2. Wind power. Wind power also became a standard European power source, although its distribution was necessarily more restricted than that of animals. The coastal areas of nearly all Western Europe, where the westerly winds blow with great regularity, as well as many interior areas where height and exposure make wind a constant feature of daily weather, were well suited to the development of wind power. As the rather complicated technologies of gearing, power transfer, and applications were gradually solved (largely by classical times), windmills spread throughout the continent. Although wind power was harnessed to perform a variety of simple mechanical functions (sawing, grinding, etc.), it was as pumping devices that it gained its greatest importance. The windmills of the Netherlands were used to drain the original lands reclaimed from under the sea, and have become such a powerful symbol of the Dutch struggle against the forces of nature that they are preserved as national monuments today, although their practical use as pumps has almost entirely vanished.

    3. Water power. The most important of these early power sources in Europe was water. The combination of generally hilly topography throughout much of the continent, and the even flow throughout the year of most rivers and streams (except in the dry-summer Mediterranean lands) meant that water power could be a more reliable and larger-scale energy source than the wind. Water power is more concentrated than other primitive energy sources, and on the eve of the industrial revolution multiple combinations of water mills were powering industries of truly impressive dimensions. The "mill," which most often meant a water mill, became the center of community life in many parts of Europe; "Miller" (in many languages) became a very common surname; and much economic activity revolved around the mill site, which often became the nucleus of town or city growth. In literally thousands of European places today, origins may be traced to a mill site, and even though the town may have developed more recently in quite different ways, the original mill building, often with its characteristic pond, may still be found near the center.

    4. Wood as the traditional fuel resource. The other energy component of economic growth, beyond power, is fuel; and the basic European fuel down to the eighteenth century was wood (as it remains today in many less developed areas of the world). Wood as a fuel had many advantages. Its combustion is perhaps the simplest of all technologies. As an almost entirely forested continent, Europe had an abundant supply; and as a metallurgical fuel, wood (in the refined form of charcoal) provided the higher temperatures required for smelting iron of better quality. In addition, of course, wood has always been a favored building material, and as the cities, ships, and other constructions of Europe grew and multiplied, the continent's forests began to shrink at an alarming rate. Moreover, a vast acreage of forest was destroyed simply to clear the land for agriculture. The sort of adversary relationship that enters into so much of Europe's folk literature is no accident. The woods and their dangerous inhabitants had constantly to be forced back; otherwise they would surely have reclaimed humans' hard-earned lands.

By the end of the seventeenth century, the shortage of wood was beginning to create what has been called an "early energy crisis." Deforestation, which had stripped the Mediterranean basin of its original cover as early as classical times, spread to all corners of the continent; only the most remote or sparsely inhabited areas were spared. As the crisis deepened, the price of wood rose steeply. Having used all locally available wood for charcoal, iron forges faced with steadily increasing demand found it necessary to import wood from farther and farther afield. One estimate of production costs in England about 1750 noted that some 90% of the cost of making iron was accounted for by the need to import wood from vast distances!

At this juncture there emerged on the scene what many authorities have considered the most significant single invention (in terms of its total technological, economic, and social impacts) in human history: the perfected steam engine. This device is largely attributed to James Watt (1769), who dramatically improved existing engines (largely by substituting metal for wooden parts) and thus launched the industrial revolution with the greatest quantum leap in energy availability since animal domestication took place so many centuries before. If the steam engine put existing animal, wind, and water power technologies into the backwaters of European development, it also posed an agonizing question: what to use for fuel? As an external combustion engine (burning the fuel outside the engine; in effect boiling water to produce steam), the steam engine could of course use wood; but there was virtually no wood left! Into this gap stepped the next major component of the European resource picture: coal.

Coal

"The rock that burns" had been known to man for many centuries, but its use on a large scale began only with the advent of the steam engine and the demands of the industrial revolution. Its rapid rise as a home heating fuel combined with rapidly escalating industrial demands to expand consumption prodigiously. Moreover, coal, in the purified form of coke, also displaced wood (charcoal) as a metallurgical fuel, and as the demand for iron and steel grew, still more coal was called for. The question rapidly became: who has the coal?

It soon became clear that, for the first time in the history of European power and fuel needs, the Europeans were dealing with a resource that was not at all evenly distributed. While animals, wind, and water were all relatively common (every nation had some), coal was, by the geologic nature of its formation, restricted to a relatively few regions where conditions had been suitable. Thus, in the midst of the rush to find coal reserves, Europe for the first time began to discover national inequities based on the presence or absence of a vital resource, a reality that was to prove extremely troublesome in future years.

It soon became apparent that some countries were much better endowed with coal than others, and in the most general terms it is reasonable to say that those nations (or eventual nations) that had the best quality and most easily worked coal quickly became the most powerful economic nations on the continent (Great Britain, Germany, France, Belgium, and later Russia and Ukraine); while those nations that came up relatively empty (Italy, Sweden, Portugal) tended to see their industrial development lag behind, and often to decline from the roles of influence they had enjoyed in previous centuries. While many other historical, social, political, and economic factors certainly contributed to the rearrangement of power relationships in eighteenth and nineteenth century Europe, the unequal distribution of coal clearly played a crucial role.

Coal is the result of a geologic process that converts vegetation into various qualities of combustible material by applying heat and pressure. During periods of earth history when vast areas of Europe and other continents were covered with shallow swamps, dead and decaying trees accumulated under the waters over millions of years (the underwater, low oxygen environment preserving much of the wood). As climates

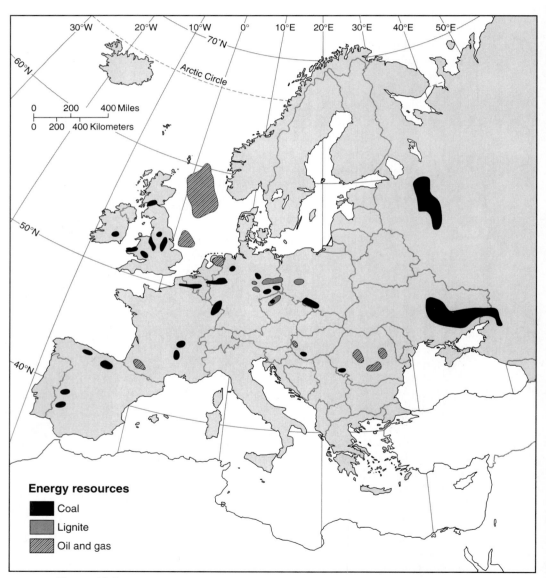

Figure 10.1 Europe: Energy resources.

Figure 10.2 Coal was the energy source for the Industrial Revolution, and remains vital to many European countries today: Forbach, France.

changed, other materials—sand, gravel, mud—were layered on top of the vegetation, eventually forming sandstones, shales, and other rock types. Except where erosion stripped away these formations, the coal was thus preserved for future human use.

The quality of the eventual product depends largely on the length of time over which the process of formation took place, and consequently on the purity of the resulting carbon. The first stage is peat, which is loosely consolidated material with very low heating values. It can, however, be dried and used as household fuel, or on a larger scale in power generation. Ireland and Russia, with large areas of peat bog, are the leading European consuming nations.

If the process continues, the next product is lignite, or "brown coal." This low-quality material is a common resource in eastern Germany, Poland, and the Czech Republic, and its widespread use as a fuel for heating and power production in this region has been one of the leading causes of environmental stress (see Chapter 13).

At the higher end of the quality scale are the resources resulting from the culmination of the coal-forming cycle. The most important of these is bituminous coal, various grades of which were (and to some extent remain) the prized fuels for both household and industrial use during the "age of coal" (roughly 1750–1950). Good quality bituminous was also used in the manufacture of metallurgical coke. The best and most valuable of all is anthracite, a nearly pure form of carbon that burns with almost no smoke. Unfortunately, despite its many advantages—higher heat values, less pollution—anthracite is comparatively rare and thus often excessively costly.

Oil

Toward the end of the nineteenth century, the technological wheel turned yet again in Europe with the invention of the internal combustion engine. Generally credited to the Germans Daimler and Benz in 1885, this new power source came to revolutionize many aspects of human society (most notably transportation) within a surprisingly few years. Although the internal combustion engine did not represent much of an absolute power increase as compared to that of the steam engine, the fact that fuel was burned within the engine itself eliminated the external boiler, water supply, and solid fuel source which had made the steam engine generally a large, cumbersome device. The internal combustion engine offered the advantages of small size, portability, and extreme diversity which, as particularly evidenced by the automobile, have made it the preferred power source of the modern world.

After a variety of experimentation (the internal combustion engine, with proper carburation, will run on a wide variety of liquid fuels), gasoline refined from petroleum became the standard fuel for these new engines, particularly in the United States and Europe, where petroleum had been used as a lubricant and in other minor applications for many years. Once again, therefore, as the demand for oil began to grow and its potential uses to multiply, the cry went out: "where can it be found?" Although its formation as a fossil fuel is similar to that of coal, oil is a much more rare commodity because of its liquid or (in the case of often-related natural gas) gaseous nature. Formed again under shallow seas, in this case the product of microorganisms, there must be the proper kinds of geologic structures to trap the oil and/or gas, or they will simply vanish into the rock formations or evaporate into the air as rapidly as they are formed. Finding oil in large quantities, therefore, was, in the days before sophisticated exploration techniques were developed, a sort of geologic lottery, with large rewards going only to a few winners!

Unfortunately for the European nations, the initial rush for oil supplies came up very short. None of the major European powers, including all of those who had gotten rich on their coal reserves, found enough land-based, easily acquired oil to meet more than a modest share of demand beyond the first few years of the internal combustion era. The only European nation where important oil deposits were found was Romania, which actually led world production for a few years in the mid-1800s, until Russian and then American production became dominant by the close of the century. The historic importance of the Romanian fields at Ploesti was perhaps most vividly illustrated by their role as the target of one of the first major Allied air raids launched against German-occupied Europe in 1942. Although terribly costly in terms of both planes and crews, the raid was seen as a strategically useful investment designed (unsuccessfully, as it turned out) to disrupt production at one of the only major sources of petroleum then available to the German war machine.

European Resources 159

Figure 10.3 Oil beneath the North Sea brought unexpected wealth to parts of Norway and Great Britain: Aberdeen, Scotland.

Beginning in the 1960s, the march of technology made available equipment capable of undertaking drilling beneath the relatively shallow waters of the North Sea. Geologic investigation had revealed the existence of oil-bearing formations some time earlier, but not until stable (and very costly) drilling rigs could be developed, towed into position, and braced on the sea bed to resemble artificial islands, could detailed exploration and eventual production actually be contemplated. Large-scale oil and gas production from under the North Sea territorial waters of Great Britain and Norway has been taking place since the mid-1970s, and although production is expected to peak and decline early in the next century, North Sea oil has brought welcome energy reserves and income to the two participants. Since the shallow continental margins are essentially parts of the mainland, oil-bearing formations may be found at other points under Europe's surrounding seas. Among other initiatives, exploration is now underway off the Mediterranean coast of Spain, and in the Aegean Sea between Greece and Turkey. While the search may prove fruitless (looking for oil is about equivalent to high-stakes gambling), it is hard for Europeans to resist the slimmest chance.

The impacts of imported oil. Even the discovery of North Sea oil, however, has not done much for the prospects of the majority of European nations. Oil consumption in such important industrial, automobile-oriented countries as France, Germany, and Italy has reached dramatic levels. Without domestic reserves, some European nations have found themselves in the position of relying on imported oil for as much as 80 percent of their total energy needs, a level considered dangerous to national security. This has led to a variety of economic patterns within Europe, as well as some major efforts to reach more realistic levels of energy self-sufficiency.

Oil import, for nearly all of western Europe, means turning to the oil producing nations of the Middle East. The world's single greatest commodity flow is the movement of oil from the Persian (or Arabian) Gulf to Europe. Tankers from these waters move continuously through the Suez Canal or (the larger ones) around the Cape of Good Hope to Europe. Since the closest European shores to the source of oil are therefore on the western extremities of the continent, a number of large, oil-oriented port cities have sprung up at what would otherwise be unlikely sites. Bantry Bay (Ireland), Milford Haven (Wales), and La Coruna (Spain) are hardly household place names, yet these exclusively oil importing and processing towns are among the most important ports in Europe. These developments have attracted a number of other oil- or energy-dependent industries as well, and the general result of large-scale oil importing has been to shift the balance of economic growth in Europe somewhat more to the edges of the continent than was the case in a pre-oil era. During the Soviet era, the nations of eastern Europe relied on oil from Russia, often provided at low cost. The collapse of this system has meant that these countries must acquire their imported supplies at market prices, and this has been one of the most difficult economic adjustments to post-Soviet reality.

Oil dependence has also heightened European concern for and interest in the Middle East. Any peacekeeping initiatives or prospects of stability in that troubled region will be certain to have warm European approval and support (note for example, French participation in the "Gulf War" of 1991). In general, however, the European nations, with their heavily energy-dependent economies and energy-consumptive societies, are perhaps most notable for their interests in energy conservation and in the development of alternative energy sources.

Conservation and New Energy Sources

Since most European states have been short of oil almost since the dawn of the age of petroleum, and since this fuel has become the fashionable energy source of the twentieth century, Europeans have been struggling to cope with the question of reliable supplies of energy. Oil has, not surprisingly, been viewed and priced as a scarce commodity in Europe, as opposed to in the United States, where abundant (and presumably "inexhaustible") supplies of oil led to its being grossly undervalued. Very high gasoline prices in Europe ($3.00 to $4.00 or more per gallon in most countries) continue to be the rule in the 1990s, and these have served very effectively to conserve oil by reducing mileage driven, encouraging public transportation, and promoting development of smaller, more fuel-efficient vehicles. While the United States, in fact, may be viewed as an economy founded on cheap energy, European nations without exception have much lower levels of per-capita energy—especially oil—consumption.

The search for new supplies of conventional oil has been exhaustive in Europe; but, except for the North Sea, improvements in prospecting and recovery technology have provided few major finds. Natural gas, related to oil in its formation, is also scarce. The largest onshore field was discovered at Groningen, in the northernmost province of the Netherlands; revenues from the sale of this gas to energy-hungry neighbors has permitted the Netherlands to support an extensive program of social services. A more modest field at Lacq in the Aquitaine region of southern France has been important in regional development; and of course the North Sea has also been a natural gas—as well as oil—bonanza for Great Britain and Norway. These supplies may already be dwindling, however, and the shortage of this clean, versatile fuel explains the intense European interest in developing and acquiring a portion of the vast gas reserves in Russian Siberia.

European Resources

Figure 10.4 Europe: Basic energy data*

	Production 1994 mil. metric tons	World Total (%)	World Rank	World Leaders		World Total (%)	Rank
PRODUCT: COAL							
Europe	287	8		China	1,100 m.t.	30	1st
Poland	134	4	7th	USA	833 m.t.	24	2nd
Germany	58	2		ex-USSR	398 m.t.	11	3rd
PRODUCT: LIGNITE							
Europe	513	53		ex-USSR	145 m.t.	14	2nd
Germany	207	23	1st				
Poland	68	4					
PRODUCT: OIL							
Europe	284	9		Saudi Arabia	428 m.t.	14	1st
Great Britain	127	4	8th	USA	389 m.t.	122	2nd
Norway	119	4	9th	Russia	352 m.t.	11	3rd

Note: Great Britain and Norway alone (North Sea) account for 87% of European production. Despite having only 9% of world crude production, Europe has some 22% of world refining capacity. Except for the former USSR, Europe has only an estimated 1.7% of world reserves.

	Production 1994 mil. metric tons	World Total (%)	World Rank	World Leaders		World Total (%)	Rank
PRODUCT: NATURAL GAS							
Europe (bill cub mtr)	264	12		Russia	607	27	1st
Netherlands	78	4	4th	USA	539	24	2nd
Great Britain	71	3	5th	Canada	149	7	3rd
Norway	30	2	11th				

Note: These three countries produce about 60% of Europe's total. The significance of the North Sea is again obvious.

	Production 1994 mil. metric tons	World Total (%)	World Rank	World Leaders		World Total (%)	Rank
PRODUCT: NUCLEAR ELECTRICITY: (000 MWe)							
Europe	132	38		USA	101	29	1st
France	59	17	2nd	Japan	40	12	3rd
Germany	23	7	5th	ex-USSR	35	10	4th
Great Britain	13	4					

Note: In terms of the overall importance of nuclear energy to the national economy, France is clearly the world leader.

*Russia and other ex-USSR nations not included in European totals.
Source: *Images Economiques du Monde,* 1995.

Despite the fact that relatively low oil prices throughout the 1990s have made research into alternative energy forms rather an expensive proposition, futuristic energy sources have also strongly tempted the Europeans. Nuclear fusion research is by 1996 more advanced and better funded in Europe than in the United States, with the European Center for Nuclear Research (CERN) facility near Geneva, Switzerland and other sites attracting research funds from several nations. Solar energy is also an attractive promise, particularly in the relatively sunny regions of the Mediterranean; and research on the concentration, collection, and distribution of this boundless source is well advanced throughout Europe. In the early 1980s, solar energy from the experimental French station at Odille, high in the Pyrenees, was fed into the nation's main distribution system, thus permitting an evaluation of large-scale production costs and problems, and making France arguably the first nation where solar energy has gone beyond small solar cells and individually solar-heated buildings.

For most European countries, however (as indeed for most of the world), the crucial question in the field of energy is how to bridge the gap between the decline of conventional sources (especially oil and in some regions gas) in the early years of the twenty-first century and the arrival in something more than experimental quantities of the energy sources of the future (fusion and solar), a prospect not generally considered realistic before about the middle of the twenty-first century. To Europe, accustomed through long years of coping with energy shortages, the options seem more clearly laid out than they do to the United States, which still seems to pin extravagant hopes on continued supplies of cheap oil. Most European countries, accordingly, are heavily into development of intermediate energy technologies; using resources already available and ideas already proven feasible, they are making great efforts to be prepared for the next energy crisis.

A major difference in European and American approaches to energy is found in the approach to conventional nuclear power (fission). France is today the unquestioned world leader in the technology and utilization of nuclear electricity, and many other European nations had been expanding their nuclear generating capacity steadily and deliberately, despite the substantial environmental questions that such a policy raises, until these were brought into dramatic focus by the 1986 Chernobyl disaster. Most countries have now decided that nuclear risks exceed benefits and have accordingly scaled back or abandoned their programs in this area. Not so, however, in France (see perception box, Chapter 17).

Beyond this, Europe still has abundant coal supplies, even if much of the remaining resource is of inferior quality and/or expensively acquired. The technology of

Figure 10.5 Solar energy collectors for modest uses are common across the Mediterranean region: Greece.

Figure 10.6 A modern generation of windmills generates electricity in many European countries: Denmark.

converting coal into other forms of energy (even into gasoline, which not surprisingly was pioneered by German scientists during World War II) is well known in Europe, and as the price of oil inevitably returns to its historic high levels of the early 1980s, the costs of this technology will come to seem more reasonable, even given the environmental price tag that goes with it (see environmental discussion, Chapter 13).

Other energy-related technologies of interest to Europeans include geothermal power (widespread on Iceland and at Bilardello, Italy); new forms of wind-generated power that are now under development (Denmark is a world leader in this technology); and waste heat recycling in many industrial processes. Europe was also the first continent to generate electricity from the power of the tides. The tidal power station on the Rance River in Brittany, France, has been operational since the 1960s, and has provided useful data on the feasibility of this process. While none of these unconventional sources generates enough energy to make a real impact on Europe's overall needs, many are of critical local importance, and together they emphasize Europe's determination to retain a leading role in the development of new energy alternatives for the twenty-first century, and to free the continent to as large a degree as possible from the hazards of heavy dependence on foreign energy sources.

As a final point, it should be noted that hydroelectric power, although widely appreciated because of its environmentally "clean" character and its ability to revitalize remote mountainous regions, has little chance of expanding beyond its current rather modest share of the overall energy package. Most European countries have developed their hydroelectric potential to the limit (Switzerland is the classic example of an Alpine nation that can no longer meet its needs by water power alone); and although a few nations have surplus hydropower to sell (Norway, for instance), the general picture is one of declining, rather than increasing, shares of national energy supplies.

METALLIC AND NONMETALLIC MINERALS

If the development of European energy resources, as noted above, contains a lesson, it is that Europe was generally well provided (although in a poorly distributed way) with coal, and has been generally deficient in petroleum; both factors have done much to determine the economic and political history of the continent. It is worth remembering, however, that no resource has value unless some human group covets it; coal was scarcely of interest before development of the steam engine, while oil was only a lubricant (or a medicine!) well into the nineteenth century. Much human history, on the other hand, has been based on the exploitation of metals, and this technology extends much farther into the past. When animals for power and wood for fuel were still the limits of human energy sources, people were already learning the process of smelting bronze from copper and a bit of tin to make a harder, more useful product. Bronze technology seems also to have been concentrated in the Middle East, where plant and animal domestications have been shown to play such an important role in human affairs, and metal working diffused into Europe at a fairly early date. The great ages of early human culture are in fact named for the metal-working skills that characterized them: if the Bronze Age is dated at about 9000 years BP, the Iron Age which followed had become generalized in Europe by about 2000 BP.

What this antiquity of metal fabrication means is that, combined with the relatively early settlement and beginnings of technological sophistication found in Europe, great pressure has been placed on Europe's metallic mineral resources for a much greater time than upon its fuel resources. There is some evidence that the tin mines of Cornwall—in what is now western Britain—may have been known to Bronze Age people in the Middle East as long as 9000 years ago (copper was abundant but tin rather scarce in that region), and that sea voyages may have been undertaken to acquire the precious metal (a trip roughly equivalent to bringing back materials from the moon in our times). This, in turn, means that Europe, which originally had a reasonable abundance of most metals, today has important amounts of very few—not that they were never present, but that they have been so heavily exploited over all the centuries of European development.

Metallic minerals differ sharply from fuel resources in the nature of their formation. Coal, oil, and gas are related to processes of sedimentation; they are organic materials laid down under shallow seas and broad swamps. Metals, on the other hand, are related to the geologic processes of mountain building. The tectonic activity associated with crustal stresses also brings to the surface a variety of minerals with the molten material from the earth's interior. Modern mineral deposition is known to be occurring, for example, along the subocean seams in the earth's crust that mark the divisions of the continental plates. Since, as has been noted, Europe is largely a mountainous continent, areas of important mineral deposits have been fairly widely distributed, with most countries enjoying at least a modest share of such valuable preindustrial minerals as gold, silver and especially iron ore. Even during the early centuries of the Iron Age, the common occurrence of this mineral in forms that could be utilized by primitive iron-making technology (iron is one of the most common components of the earth's crust) meant that smelting was a very widespread activity.

Iron ore. As the pace and scope of industrialization began to increase, however, culminating in the extravagant resource demands of the industrial revolution, larger bodies of iron ore, especially those near coal fields, were rapidly depleted. This was

European Resources

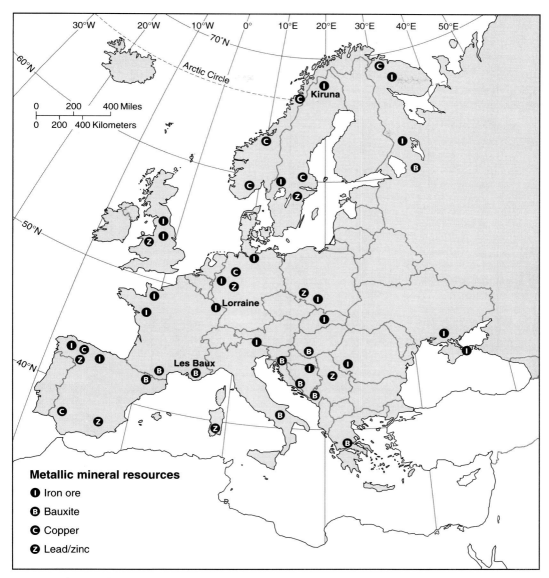

Figure 10.7 Europe: Location of selected metallic minerals.

notably the case in both Germany and Great Britain. As a result, there are only two major iron ore-producing regions in Europe west of Russia and Ukraine today (neither very significant on a world scale), and a number of other areas where modest production can at best reduce the level of imports. The survival of even two important producing regions must be explained by exceptional factors.

The most important region of iron ore production in northern Europe today is in extreme northern Sweden, where the towns of Kiruna and Gallivåre are the center of a district producing very high-grade ore (up to 60% iron content). The ore is shipped from Luleå on the Baltic (where a small steel mill has been established) in the summer,

Figure 10.8 Europe: Other mineral resources*

	Production 1993/1994 (Million Metric Tons)	World Total (%)	World Leaders	Total	%
PRODUCT: IRON ORE					
Europe	18	3	Brazil	100	18
Sweden	12	2	ex-USSR	93	17
Norway	1.6				
PRODUCT: STEEL					
Europe	183	25	Japan	96	13
Germany	41	6	ex-USSR	95	13
Italy	26	3	China	92	13
France	18	2			

Note: With 25% of world steel production and only 3% of the iron ore production, it is clear that Europe is a major importer of this commodity. Nineteen European countries now produce steel.

PRODUCT: BAUXITE					
Europe	6	5	Australia	42	38
ex-Yugoslav.	3				
Greece	2				
PRODUCT: ALUMINUM					
Europe	5.8	30	USA	3.3	17
Norway	1		ex-USSR	3.2	17
Germany	0.5				

Note: Although Europe is an importer of bauxite, this is one of the few minerals whose production is even close to demand. This reflects the fact that aluminum is a 20th century technology; thus Europe's reserves were not used up.

PRODUCT: COPPER					
Europe	.7	8	Chile	22	24
Poland	.4		USA	18	19
PRODUCT: LEAD					
Europe	.3	11	Australia	.5	19

Note: Copper and lead are typical of the European pattern as regards other industrial metallic minerals: modest supply and a heavy import requirement.

*Russia and other ex-USSR nations not included in European totals.
Source: *Images Economiques du Monde*, 1995.

from Narvik in Norway in the winter as the Atlantic remains ice-free while the Baltic freezes over. This ore remains for use today because of the remoteness and isolation of its setting (above 67 degrees north latitude), and the fact that the small deposits of iron ore found in southern Sweden were adequate for national needs until fairly recently. Only with the depletion of other European sources and the rise of a world trade in iron ore did the deposits at Kiruna become commercially attractive.

The second major iron ore producing region is in Lorraine, in eastern France and Luxembourg. Ores in this field are much lower in iron content (some 20%), but their location in the heart of Europe's greatest industrial zone would seem to have made them a logical choice for early utilization and exhaustion. Why do they still remain? The answer here is not physical isolation but rather chemical composition. Most of the Lorraine ores contain phosphorous as a contaminant, which meant that while they could be used in very primitive ironmaking, they were virtually useless in the age of the blast furnace and the growing demand for quality iron and steel (phosphorous renders the iron weak and unstable, and may also contaminate the furnaces). It was not until development in the 1870s of the Gilchrist process, a method of using ceramic bricks on the inside of the furnace to remove the phosphorous chemically during firing, that the Lorraine ores could be fully utilized in modern steelmaking. Thus, some of the ore remains today.

Other metals. Since Kiruna, Lorraine, and the smaller fields are entirely inadequate to meet modern demand, Europe, with the exceptions of Russia and Ukraine, is in general a major iron-ore importing region. Ore from Canada, South America, and western Africa is a significant imported commodity, and reliance on these overseas sources has been a major factor in relocating the European iron and steel industry to coastal locations. Other industrial minerals show a similar pattern: Europe as a have–not continent, with often even less locally available resources. Copper, lead, zinc, tin, ferroalloys, and indeed nearly the entire range of metal ores must be almost entirely imported. As a generality, the countries of southeastern Europe, where industrialization was late arriving, have the largest remaining European reserves of these minerals west of the former Soviet nations, but none of them rank among world leaders. Russia, with its vast Siberian resources, is a world leader in the production of nearly all metallic minerals.

One exception to this general rule is bauxite, or aluminum ore, which is found in France (taking its name, in fact, from the village of Les Baux, near the Mediterranean), and in moderately important quantities in several southeastern European countries. Although supply is far from equalling demand, Europe is closer to self-sufficiency in this commodity than nearly any other mineral, while Russia is a world leader. This reflects the fact that aluminum is strictly a twentieth-century product, requiring advanced technology in its fabrication; thus there was no early interest in its ore, which in fact resembles a clay material more than a recognizable metal.

Nonmetallic minerals. The other major category of minerals consists of the nonmetallics: items such as sand and gravel, building stones, and related products. Although locally important, all European countries have generally adequate supplies of these materials which, since their value is small, rarely enter into long-range commerce. One element of variety in Europe stems from the fact that older buildings were generally always built of locally available materials. Stone was naturally often used, and so differences in color and texture may distinguish villages and give the buildings a uniform appearance quite different from those of their neighbors.

A few types of nonmetallic minerals used in building are, however, of sufficient value to travel across the world and bring prosperity to their producing regions. Significant among these is marble, a metamorphosed, resistant variety of limestone which is hard enough to take a fine polish and resist wear, and may also contain attractive chemical veining in different colors. Some European marbles, such as those produced at Carrara in northern Italy, are internationally prized.

Finally, the nonmetallic mineral category also includes a large variety of natural chemicals, or "salts," such as potash, gypsum, rock salt, and many others, which are used as fertilizers and as the basis of very significant chemical industries in several countries. These minerals have historically been particularly important in central Europe, where many German-language place names containing "Salz" (for example Salzburg, Austria) or "Hall" are evidence of their longstanding importance.

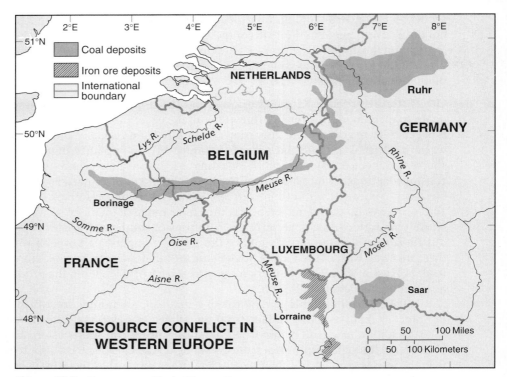

Figure 10.9 Resource conflict in Western Europe.

PERCEPTION: RESOURCE COMPETITION AND EUROPEAN HISTORY

As the increasing demand for unequally distributed mineral resources became more intense in Europe following the industrial revolution, and the inadequacy of national boundaries to ensure reasonable distribution of these advantages became apparent, controlling the requisite minerals became an important item of policy in many European states. The result of this was the addition of an extra element of tension in an already strongly nationalistic age. Unequal resource distribution was clearly a contributing factor to the political pressures and outright warfare which characterized Europe until 1945. The portion of the continent near the North Sea containing northern France, Belgium, Luxembourg, the Netherlands, and western Germany is a striking object lesson in the positive and negative results of the irregular distribution of vital mineral resources.

Although some national boundaries in this region, such as those of Belgium, date only from the early nineteenth century, others, such as those of Luxembourg, trace back directly to the time of the early Frankish kings. In general, the borders were well established before increasing interest in minerals led to any systematic geologic investigation of their location. As the importance of coal and iron ore resources began to become clear during the spread of the industrial revolution across the continent in the early nineteenth century, tensions and competition began to increase across many of these ancient boundaries, leading in some cases to disastrous outbreaks of warfare as the various countries sought to manipulate their destinies and ensure their success by controlling the resource base.

The coal fields shared by Belgium and France, or by Belgium, the Netherlands, and Germany have posed only minor international problems; but the resources shared by Germany and France have been a bitter bone of contention between these two major powers. The major coal fields of the Ruhr—Europe's largest and finest deposit—are very near this border, while those of the Saar, the third most important continental field behind the Ruhr and Silesia (excluding Russia and Ukraine), actually straddle the boundary. Moreover, the Lorraine iron ore field, of increasing significance with development of the Gilchrist process, is temptingly close to the German border. Thus, among the complex war aims of both rival nations in 1870–71, 1914–18, and 1939–45 was effective control of both coal and iron ore, and a consequent grip on the regional iron and steel industry. When Germany (or Prussia, in 1870) gained the upper hand, as in 1870 or 1940, the Lorraine fields were annexed to Germany (both Lorraine and the neighboring province of Alsace in fact remained part of Germany from 1871 to 1918). When France was triumphant, as in 1918 or 1945, the entirety of the Saarland was annexed to France, the German portion only being returned after several years. Given this bitter history, with its grim toll of human misery and economic disruption, it is small wonder that the first goal of the European Economic Community after World War II (see discussion of international organizations, Chapter 14) was to internationalize the coal and steel industries, making the resources equally available to all, thus removing at least one temptation to further aggression.

On the brighter side in the same region, two national advantages should be noted. By a casual accident of history, the Netherlands acquired a portion of the province of Limburg , which extends southward between Belgium and Germany. Barely 20 miles (32 kilometers) wide at its narrowest point, this strip

Figure 10.10 The luck of history gave Luxembourg a corner of the Lorraine iron–ore field: the steel mill at Esch from the French frontier.

of the Netherlands—almost a geographic afterthought—proved to be above an extensive coal field, extending from northern Belgium to the area around Aachen in Germany, which became critical to the economic development of the nation. Although largely abandoned today, this small Dutch coal field is a good example of the right resource being available at the right time to have a major impact on a national economy.

Even more spectacular is the case of little Luxembourg, whose ancient southern boundary with France intersects a small corner of the Lorraine iron ore field. This resource, as might be imagined, has been of enormous importance to a nation of 999 square miles and less than 400,000 population! In fact, it seems reasonable to assume that without this advantage the Grand Duchy would long ago have been absorbed by Belgium (with which it already shares a close economic union). Within a half mile of the French border, one of the largest integrated iron and steel mills in Europe was built to take advantage of the ore. Throughout the 1960s and 1970s, this factory was the single leading support of the Luxembourg economy. With the decline of steelmaking as a key activity throughout Europe, it has now been replaced in importance by the banking industry and the international institutions to which Luxembourg is home; but it remains a striking testament to the economic advantages that mineral resources can bring when they are found in otherwise disfavored countries.

ADDITIONAL READING

ALDCROFT, D.H., *The European Economy, 1914–1970.* London: Croom Helm, 1978.

GORDON, R.B., et al., *Toward a New Iron Age? Quantitative Modeling of Resource Exhaustion.* Cambridge, MA: Harvard University, 1987.

HOFFMAN, G., with L. DIENES, *The European Energy Challenge: East and West.* Durham, NC: Duke University, 1985.

LUCAS, N., *Western European Energy Policies: A Comparative Study.* Oxford: Clarendon, 1985.

MAULL, H., *Europe and World Energy.* London: Butterworths, 1980.

DISCUSSION QUESTIONS

1. How does Europe's production of petroleum relate to its consumption? What adjustments have European countries made to at least partially cope with the problem?
2. "No mineral has any value unless it is assigned one by a human group." What does this mean? Why has gold, of all metallic minerals, been so consistently prized over time?
3. Given its small size and population, Luxembourg has probably the most impressive resource base of any European country. Why is this?
4. Since the 1986 Chernobyl nuclear disaster, most European countries have eliminated or cut back substantially on their nuclear energy projects. The exception is France, which has proportionately the world's most ambitious program. Why are the French so committed to the nuclear option?
5. Most of the countries that emerged to become the leading industrial powers of Europe in the nineteenth century shared one vital resource: coal. What is the connection?

11
Industrial Location in Europe; Evolution of the European Industrial Model

Economically, the most significant aspect of Europe is its industrial strength and the vital role of this activity in all European nations. This is true for at least three basic reasons. First, although agriculture remains a highly visible component of the economy in most countries, its role is minor compared to industry in all but a few, as nearly all earn by far the major share of their national income from the production and export of industrial goods. While employment in the tertiary sector (provision of goods and services) is rapidly increasing, industry remains the motor of the European economy. Additionally, the European model of industrial operation—factory design and operation; labor–management structures; and the strongly rooted colonial pattern of importing raw materials, adding the value of manufacturing in Europe, and re-exporting the finished products—remains the only viable industrial model throughout the world today. Improved on here and there (as in the United States or Japan), and widely condemned (although avidly copied) by the socialist and "third" worlds, the European model has never really been replaced. The current rush of formerly communist countries in eastern Europe to restructure their industrial base along more traditional nonsocialist lines is ample evidence of the success of the basic western model. Finally, traditional European industrial strength and the existing infrastructure of industrial development have enabled many European nations to remain in the forefront of modern industrial change (especially including the application of vital new technologies) despite the handicaps of modest size, resource deficiencies, and often antiquated facilities.

There are numerous ways of viewing industry, both in Europe and elsewhere, but for our purposes two approaches may be emphasized: the reasons for industrial location and the evolutionary development of industry which has enabled Europe to become so dominant in this area. Following these discussions, it will be useful to look at at least one European industrial region (the Ruhr) in some detail, with a view to seeing how the considerations of locational constraints and the historical process have interacted to create the modern picture.

The importance of industrial location may be understood by a simple glance at the map of Europe. Industrial activity, as so many other factors whose distribution can be analyzed geographically, is by no means evenly scattered over the landscape. Some areas seem to attract far more than an average share of industry, while others suffer decline or stagnation as a result of their inability to recruit or retain industry. What are the reasons for this? Looking more closely, it is possible to recognize a general zone of industrial (and in fact overall economic) concentration roughly 200 miles (320 kilometers) wide extending in a curving line from the Midlands of England through northern France and the Benelux states, into central Germany, down the Rhine corridor to Switzerland, ending in northern Italy: a distance of some 1200 miles (1760 kilometers), with an eastward extension into southern Poland and the Czech republic (figure 11.2). This region has a remarkable concentration of industry, transportation routes, population, urban centers, and, for that matter, quality agriculture. It is, in effect, the economic "heartland" of Europe.

Those countries and regions within or close to the "heartland" generally have the most impressive statistics reflecting economic growth, living standards, and social conditions in all of Europe. Increasing distance from this zone normally implies greater difficulty in keeping up with the economic progress of the continent. It is possible to do this, through some combination of technological innovation, political intelligence, and social cohesion, as the Scandinavian nations abundantly demonstrate. More commonly, however, those European nations or regions on the fringe of the economic "heartland" have had the most difficulty coping. Countries on the bottom of the European economic list, such as Portugal, Greece, Albania, Bulgaria, Ireland, Slovakia, the former Yugoslav states, and many of the European pieces of the former Soviet Union, owe at least part of their difficulties to pure geography: the remoteness of their locations. Even in more centrally located nations, the same phenomena can be observed. "Backward" regions in otherwise highly developed countries are usually those located farthest from the development axis. Examples would include the Italian south, remote regions of western and southern France, Wales and Highland Scotland in Great Britain (see discussion of minority languages, Chapter 5).

The general discreditation of the socialist economic model since the late 1980s has put the eastern European nations at a further disadvantage as they seek to make their industrial base more competitive with the west. The revealed deficiencies of the planned economy and the heavy weight of centralized planning and governmental bureaucracy have made it difficult for these countries to profit quickly from their new political freedoms, while at the same time rationalizing their industrial production along western lines often generates considerable hardships in the form of increased unemployment and inevitable plant closings.

Figure 11.1 Basic types of european industry

Type of Industry	
Heavy	Nearly all producer goods. Usually features coal mining, steel manufacturing, shipbuilding, heavy chemicals, etc.
Medium, diverse	Producer and consumer goods. Features engineering industries (automobiles, machinery, metal products, tools), and may add textiles, rubber, and similar industries.
Light, very diverse	Nearly all consumer goods. Sometimes called "capital city" industrial model, since it depends heavily on market location and thus grows rapidly around expanding cities. Manufacturing extremely diverse but might include tailoring, housewares, scientific instruments, luxury items, light machinery, etc.
Port-associated	Medium, diverse to heavy industries related closely to the import of bulky raw materials from overseas. Typical activities include oil refining, flour milling, ship maintenance and repair, and processing of wood and tropical commodities. Transshipment or "break-in-bulk" activities are also very important functions of most ports.
Dispersed	Includes widely scattered activities based either on the perishability of the product or raw materials (agricultural processing, baking, printing), or on the distribution of low-cost power sites. This category would also increasingly include industries not bound by normal location factors (high-technology, high-value, or small-scale industries).

FACTORS OF INDUSTRIAL LOCATION

Setting aside the negative aspects of industrial distribution, however, what are the factors which can be identified as crucial to the location of industry, and thus to the establishment of modern economic strength in Europe? It is perhaps possible to recognize nine of these, some of which have been vital in any age, some of which have declined in their importance over time, and some of which are probably of greater significance today than at any time in the past. Taken together, they enable us to understand the apparently complex pattern of European industry with some clarity.

Market. Of greatest importance in explaining the location of most industrial activity anywhere in the world is the interaction of three closely related factors: market, materials supply, and transportation. Analysis of these factors, in fact, forms the basis of most practical work in industrial location, both in Europe and elsewhere. Of these, market is usually the most significant. In the absence of other compelling reasons, industry will tend to locate to its market outlets, thus both reducing shipping costs and being in a position to adapt quickly to changes in the market by being close at hand. For many types of manufacturing, "market" implies large concentrations of people, hopefully with some purchasing power, and this in turn translates into cities. Many kinds of industries, notably those producing consumer goods, have thus located in urban areas; and as these have expanded so has their attractiveness to still more industry. This sort of "snowball" effect, whereby cities attract industry which in turn attracts additional labor and services, creating a larger city which attracts more industry, and so on, is one of the critical concepts in modern economic development, especially in the sense of revitalizing backward regions.

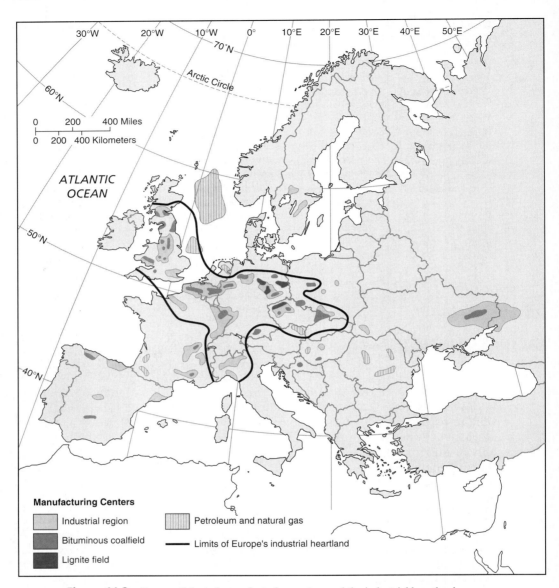

Figure 11.2 Europe: Selected manufacturing centers and the industrial heartland.

With their larger concentrations of people, cities are also more likely to have a greater variety of specialized markets, and to represent greater wealth than smaller communities or rural areas. This attracts a further range of manufacturing activity: those whose markets are small but exclusive, or whose products are expensive. The concentration of high-fashion clothing and other luxury goods as manufactured items in Paris, London, Milan, and a few other European cities is a good example of this pattern. Given these relationships, it is clear that cities and manufacturing are closely related, and it is thus no surprise to learn that the growth of industry and the rapid expansion of cities in Europe have been contemporary phenomena.

For many industries, of course, "market" represents something entirely different than simple concentrations of people. Particularly for manufacturers of "producer" goods, or those that are generally used to make other things and have no appeal to an ultimate consumer, the "market" might be another manufacturing plant or industrial complex which may have nothing to do with an urban location. Basic iron and steel making, chemical manufactures, or the production of machinery of all kinds—all vitally important European industries—are examples of activities which will often locate to market, but for which the market is a vastly different concept than for the maker of "consumer" goods.

Materials supply. The acquisition of whatever is used in the manufacturing process is also of vital interest to industrial location analysts, and may be considered the opposite pole of location from the market. Sometimes the locations of market and materials are close together, sometimes they may be half a world apart; but somewhere along a hypothetical line connecting them will be the optimum site for the manufacturing activity. This factor is sometimes identified as "raw materials," but this term is a bit misleading, as it implies strictly unprocessed materials (wood, coal, iron ore, etc.), whereas in reality the materials used by a majority of manufacturing plants are already at least semi-finished (steel plates or cables, refined chemicals, etc.).

In general terms, if the materials used in a manufacturing process are heavy, bulky, and/or low in value, it will be advantageous to locate the industry as close as possible to the materials source, in order to reduce the excessive transportation costs of moving this sort of item. The classic example of this type of location is the manufacture of cement,

Figure 11.3 A coastal site is a vital locating factor for many industries utilizing heavy, bulky raw materials: Dunkerque, France.

which is nearly always located directly on the rock (usually limestone) from which the product is made. In fact, when quarrying of the rock moves a relatively short distance from the factory, it is often cheaper to build a new plant than to transport the rock an extra few hundred meters! This principal also accounts for the location of so much European industry of this type at coastal locations. As a continent which must import such a large share of its basic, unprocessed materials, Europe has seen a major share of its industry dealing with wood, grain, oil, and metal ores gradually shift over time to become concentrated in major port cities around the extensive coastline.

On the other hand, if a manufacturing activity is concerned with the utilization of materials which are small, lightweight, and/or of high value, there is absolutely no reason to consider a location on the source of supply, since any transportation costs are likely to be a minor item, given the character of the goods involved. Many modern "high tech" industries, for example, which might use such "raw" materials as electronic components or silicon chips, can quite literally transport both their materials and products by air freight without considering the cost excessive. Thus freed from basic locational constraints, such industries will locate to market, to other locally important factors, or even to amenities and an attractive life style. The "sun belt" phenomenon is by no means unknown in Europe!

Transportation. The link that bridges materials and markets, and the third factor of particular importance in determining industrial location is transportation, or the infrastructure and cost of moving goods to and from the manufacturing site. At local scale, industrial plants are often sited where there is an abundant variety of readily accessible transport choices available. Rail connections, freeway interchanges, airports and/or waterways close at hand are very persuasive considerations for many industries. This means, of course, that those parts of Europe where the transportation network is most dense have a built-in advantage in attracting new and expanding industry. Thus the relatively flat lands of the North Sea plain, so useful for agriculture, urban growth, and the easy construction of transportation routes, also contain an impressive share of the continent's industry, as do such important transport corridors as the Rhine and Rhône valleys. Conversely, mountainous regions and the remote corners of Europe are even more developmentally disadvantaged because of an inadequate transportation infrastructure.

At a larger scale, the fact that not all methods of transportation are equally costly has a strong effect on the distribution of Europe's industrial activity as manufacturers attempt to minimize the costs of acquiring their materials and reaching their markets. The basic reality is that water transport is almost invariably the cheapest means of moving goods around, particularly where time pressures are not an important consideration. This, combined with the fact, stated previously, that Europe relies to an extraordinary degree on imported basic materials for its industry, accounts for the exceptional importance of the continent's rivers, lakes, canal systems, and coastal sites in industrial location strategy. Especially where low-value, high-weight materials are involved, it is vitally important to keep transportation confined as much as possible to the water, and even over relatively short distances, waterways will be used wherever practicable to move coal, rock, sand and gravel, mineral ores, non-perishable agricultural produce, and other materials.

Only pipelines can compare with waterways in transport cost terms, and while these are quite important to Europe, they cannot handle the volume or variety of materials found on the fleets of barges that seem to be ever present on the waterways. Rail and road transport costs more than water transport, but these have the advantages of greater flexibility and faster service. Rail transport is generally advantageous over longer distances, road transport within a shorter radius, although these distinctions are often minor in Europe. Manufacturers of high-value, low-weight products (such as many consumer-oriented goods) will use either of these transport possibilities, dependent on local factors. Air freight, finally, is by far the most expensive means of transport; but where the materials transported are very valuable and time is a critical factor, it may be employed.

Labor. Although, with the general decline of labor intensity in European industry over the past half century or so, labor supply as a factor of location has seen its role diminish, its former significance helps explain the location of an important share of the industrial plant. In particular, from the time of the industrial revolution the larger scale of industry demanded more workers, and location in growing cities was thus a necessity; the city was at once a market and a source of labor, and the relationship between urban growth and industrial expansion was thus reinforced. Moreover, if particular skills were required, it was much more likely that suitable individuals would be found in cities than in the countryside.

In modern times, however, the picture has changed. The percentage of the labor force employed in manufacturing is declining in nearly every European country, especially compared to the that of employment in the services sector. Moreover, the prevalence of national newspapers, radio, and television means that news of job opportunities reaches qualified workers wherever they may be: labor will move to industry rather than industry being forced to seek out labor. Throughout much of the twentieth century, the result has been a decline in labor as a factor of location. Employers are still anxious, however, to reduce labor expenses, which are usually a major part of their overall cost package. This gives a potential advantage to poorer countries and regions where wage scales are lower. Thus, provided that labor skills are comparable and other location factors at least neutral, many industries have successfully relocated to cheaper manufacturing sites.

In an era of widespread unionization, labor costs have to some extent evened out from one region of a country to another; but costs for the same amount of labor remain rather higher in cities than in smaller places, and many depressed regions within several countries seek to lure industrial jobs with the promise of cheaper labor. This process can have unexpected results, however. When the French government persuaded the Citroën automobile company (now part of Peugeot) to build a large new assembly plant in remote Brittany in the early 1960s, it was embarrassed to discover that few people in the local region knew much about putting cars together! Forced to attract labor from the Paris area with expensive inducements until proper training programs could be initiated, the company learned a harsh lesson about the value of suitable labor as a basic factor of location.

Capital. As with labor, the significance of capital availability as a major factor of industrial location has diminished with time. Its historic importance, however, still helps explain the concentration of much European industry even today. For most of European history, capital was a fairly rare commodity, concentrated in the hands of a few governments and banking families, all with urban interests and locations. Thus, the provision of financial resources for the establishment of new industry was a very restricted possibility, and since many conservative lenders were reluctant to see their invested capital move very far beyond their sight and daily control, early industry was further concentrated. The result: one more way in which the processes of urbanization and industrialization proceeded in tandem in Europe from the eighteenth through much of the early twentieth century.

At the present time, this pattern has changed in several ways. First, there is a great deal more investment capital available to European industry than has previously been the case, which means that its geographical distribution is much less restricted. Second, the flexibility of lenders has increased and their expectations have changed considerably. Nowadays, any prospective borrowers would no doubt be asked to produce figures showing that they had done their "homework," that is, that the proposal for a manufacturing facility had been fully analyzed in accordance with modern location theory (markets, materials supply, transportation, and all other relevant factors). The optimum location, in other words, may have nothing to do with the location of the lender. Nonetheless, this idea is a recent innovation, and a fairly large share of Europe's older industry owes its location at least partially to where the needed money could be found.

Power and fuel costs. As a factor of industrial location, the cost of power and fuel is a relatively minor consideration for most industries. Where modest amounts of these resources are required, costs are relatively similar throughout Europe, and differences would rarely be so great as to influence the location planning process. For a few industries, however, power costs are such an overwhelming consideration that they will locate to reliable supplies of cheap power regardless of the distance from markets or materials, or of any other considerations. The location of such industries defies analysis unless the role of power costs is understood. A notable case, in Europe

Figure 11.4 Aluminum refining is a classic example of an industry that invariably locates to cheap energy: Tyssedal, Norway.

as elsewhere, is the refining of aluminum. Because of the enormous power demands of the process, aluminum refining is carried on, for example, at remote points on the coast of Norway (a country with abundant supplies of cheap hydroelectric power), even though there is no bauxite and only a very small market in that nation.

Space requirements. One of the ways in which industrial location in Europe has been changing most rapidly in recent years is in response to the need for greater amounts of space in many manufacturing processes. The growth of assembly line operations and the relatively higher costs of vertical movement within industrial buildings has placed a premium on single-story structures of large size. This modern shift in plant design, however, sharply contradicts the traditional values of locating industry in cities, where market, labor, and capital may have been present, but space was usually restricted. Moreover, as we have seen, the relationship of industrialization and urbanization was such that many European cities actually grew up around manufacturing plants in the nineteenth century, and it is still in fact common to find an unusual variety of basic industry located in the very heart of these cities.

With the need for more efficient design and greater amounts of space now becoming a clear concern of industrial construction or expansion in the late twentieth century, many European manufacturing operations find themselves forced to make difficult decisions. Since their present sites are now heavily impacted by dense urban growth, acquiring adequate land for growth in their present locations is often impossible, or at least prohibitively expensive. To solve this problem, many industries have moved away from their original locations altogether, often selling the valuable urban land at a substantial profit, and have built new plants in rural areas, where land costs are less and local governments may provide location incentives. The problem with this strategy is that by moving away from the city, with its markets, transportation connections, communications base, and other advantages, industry risks losing its competitive advantage. A good compromise is to find a location far enough from the city to provide adequate space, yet close enough to retain many of the other advantages; thus, the growth of smaller cities and towns within a radius of 120 miles (200 kilometers) or so of larger centers is one of the most striking features of European population change at the present time.

An additional interesting process that occurs as a result of this shift in industrial location is the freeing of large tracts of former industrial land in the central districts of many European cities. At a time when these places are attempting to cope with modern problems of transportation, housing, recreation space, the provision of services, and general "livability," while at the same time retaining the tourist-attracting antique charm of their older quarters, such land can be a most valuable asset in urban planning and redevelopment. In numerous cities, what were industrial sites as recently as the 1960s have now become apartment complexes, commercial developments, office space, or parks and pedestrian malls.

Government decision and planning. The role of government in determining the location of modern industry is vitally important in nearly all European countries.. This is in sharp contrast to the United States, for example, where this factor plays almost no part in determining the pattern. As background to explaining this difference, it must be noted that European governments in general have until recently exercised a much greater degree of ownership and control over basic segments of their economies than is the case in North America. The former Soviet Union was the classic case: all means of production were controlled by the state, which made all relevant decisions in a highly centralized fashion. This same principal also applied to the formerly socialist states of eastern Europe, although most of these, as well as most of the former Soviet republics, are now moving away from excessive concentration of the decision-making process (see discussion of eastern European regions, Chapter 23).

Even in the parliamentary democracies of western Europe, however, government influence remains surprisingly pervasive. In general, nearly all public utilities are nationalized, as are many transportation systems (notably railroads and airlines). Power production (coal mining, nuclear plants) is nearly always a government monopoly, as is, in some countries, iron and steel manufacturing. Financial institutions have in some cases been nationalized, as have some key industries across the continent (Renault, the major French automobile maker, is a good example). In perhaps an even greater number of cases, governments may fall short of outright control of an industry, but still hold an important financial stake, thus giving them considerable leverage in decision making. Since the mid-1980s, the advent of many conservative governments in western Europe has led to a trend toward de-nationalizing (or "privatizing") many of these holdings; but while the system may indeed be changing toward one of less government intervention, the location of all industry established up to this time by government initiative remains the same.

What all this means, broadly, is that 1) by controlling so much of the necessary infrastructure and basic industrial plant, European governments have been in a strong position to influence the location of new or relocated industries, and 2) they may have made these decisions for social, as well as economic, reasons. Industrial policy has thus been a useful tool that governments have used to fight unemployment or equalize regional imbalances. Unfortunately, it also means that some of the more pragmatic factors of location (market, materials supply, transportation, etc.) have been given a lower priority, so the economic success of these industries is often marginal at best.

Basically, western European governments have influenced industrial location in two ways. One is by providing incentives for privately controlled industries to locate or relocate in selected parts of the country, which can be identified (by living standards, unemployment rates, and other measurements) as lagging behind national standards. Nearly every country recognizes "development zones," and industries locating in these areas might expect to profit from such incentives as tax abatement, provision of utilities and transport access, free or low-cost land, low-interest loans, etc. In many cases, a second group of somewhat less disadvantaged regions may qualify for a certain part of the industrial incentives. Industries insistent on locating or expanding in areas already highly developed (or even overdeveloped) will receive none of these benefits; but the fact that many industries still choose such locations means that the attractiveness of market proximity and other factors is hard to overcome. No amount of government encouragement can make an industry profitable if its location is economically and geographically unsound.

The second way in which governments may influence the location of manufacturing is through outright decisions involving industries they control. This leads to even more arbitrary judgements than simply providing incentives to private capital, and is a device normally used where there is some compelling social or political reason to locate an economic activity in a certain place, or where there seems little hope of developing a given backward region without some positive action. Southern Italy, for centuries the poorest of Italian regions, has seen a remarkable amount of industrial growth over the past twenty-five years as a result of the forced siting of plants in sectors then under government control (iron and steel, oil refining, petrochemicals). Although this process has not come close to solving all the problems of the region (see Italian regional discussion, Chapter 21), and although many of the industries involved are only marginally profitable, if at all, it reflects the government's determination to use every means at its disposal to revitalize the South. Another example of this process is in Norway, where the small coastal steel mill at Mo-I-Rana is maintained as a government attempt to provide some economic stimulus and stem depopulation in a remote, inaccessible region.

The human factor. If we analyze all of the factors listed above and apply the results to a map of industrial location in Europe, many patterns can be explained and the broad outlines of the continent's manufacturing can be understood. Nonetheless, a very large number of individual operations cannot be explained by this process, and in fact seem to contradict most of the rules of location that we have identified. These are the industries—in Europe as elsewhere—that owe their locations to nothing more than a personal (often uninformed) decision on the part of the founder. These are often smaller industries; sometimes they have been run as family enterprises for generations. Since their location sites are simply the village or town of the founder, they are often poorly adapted to modern demands, and may in fact be blown away by the increasingly harsh winds of competition. In some cases, such industries may once have been properly located, given the conditions of the eighteenth or even nineteenth century (a small textile mill on a water power site, for example); but the world has changed rapidly around them, and their location now may be hopelessly isolated. Literally thousands of these industries survive in Europe, based on habit, inertia, low expectations, a sense of family or community loyalty, and perhaps some receding image of better times to come.

THE EVOLUTION OF EUROPEAN INDUSTRY

Europe's relative economic success in the modern world has been largely attributable to the power, variety, and technological expertise of its industrial establishment. Yet, even if the location of modern industry conforms largely to a certain set of geographic criteria (see above discussion), how is it possible to explain why Europe, of all the world's possibilities, should have come to be the industrial continent par excellence? The answer lies in four related phenomena: the dense and relatively skilled populations that began to evolve in Europe at a comparatively early date; the steady (although with many lapses) growth in agricultural productivity that generally enabled Europe to free labor for nonagricultural pursuits; the rich and diversified mineral wealth (now largely exhausted) of the continent; and the early realization that importing raw materials and exporting finished products, complete with the value added by manufacturing, represented the surest way to national wealth and the best hope of raising living standards in a narrowly circumscribed space.

What is most remarkable about the European industrial model is that it has been translated virtually intact into all parts of the world. Much of this is the result of the outpouring of European peoples and ideas during the great waves of emigration and colonialization in the eighteenth, nineteenth, and early twentieth centuries: all European colonies were closely linked to European economic strategies. Yet the virtue of the model continues to be reflected in the American and (more recently) Asian translations which have lately put such pressure on European industrial production.

It is convenient to consider the evolution of European industry in four phases. Importantly, each successive phase built on its antecedents, and each was successful in its own place and time. The modern wonder of European industry did not spring full-blown from a nineteenth century agricultural landscape; but was rather foreshadowed by a long history of incremental and difficult, yet invariably focused, change. These stages of industrial evolution may be summarized as follows:

Primitive household industry. This is essentially the industry of the stone age. Without power assists, without sophisticated philosophies of material supply or trading possibilities, the "making" of things was essentially primitive, at individual scale, and distinctly ad hoc . Yet the genius of people—notably their thought processes and manual dexterity (featuring the opposable thumb)—was sufficient to create a vast range of "industrial" products, including weapons, tools, religious objects, jewelry and ornamentation, and the pots and cups of everyday life. There were no considerations of far-off markets, industrial materials (stone, bone, wood) were generally those that came immediately to hand, and the workplace was almost invariably the living place; yet these early artisans produced a volume and variety of goods that continue to amaze later generations.

Primitive household industry survives in a sense today as the simpler forms of "do-it-yourself" handicrafts. The modern person who nails together a birdhouse in the garage is in effect a distant cousin of the stone age industrialist who chipped a hunting arrow out of a handy piece of chert in front of the family cave.

Simple powered household industry. Just as the addition of power drills, saws, and other sophisticated equipment broadens the creative horizon of the modern "do-it-yourself" enthusiast, so the evolution of enhanced power sources changed forever the parameters of industrial production and led Europe toward new qualitative and quantitative levels of output. The arrival of more reliable and especially more powerful sources of energy (see Chapter 10)—notably animal, wind, and water power generation—opened dramatic new industrial vistas of which Europeans were quick to take advantage. From the gradual introduction of these innovations, beginning perhaps 9000 years ago, until the energy revolutions of the nineteenth and twentieth centuries, simple-powered household industry spread across the face of Europe, bringing not only enhanced levels of production, but also new ideas about acquiring and stockpiling raw materials from remote sources and profiting from the demands of more distant markets.

Two aspects of this form of industry deserve attention, as implied in its name. First, the addition of power to the manufacturing process meant inevitably (since simple power sources, while widespread, were not universal) that industry on this new scale became more geographically concentrated and that specific industrial sites began to be identifiable. The mill is a good example. Whether powered by wind or (more significantly) water, the mill where grain from the surrounding farmlands could be ground into flour, or where wood, textiles, and other basic materials could be processed using the simple power adaptations available, became the focal point of its community, quickly attracting additional services and population in a familiar pattern of growth. The miller almost always became a figure of importance in medieval and Renaissance Europe ("Miller" or its equivalent is one of the most common family names in any European society). Controlling the contemporary levers of industrial power, millers were in a sense the first successful industrialists, and as such often became objects of envy and mistrust. Songs and stories from these periods commonly portray the miller as someone to be tricked or cheated if humanly possible!

The second feature of interest of simple powered household industry is the word "household." This correctly implies that at this time living space and productive space remained undivided. The miller and his family lived above the mill, and similar arrangements existed for other sorts of industrial activities. The separation (see next section) of living and work space marked one of the great cultural and geographic watersheds of European history.

Although rare, survivals of this type of industrial model can still be found in Europe today. The small cheese factory in a Swiss village, with the family's quarters upstairs; the renovated local hand-weaving industry in an English cottage; the hand-carved wooden souvenirs made in and sold from the home in countless tourist-attracting towns: all are relevant examples. Their modern success reminds us that this form of industry was the European standard for many centuries and that the wealth generated by the great volume of these individually modest activities was a key factor in enabling Europe to move into more modern modes of production.

Community workshop industries. In about the thirteenth and fourteenth centuries, as Europe emerged from the period of relative stagnation known widely (if somewhat inaccurately) as the "Dark Ages," the continent began to experience renewed demographic and economic growth and especially started to develop more elaborate and profitable trading patterns both at home and throughout the world. In this setting, the shortcomings of the simple-powered household industrial model became increasingly clear. In particular, the inability of the myriad small producing units to generate sufficient volume to meet demand, to assure overall quality control, or to group into viable marketing organizations, meant that many chances for expansion were lost. Into this void—powered by the seemingly endless human urge to recognize and profit from opportunities—began to emerge a new concept of industrial organization, one that would have profound implications for the social structures and geographic patterns of Europe, as well as for the continued economic success of the continent. Beginning on a serious scale in about the thirteenth century, evolution of what is often termed the "community workshop" industrial model changed Europe's industrial picture substantially, and introduced as well such potent social forces as the guild system.

From the standpoint of industrial theory, the growth of community workshops marked a crucial change from home-based to workshop-based production. Rather than working independently at home, workers in a particular trade would assemble at a single place to be part of a manufacturing process which thus lent itself to greater volume, more consistency in quality, and the scale required for long-term and long-distance trading. The innovation of moving from home to work, to return at the end of the day, marks the beginning of the commuting society which continues to dominate most developed economies today. It also, since industrial units became much larger and since transportation limited to a horse's speed did not permit workers to live far from their workshop, had much to do with the early growth of towns and cities.

The guild system was perhaps the most important social structure associated with community workshops. Organized in nearly all European cities, for centuries the guilds exercised dominant control over both people and processes involved with nearly all forms of economic activity; only a very few professions were not part of the guild system. Typically, boys in their early teens or even younger (there were a few female guilds, but most European societies were male-dominated) would be apprenticed to a local guild (normally but not invariably one for the activity of which they showed a certain aptitude). The guild provided room, board, and a basic education, while the child learned his craft by observing, helping, and running errands for the guild's professionals.

If the period of apprenticeship (variable, but often six years) went well, the child might become a "journeyman," or a member of the largest component within the guild, who were responsible for most of the group's industrial production. This promotion gave a solid social and economic standing to the individual. At the top of the pyramid were the masters of the guilds. These were the people who were in theory (and often in reality) the best craftsmen in the guild, and who had risen to become powerful figures in the political, social, and economic life of their communities. The journeyman who aspired to become a master suffered a testing time. Having to endure intensive scrutiny by the existing masters, he was favored by political acceptability and "correct thinking." Highlight of the process was the presentation to the full array of masters of a specimen of the candidate's best work, which he hoped would be of sufficient quality to assure his election to the elite: hence the modern term "masterpiece."

Figure 11.5 The authority and wealth of the guilds is reflected in the architecture of their buildings: Bruges / Brugge, Belgium.

As it dominated the European economic scene for centuries, the guild system, like most institutions, had positive and negative impacts. Negatively, the guilds, by promoting conformity, stifled creativity: original conceptions were rarely well received. On the other hand, the guilds, through their structure, had a considerable degree of built-in quality control: shoddy products damaged the reputation of the entire guild. To this day, European industrial products have an enviable (if sometimes misplaced) reputation for good quality: an enduring legacy of the guild system.

In the seventeenth century, therefore, on the eve of the industrial revolution, Europe was already the most industrialized of continents. The successful combination of simple-powered household industry in the countryside and the guild-dominated community workshops in the growing cities produced a volume and variety of goods largely adequate to meet Europe's needs. Looming events, however, were to shatter this comfortable balance and alter the geographic pattern of European industry beyond all recognition.

Corporate urban industry. In the seventeenth and eighteenth centuries, Europe underwent an astonishing series of social and economic transformations that resulted in the industrial model that continues to characterize the continent (and to a remarkable extent the world) today. This complex, pervasive, and ongoing process is generally summarized as the "industrial revolution," and it is widely considered among historians to be one of only a handful of events that have dramatically altered the parameters of human experience. Since countless volumes have been written about the industrial revolution, and since it remains today a topic of substantial interest to historians, there is little point here in providing an exhaustive summary of the topic. Nonetheless, the process has had such profound implications that it seems necessary to at least briefly review its geographic patterns.

Basically, the industrial revolution was the creation of Great Britain, more specifically of the English Midlands (figure 11.6). In the eighteenth century, this region had already become the nation's most significant manufacturing area, based on a combination of simple-powered and community workshop industries. This was also, however, the time at which Britain became the dominant world power and built a massive colonial empire. The labor demands thus placed on a relatively small population (perhaps no more than an average 30 million) became severe. The army, the colonial administrations, and especially the navy ("press gangs" scoured the land for men who could literally be seized and forced into service for several years) represented enormous demands. Who was left to produce the goods increasingly required not only at home, but also in the colonies?

Industrial entrepreneurs in the Midlands realized that the only way to meet increased demand was to replace scarce labor with machines: a machine capable of providing the output of ten workers would do much to overcome shortages of human hands. Throughout the nineteenth century, therefore, invention and technological innovation steadily transformed industry from a human-based to a machine-based activity. Each year saw some new improvement that added to the quality, volume, and efficiency of output. Although all industries were eventually affected, it is scarcely surprising that most of the innovations were involved with textile manufacture, since this was both the industry whose products were most in demand in a colonial world and the one whose operations lent themselves most easily to mechanization.

Figure 11.6 The English Midlands were home to the Industrial Revolution.

As the century wore on and the colonial age reached its zenith (1885, when the European powers met at Berlin to literally carve up Africa into spheres of influence, is perhaps a representative date), demand for industrial products continued to increase geometrically. At this point, however, a new factor came into play.

The products of the textile mills, with their higher quality, smooth finish, and (thanks to increased productivity) lower cost, rapidly put the workshops and simple-powered household industries out of business. Entire categories of rural occupations vanished virtually overnight—certainly within a generation. Weaving and its related activities is the most obvious example, but all industries that had been similarly dispersed eventually suffered the same fate. The process spread across Europe from the Midlands like the ripples from a stone dropped into a pond. In 1700, the European countryside was a diverse economic and social landscape, with countless modest industrial activities complementing the farming community. Two centuries later, nearly all the alternatives to farming had vanished, as well as most of the mills and small communities supported by dispersed industry.

One final process completed the transformation of Europe's industrial population from rural to urban, especially in Britain. In traditional settlement patterns (see Chapter 6), each village had a "commons," on which any citizen enjoyed the right to pasture animals. Many people, too poor to own land, were thus able to eke out an existence in their village. Reacting to the increased demand for food as cities began to grow rapidly, as well as to the growing need for labor in the large new urban mills, the British parliament passed a series of "enclosure acts," affecting many counties, which specified that "common" land was to be sold and put into the production of crops. This instantly impoverished thousands of people who had been subsisting on the marginal grazing, forcing them as a matter of survival to migrate to the growing cities where jobs in the new industries were their only choice.

It is because of its impact on population patterns, social conditions, and the landscape, as well as on economic development, that the industrial revolution is widely considered to be one of the most significant historical processes. It is little wonder that shortly after 1800 London became Europe's first millionaire city since ancient Rome; or that Great Britain became the first country to have a majority of its citizens living in cities (about 1850).

PERCEPTION: THE RUHR

German industry, traditionally heavily concentrated in the Ruhr district of the state of North Rhine–Westphalia, is today more spatially diversified. The Ruhr, however, remains a remarkable region of industrial might, grouping within a rough square some forty miles (sixty-five kilometers) on a side about 10% of the country's total population, and forming the most important and most impressive single industrial concentration in all of Europe.

The Ruhr owes its significance to a combination of locational and human factors. Its physical setting is ideal. Lying at the western end of the contact zone between the North Sea Plain and the Hercynian formations to the south (known as the Börde), as well as directly on the Rhine, it is thus at the crossroads of what have for centuries been the major east–west and north–south routeways in central Europe. Moreover, its surrounding region, featuring rich soils formed from windblown deposits of loess, is one of prosperous agriculture, always an important support for industrial development. Finally, and certainly

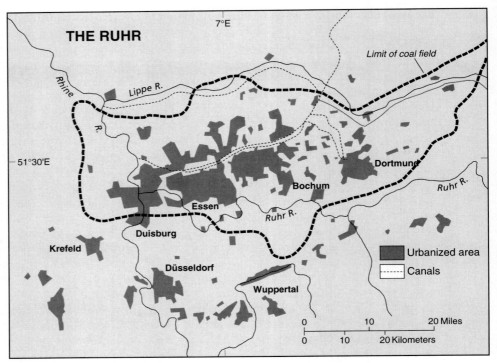

Figure 11.7 The Ruhr is Europe's most concentrated industrial district.

most importantly, high quality coal reserves were found near the surface at many points along the Börde, and these were particularly valuable in the Ruhr, which became Europe's richest coal mining district.

Human factors, however, also played a major role in the great concentration of industry in the Ruhr. German industrialists have long been fascinated by the concept of vertical integration, which implies direct control over all phases of a manufacturing process, from acquisition of raw materials to generation of the final product. Thus, in the Ruhr, giant corporations such as Krupp or Thyssen would control the coal and iron ore mining, the basic iron and steel production, and the subsequent manufacture of all related products, each operation receiving its materials from an adjacent plant, and passing its product to yet another. This type of monopolistic philosophy (much admired in the United States by Henry Ford) naturally led to the massive concentration of heavy industry in the Ruhr.

The particular modern genius of the Ruhr, however, has been its remarkable ability to maintain its economic vitality in an era when its traditional strengths have been substantially eroded. Coal and steel, for example, are static or declining industries in Germany, as they are in most of Europe and much of the world. In the Ruhr, however, more efficient steelmaking has simply shifted away from the coalfields to new locations along the Rhine , where the now most important raw material, iron ore, can be cheaply imported by water transport. The "classic" steel town of the Ruhr, Essen, although no more than twenty miles (thirty-two kilometers) from the Rhine , now produces no steel at all!

Figure 11.8 Resources and location combine to make the Ruhr Europe's most concentrated industrial region: Steel mill at Krefeld on the Rhine.

Moreover, the Ruhr has been successful in attracting a wide range of new technologies and related industries, which are drawn by the skilled labor pool, the investment possibilities, and above all the still-outstanding location at the heart of Europe, which continues to be the single greatest appeal of the Ruhr.

ADDITIONAL READING

BLACKABY, F., *De-Industrialization.* London: Heinemann, 1979.

DICKEN, P., *Global Shift: Industrial Change in a Global World.* London and New York: Harper & Row, 1986.

POUNDS, N.J.G., *The Ruhr: A Study in Historical and Economic Geography.* London: Faber, 1952.

RILEY, R.C., *Industrial Geography.* London: Chatto and Windus, 1973.

WARREN, K., *The British Iron and Steel Industry since 1840: An Economic Geography.* London: Bell, 1970.

DISCUSSION QUESTIONS

1. Market is generally considered to be the most important single factor governing the location of modern industry. Why is this true?
2. Coastal location is of great importance for many European industries (e.g.: iron and steel, oil refining). What is the significance of waterside sites?
3. Many older, smaller European industries are found in small, remote towns and seem to defy locational analysis. Why are they there?
4. Government decision has traditionally played a greater role in the location of European industry than is the case in the United States. Why has this been true?
5. Cement manufacturing is considered a classic case of an industry locating to its raw material supply. Why?

12
European Tourism

Tourism in the modern sense began in Europe in the early nineteenth century. Since that time, and particularly over the past half century, tourism has emerged as the world's leading growth industry: currently expanding at a sustained rate of 4% to 5% annually, it has become both the main economic hope of much of the "third world" and often the only dynamic sector of more prosperous and mature economies. Moreover, several trends of the 1990s indicate that tourism is likely to expand even more as a key economic support. In Europe, for example, the dramatic and swift removal of barriers to international movement in the Eastern nations (as well as, to some extent, in the nations born of the former Soviet Union) seems certain to substantially increase tourism, in the senses both of opening attractive new destinations to outside visitors, and of releasing the pent-up travel aspirations of millions of eastern Europeans.

In the context of world tourism, Europe remains by far the most important continent by almost any measure. Europeans spend more on tourism than any other group (Germans are clearly the world leaders in per-capita expenditure), yet at the same time earn more from this activity than the rest of the world combined. In 1987, Spain at least briefly surpassed the United States as the single nation with the largest gross tourism receipts (note that Spain's population is less than 20% that of the U.S.), while Austria claims the highest per-capita income from tourism. Much of this activity is of course intra-European; but Europe nonetheless remains the favored destination of travelers from most other corners of the world.

Not all this is good news. Some parts of Europe have become almost pathologically dependent on tourism (in Venice, Italy, for example, it is difficult over much of the year to find anyone who is neither a tourist nor supported directly by tourism). The time- and space-concentration that characterizes much tourism also takes a heavy toll on the environment. The Mediterranean basin is by far the single most visited region on earth: some 35% of the total world tourism market; and the annual arrival of an extra 100 million visitors (nearly all in July and August) has predictable consequences for what is already one of the earth's most endangered ecosystems.

Thus, in many senses, an understanding of the challenges and opportunities presented by the tourism industry both begins and ends with consideration of the European scene.

Europe: The Continent Where Tourism Began

As one of the most adaptable of species, humanity has moved successfully into nearly all of the world's environments, blending into certain landscapes while modifying others according to need and to the technological forces available. Movement has thus been a continuing theme in human history; but such activity has normally been motivated by the hope of gaining economic advantage, by cultural pressures, or by fear. Demographers and geographers who have studied human migrations have evolved various models to explain the phenomena, but all of these have stressed the practical and immediate rewards of movement (escape from enemies, the search for more or better land, urban employment prospects, etc.).

The concept of movement for some trivial or purely pleasurable end—change of scene, recreation, curiosity—is a distinctly uncommon theme throughout most human experience. People lived and died in the precincts of their village, and whoever ventured beyond the limit of the village fields, except possibly to a periodic market, was an object of wonder, concern, and possibly pity.

Figure 12.1 Newly opened frontiers in Eastern Europe will mean increasing tourism in both directions: Budapest, Hungary.

Figure 12.2 Tourism balance of payments, selected European countries, 1989.

Country	Receipts ($ Billion)	Spending ($ Billion)	Balance ($ Billion)
Spain	16	2	+14
France	17	9	+8
Italy	12	5	+7
Austria	9	6	+3
Switzerland	6	5	+1
Great Britain	11	13	−2
Netherlands	3	6	−3
Germany*	9	24	−15

Source: OECD.
*Data for former West Germany only.

Exceptions to this rule are of course numerous. Certainly from Greek and Roman times onward, accounts of travels and voyages form a large part of the body of surviving literature. In this setting, the elitist nature of most early travel was clearly established, voyagers being by definition wealthy enough to provide for their needs and educated enough to record their adventures and impressions—a rare combination in an era when the average citizen was unable to envision life above the level of the plowed furrow, let alone escape the immediately pressing demands of providing daily food and other necessities.

Figure 12.3 The Mediterranean coast in August summarizes the problems of time and space concentration: Frejus, France.

Even among these few early voyagers, travel strictly for its own sake or out of curiosity was unusual. Some other basic purpose was most often involved, whether commercial trading, military campaigning, sponsored scientific investigation, or exploration under royal charter; and early "travel journals" are normally by-products of more purposeful movement, satisfying the creative urges of their authors while filling idle moments in the larger enterprise. Nonetheless, human curiosity and the restless desire to experience something different—still probably the bottom line of much modern tourism—appear to be the threads that hold together unforced human movements across the ages.

Although colorful and adventurous, therefore, and the inspiration for a considerable body of literature and for geographic descriptions which often formed the major bases of understanding remote nations or regions, travel of any kind—and especially travel for any but strictly utilitarian motives—involved only a minute percentage of humanity. Although curiosity may be a basic human trait, it was not until the three other basic prerequisites of tourism: money, time, and infrastructure, began to be more widely distributed that the patterns of tourism as we know them started to emerge. Most of this process took place in Europe, or with the impetus of European ideas, and it is thus possible to consider Europe the continent where tourism began.

Transportation and Tourism

Two seemingly unrelated processes, each derived from the industrial revolution, began interacting in Europe (as well as America) during the last half of the nineteenth century, accelerated the process of change during the 1920s and 1930s, and led eventually, in the years following World War II, to an expansion of tourism on a scale undreamed of a century earlier. One of these processes was technological, the other, social.

The technological innovations involved many of the dramatic changes which may be grouped as the transportation revolution. Since its domestication—perhaps some 7000 years ago—and subsequent association with wheeled vehicles, the horse had reflected the ultimate speed of human movement: an average of about eight miles per hour under good conditions; occasionally more, quite often less. This meant that any travel was tedious and time-demanding: a twenty-mile trip from farm to market town was a lengthy and infrequent undertaking; a visit to a city one hundred or two hundred miles away could be the adventure of a lifetime.

The industrial revolution changed all that, of course, as it changed so many other things. Through the specific medium of a steam engine connected to wheels and designed to run on a fixed road or track, the tyranny of the horse's pace was broken at last, and humanity's obsession with speed was given new dimensions, launching a period of a century and a half in which innovation succeeded invention at a dizzying rate. The dawn of the railway age is conventionally dated from 1825, when the Stockton and Darlington Railway in the English Midlands brought together a functional steam locomotive, a smooth-railed track, and a specific commercial vision (hauling coal) to form the first specifically constructed public railway. Scheduled passenger services began in both England and the United States in 1830; and from that point the race was on, as British ingenuity, American enthusiasm, and European colonial expansion spread rail lines and thus rail travel around the world.

As the industrial revolution created marvels of transportation technology, it also created a more extensive moneyed class with both the desire and the means to contemplate travel for the sake of change, out of curiosity, or simply to "broaden one's horizons." In 1841, Thomas Cook (at the time a missionary) persuaded the Midland Counties Railway Company to run a special round-trip excursion to a temperance meeting, charging a shilling per person, thus marking the beginning of the travel agent business. By 1856, the first Cook-arranged "grand tours" of the European continent were underway.

The second principal impact of the industrial revolution in the realm of transportation was the adaptation of steam power to movement on water. Robert Fulton is generally credited with operating the first passenger-carrying steamboat on the Hudson River in 1807. The first Atlantic crossing at least partly under steam was made in 1818, and in 1840 Samuel Cunard inaugurated the first regularly scheduled service between Britain and America, using a fleet of four specially constructed steamships. As passenger carriers, steamships were particularly associated with the great waves of emigration that shifted surplus populations from Europe to America and other destinations in the century up to 1930. Beyond this, however, travel by sea generally belonged to the province of the wealthy, and the great ships that became famous on the Atlantic run created a legend of style and opulence that epitomized "travel" in the popular imagination.

The third major component of the transportation revolution was the automobile, which not only increased the speed of land travel (once suitable road networks were in place), but also liberated the traveler from the fixed itinerary and printed schedule. The automobile's impact on human travel for pleasure has been exceptional, expanding destinations beyond the periphery of the rail lines, bringing the curious to previously inaccessible regions, breaking forever the isolation of countless rural villages, and reducing both the scale and cost of travel, thus marking a key stage in the democratization of tourism.

The functional automobile is usually considered to date from 1885, when the Germans Daimler and Benz employed an internal combustion engine to power an early model (there are many competing claims for the "first automobile"); but it was Henry Ford's assembly line techniques that standardized the product and lowered its cost to the point where any worker could aspire to own one. By 1926, the image of the motor car as a toy for the rich had been shattered, with momentous implications for, among other things, tourism.

The final major technological innovation in transportation—and the one which has clearly had the greatest impact on international tourism—was the airplane. Like the automobile, of interest mainly to the wealthy during most of its early years, air travel has done more to revolutionize the conditions of tourism over the past forty years than any other single factor. It has served to increase range, reduce costs, and decrease the proportion of time spent in movement, which only a minority of travelers consider the high point of a vacation. The Wright brothers achieved powered flight in America in 1903, and several European pioneers were close behind; but limited range and light payloads kept the cost of air travel high for many years. The North Atlantic was not crossed regularly by a scheduled commercial air route until 1939. It was not until the 1950s that larger, faster aircraft were introduced into service, reducing travel costs by carrying larger numbers, thus removing air travel from the province of the wealthy few.

Worldwide, generally lower costs and greater flexibility have largely enabled the automobile and the airplane to replace the railway and the ship as preferred means of tourism travel. The ship, however, has found new life in the increasingly popular cruise industry, which many European countries (Greece and Norway are good examples) are promoting heavily both at home and in other parts of the world. The railways of Europe, with their short distances between destinations and their downtown-to-downtown time advantages over airlines, have also continued to prosper, based to no small extent on the numbers of tourists utilizing their services.

Technology continues to evolve continuously in the realm of transportation, often with dramatic (if sometimes unintended) effects on tourism. A new generation of high-speed trains, running on specially built tracks, is spreading across Europe: by 1990, a French TGV had achieved an in-service speed record of 310 mph (500 kph)! Mileages of superhighways, often built at great expense through spectacular landscapes, continue to expand throughout Europe, while the long-awaited tunnel under the Channel, connecting Britain with France, has finally been opened (at enormous cost), thus linking Britain directly to the continent for the first time. Tourists will thus have an increasing range of destinations within reach over a fixed time.

The Paid Vacation

The technological impact of the industrial revolution, as applied to various means of transportation, led directly and in a surprisingly brief time to changes which eclipsed anything that had been seen throughout history. However, even though modern transportation began to set before humanity a remarkable variety of increasingly fast, safe, reliable, and even comfortable ways of moving about, who had the time or the money to make use of them?

In societies whose citizens were largely engaged in wresting a bare living from the land, or were chained at least figuratively to the shop floor, only the privileged had the temporal freedom and the cash to indulge whatever tastes in travel emerged and were encouraged. Even the rapid expansion of an entrepreneurial or upper-middle class in the urban industrial areas could not hope to provide the number of new customers required to transform tourism into a truly mass industry.

Another aspect of the industrial revolution, however, provided the impetus needed to complete the process. As the countrysides were drained of their labor to provide hands for the factories, in the familiar process of destroying the rural artisan class by providing only urban alternatives, pressures for social change began to build in the new cities and factory towns that sprang up as the process of industrial change spread outward from Britain like ripples on a pond into which a stone is thrown. Increasing political influence, examples of enlightened management, and above all unionization were among the factors that eventually began to bring about reform in the conditions of industrial labor. Wages, hours, and working conditions have been perhaps the most consistent issues throughout the history of the labor movement, but at various times a number of fringe benefits began to filter into the process. Among these, the novel concept of the paid vacation has had by far the greatest impact on tourism.

The notion of time off at employer expense is intrinsically alien to the capitalist mind, and it was not until the large-scale unionization of major industrial sectors in the 1920s and 1930s that the dream of "two weeks at the seashore" could become a reality for large numbers of workers and their families. The two-week paid vacation was enacted in France, for example, in June 1936. Thus August of that year marks the beginning of the massive annual French exodus to the mountains and especially the coast that some cynical observers have likened to a migration of lemmings! Following World War II and the immediate period of reconstruction, a third week was added to the paid vacations of most workers in 1956; the norm is now four, while five weeks is increasingly seen. A similar evolution in the concept and provision of paid vacations can be seen in most other developed industrial nations, although the amount of time may vary from country to country.

Early European Tourism: Setting the Pattern

Without doubt, the customs, facilities, and destinations of international tourism before World War I, and even to a large extent in the interwar period, were conditioned strongly by British taste and culture, especially by those among the English, Scots, and Welsh who were profiting most from the new industrial cornucopia. It is fair to say that these were the first "tourists" in the modern, unreliant sense of the word: demanding their accustomed comforts, patronizing their hosts, finding local citizens "quaint" in one form or another, and bringing upon themselves the coldness and condescension that only providing services to the insensitive rich can engender in a human population. Two destinations, still enormously popular, are particularly associated with the beginnings of British tourism to the continental mainland in the nineteenth century: Switzerland and the Mediterranean Riviera.

In the best traditions of early travel, visitors to Switzerland in the years before 1800 were largely motivated otherwise than by the tourist's obtrusive curiosity. Diplomats, commercial travelers, and titled gentry on various business made up the list, and wrote of what they encountered. The mountains alternately attracted and repelled, challenged and soothed. The adventure, tinged with a rather cultivated mystique, of climbing mountains "because they are there" emerged in the eighteenth century and became a veritable cult, focused on Switzerland, by 1850. The Swiss narrowly edged the British to the top of Mont Blanc in 1786, but Whymper's victory on the Matterhorn in 1865 captured the European imagination (especially since four of his party died in the descent), and seemed to reflect to the chauvinistic English the finest qualities of their race. The rush, to view if not actually to climb these peaks, was on. Thomas Cook had already brought 130 people on his first Swiss tour in 1863.

As various Swiss citizens began to discover that tourism could be actively promoted, rather than simply happening, the notion of health cures and their implantation into the Swiss environment began to spread. Among its other contributions, the industrial revolution, with its urban crowding and unregulated air pollution, made tuberculosis—already common in the damp, cool climates of northwestern Europe—the continent's major health problem. The traditional treatment, for those few who could afford it, was a prolonged stay in the warm, dry environment of Egypt; but when Swiss doctors began to discover that mountain air was almost equally beneficial, promotion of cures greatly stimulated tourist business. This process began in the 1850s, and although there were certain conflicts between treating the sick and catering to the traveler, the image of a "healthy" Swiss climate had a generally positive effect on tourism revenues.

The Mediterranean coast, although malarial in many places into the twentieth century and avoided by traditional populations of the region, also proved to have climatic attractions that began attracting well-to-do British around the turn of the nineteenth century. The French Riviera, or Côte d'Azur, as one of the most protected, sunny, and colorful parts of the coast, became especially favored as a winter retreat from the cold and damp of Britain. The British presence grew steadily, and actually contributed financially to improving the tourism infrastructure. The famous beachfront drive in Nice, still known as the "Promenade des Anglais," was built in 1822–23 through subscriptions put up by the visitors.

Modern European Tourism: Advantages and Benefits

In the 1990s, Europe has retained its position as the world's most important tourism continent for several good reasons. It is the region where the four prerequisites for tourism: money, leisure time, infrastructure, and curiosity are the most abundant, well developed, and widespread. Europeans, with Germans leading the way, are tireless explorers of their own nations, the continent beyond their national borders, and the world beyond Europe. At the same time, Europe is the destination of choice for travelers from the world's other wealthy, tourist-providing regions (essentially North America, with an increasing Japanese component), as well as for the small but growing number of citizens from the less developed world who are able to find the money and time to indulge their curiosity. Whatever their origins, Europe offers tourists many advantages unequalled by other potential (and certainly hopeful) destination regions.

1. **Access, size, variety.** The physical conditions and historical background of Europe are the root cause of many of these advantages. Since Europe is a small continent, distances are short, and travelers can easily sample the attractions of many countries and regions in a relatively short time. Moreover, European countries are generally easy to reach and easy to move around within (in contrast to many other parts of the world), so tourists are rarely forced to cope with problems of difficult or limited access. The physical variety of the continent, as described in earlier chapters, is also a major tourism advantage. Climatic change over short distances from north to south, or from lowland to highland, as well as the intricate patterns of landscape, make different scenes and varied temperatures readily accessible.

Beyond this, the lengthy history of human occupance and landscape modification in Europe, often against a backdrop of wealth, power, and worldwide influence, means that the continent is a vast showcase of art, architecture, and other evidence of the human condition that constitutes a powerful appeal to modern tourists. The great palaces and galleries are on every visitor's itinerary; but the smaller details of mankind's presence, the manicured quality of much of the European countryside, the cultural changes from region to region and country to country, and the abundant variety of goods offered for sale all form part of the richly textured human mosaic that is perhaps Europe's greatest attraction.

2. **Organization and infrastructure.** As befits the continent where tourism began, Europe has in most senses the best organized tourism operations and the most highly developed network of tourism-related infrastructure improvements to be found anywhere: outstanding transportation systems, including thousands of kilometers of superhighways, high quality networks of secondary roads, rail connections—including many high-speed and luxury trains—to most corners of the continent, internal and external air

connections to all the world's destinations, intricate ferry and shipping linkages across the seas and rivers and among the islands, and elaborate public transportation networks within and between major cities. While these systems are not designed exclusively to promote tourism, they are in themselves no small part of the European appeal.

All types and levels of accommodation are also available to the tourist in nearly all countries, backed by systems of government and private inspection and rating to ensure quality and consistency. From luxury hotels to more modest inns, apartment and farmhouse rentals, and abundant campsites, the entire spectrum of tourist affluence and preference is catered to. The same holds true of restaurants and the provision of food in general, always a key concern of travelers. Beyond this, the variety of things for the tourist to see and do in Europe is virtually endless, ensuring that every taste can be accommodated and that Europe can be a favored destination for nearly everyone. All varieties of participant summer and winter sports are available and well organized, as are hunting and fishing, spectator sports of all kinds, and in many countries more arcane specialities such as barge cruises or ballooning.

Concern for tourism also appears in the way most countries organize these activities. Many nations have specialized agencies with offices in nearly all towns to deal with reservations, tours, and accommodations, and to provide tourists with information and suggestions. Elsewhere, city or regional governments may provide the same services where a national agency is lacking. Money can be easily changed throughout most countries, efforts are consistently made to reduce the red tape that can plague tourists at border crossings, and multilingual signs and even translation services smooth the traveler's way. All of this reflects the importance of the tourist's choice of destination and the well-known truism in the tourist industry that a single unfortunate experience can easily cause customers to take their valued trade elsewhere.

3. **Safety.** To be successful, tourism must be a "feel good" industry. This accounts for the unfailingly optimistic and glowing quality of all travel brochures and promotional materials, many of which deal in half- or even non-truths regarding social conditions or

Figure 12.4 Tourism may provide income for the preservation of priceless art and architecture: Venice, Italy.

Figure 12.5 New construction in Alpine regions has been a major result of the recent boom in winter sports: Les Contamines Montjoie, France.

the economic situation of a given destination. With few exceptions, tourists—at least while removed from their normal lives and concerns—are interested almost exclusively in the amusements, diversions, and tranquility for which they have paid. Thus, the psychology of tourism is a key factor in the bitter competition for tourist revenues waged among nearly all countries and regions. The possibility, almost as much as the reality, of physical danger, unrest, or somehow difficult circumstances, is often sufficient to reduce tourist flows to trickles literally overnight.

In this context, Europe is again comparatively well-placed. The daily life of most countries is usually well organized and well regulated; crime occurs at only moderate levels and that which might be directed at tourists (robbery, petty vandalism) is nearly always nonviolent; and organized terrorism seldom strikes at specifically tourist-oriented targets. Even the dramatic demonstrations in eastern European countries in 1989–90, heralding the end of Soviet-backed authoritarian governments were, with the exception of Romania, generally peaceful and actually had the effect of dramatically stimulating tourist flows from many corners of an astonished world. Nonetheless, the sensitivity of tourists is legendary, and the slightest untimely incident (a strike, a political demonstration, a kidnapping, etc.) can mean substantial lost tourism income to the unfortunate country involved. The uncertainties caused by the Gulf War of 1991 had a chilling if brief impact on travel to nearly all European countries; while the tragic civil war in former Yugoslavia has all but completely eliminated tourism in one of Europe's most beautiful and tourism-dependent regions.

The relative stability of European societies in recent years has thus been a positive advantage to their tourism programs. Additionally, tourists are invariably (and often with good reason) concerned about health, and here again Europe as a tourism destination has considerable advantages. Health care and medical facilities, including emergency services, are widespread and of generally good quality; and dangerous

European Tourism

diseases, while not totally unknown, are quite rare. Public water supplies are safe almost everywhere (despite the fact that nearly all Europeans are major consumers of bottled water), and food is abundant and of almost always good quality, in addition to being treated as an art form nearly everywhere.

Tourism, of course, has its dark side as well, and the relative expense of European travel, along with the xenophobic reactions of many Europeans when confronted by hordes of often insensitive tourists, can make traveling on the continent a negative experience. Moreover, the increasing preference of many tourists for low-budget vacationing (utilizing campgrounds, backpacking, buying cheap food instead of patronizing restaurants) means that the golden dream of tourism income is often a bit tarnished for local citizens. In fact, in some cases, the cost of providing services for tourists may exceed the revenue they bring.

Beyond this, the extreme concentration of tourist activity at certain times of the year (especially July–August), and the heavy preference of most visitors for certain—often environmentally fragile—locations, means that tourism risks generating more environmental costs than its contribution to local economies can justify. This is notably true along the shores of the Mediterranean and in certain Alpine districts (see environmental discussion, Chapter 13).

PERCEPTION: BALANCING THE SWISS BUDGET

Although tourism is commonly perceived to be the economic salvation of many poorer countries and regions of modern Europe (notably along the Mediterranean fringe), it is in fact a vital element in the national accounts of nearly every nation. Switzerland, for example, is widely conceded to be the world's wealthiest country (discounting the fluctuating fortunes of certain oil-producing states); yet the Swiss rely heavily on tourism to produce this wealth. The process is instructive, and shows clearly why tourists and their money are so coveted by all nations.

Figure 12.6 Switzerland's natural beauty and tourism organization attract enough visitors to nearly balance the national budget: Luzern.

As do most European and other economically developed nations, Switzerland earns the vast majority of its income from the export of high-quality industrial and agricultural products, as well as services. Swiss cheese and chocolates, watches and electrical equipment, to name only a few items, command top prices around the world. On the other hand, Switzerland must import virtually all of the raw materials required by its advanced economy (notably including oil), as well as certain basic food items that cannot be produced in an Alpine setting. The result is that the nation runs a chronic deficit in its import–export balance.

It is the trade in services, especially financial transactions and tourism, that closes the gap and allows the Swiss to enjoy their high standard of living. Swiss banking and investment activities are well known; but Switzerland is one of the few wealthy countries in which receipts from visitors more than equal the amount spent by nationals traveling abroad (see Figure 12.2). Building on its long history as a tourist-welcoming nation, with probably the most highly developed tourism infrastructure anywhere, and profiting from its solid reputation as a safe, reliable place, Switzerland is easily able to market itself as a preferred tourism destination despite its relatively high costs. The fact that its winter and summer seasons are almost equally attractive also serves to balance and enhance tourism revenues.

ADDITIONAL READING

BERNARD, P., *Rush to the Alps: The Evolution of Vacationing in Switzerland.* Boulder, CO: East European Quarterly, 1978.

LABARGE, M.W., *Medieval Travelers.* London and New York: Norton, 1982.

LUNDBERG, D.E., and M. LUNDBERG, *International Travel and Tourism.* New York: Wiley, 1985

WILLIAMS, A.M., and G. SHAW, eds., *Tourism and Economic Development: Western European Experiences.* London: Pinter, 1988.

YOUNG, G., *Tourism: Blessing or Blight?* Harmondsworth: Penguin, 1973.

DISCUSSION QUESTIONS

1. Mass tourism is generally a phenomenon only of the years since the 1950s. Why? What combination of factors is necessary if tourism is to operate on a large scale?
2. The Mediterranean coast is the world's most popular tourism destination. What attractions make this region such a magnet for tourists?
3. One of the great advantages of tourism as an economic activity is its ability to provide income to otherwise disfavored regions. How is tourism different in this sense than, for example, manufacturing?
4. Why is tourism income considered to be "clean" money? Why is such income considered vital even by such a rich country as Switzerland?
5. Tourism numbers, including both European tourism and travel outside Europe by Europeans, will almost certainly continue to increase. What effects do you think the opening of eastern Europe's frontiers will have on tourism?

polluted lake are thus more than simply news items: they offend the very nature of things. In the same way, European governments have until quite recently accorded little or no priority to environmental issues. Economic matters have traditionally been of greater importance, and the environment was seldom seen as a national concern: cabinet-level ministries of environment date for the most part only from the 1970s.

Types of Environmental Stress

Air pollution. Problems of air quality have recently received considerable scientific and popular attention in Europe, especially as their impacts have become increasingly visible. "Acid rain" has in fact become such a common catch phrase that virtually no political speech on the environment is complete without several promises to control it. Air pollution is of course no stranger to Europe: the rapid eighteenth–nineteenth century growth of industry and of urban home heating systems based on burning coal (each room in a large house would commonly have its own fireplace with a separate chimney) created "smog" of appalling proportions in many cities and led to astonishing increases in the rates of tuberculosis and other respiratory problems. The simultaneous rise of tourism in the Alps and Middle East relates in no small measure to attempts by those classes with time and money to escape from such conditions, at least temporarily. By the late twentieth century, although the chimney pots no longer (in most cities) belch forth smoke from inefficient fires of cheap coal, contamination of the atmosphere is more widespread, more serious, and certainly more insidious. There are three basic reasons for this: continued expansion of both population and industry, the generation of electricity in large plants, and the advent of the automobile.

Any sort of combustion (in effect an oxidation process) adds a predictable series of products to the atmosphere. Water vapor and carbon dioxide (CO_2) are chief among these, and may have a strong influence on weather and climate patterns (see discussion of the "greenhouse" under climate, Chapter 2). Burning fossil fuels, however, also adds an additional, more threatening group of chemicals to the air. Since coal, oil, and natural gas are all formed in an anaerobic environment (that is, under shallow waters where oxygen is limited), they are inevitably linked to the natural sulphur cycle of reduction normal in this setting. Sulphur is, in fact, an economically important by-product, especially of oil and gas exploitation. Varying amounts of sulphur are thus chemically locked into these fuels, to be released as they burn—no matter how efficiently—mainly as sulphur dioxide (SO_2). In the atmosphere, this gas reacts with the abundant water vapor to produce in effect a weak solution of sulphuric acid (H_2SO_4), which then eventually falls to earth either in dry form or as precipitation with a significantly lower than normal Ph rating.

Acid rain has three principal impacts:

1. **Water bodies.** Through direct fall or runoff, acid rain raises the acidity of water bodies, notably lakes. This is not a substantial problem in southern Europe, where there are few lakes, mainly with natural buffering and high Ph values. In the north, however, especially Russia, Finland and Scandinavia, the problem is acute, since the thousands of small glacial lakes are naturally acidic in any case (a result of the granitic basins occupied by many of them and of the acidic nature of the conifer vegetation

which covers their drainage areas), and have delicately balanced ecosystems. Even a slight addition of acidified water may effectively turn them into "dead" lakes without either fish or vegetation. Sweden has thus been predictably upset by the acid rain "exports" of Great Britain and has gone to the expensive length of adding alkaline minerals to many lakes in an effort to restore their Ph balance.

2. **Vegetation.** Acid rain also has a devastating impact on many kinds of forest vegetation, not just through direct fallout, but also as a result of complex chemical reactions which are still only imperfectly understood. For example, the acid reacts with soil chemicals, releasing normally tightly bonded elements such as aluminum. These ions may then be absorbed by the trees' root systems, effectively poisoning them from within. In any case, the devastation can be substantial, the more so since damage is insidious, with symptoms often appearing only after it is too late to save the plant. Estimates of forest acreage affected by this problem in Europe are startling, and since most modern Europeans have a strong traditional attachment to their woodlands, the acid rain problem has been in many cases the primary vehicle for enhancing environmental concern. Visible threats to such world-famous regions as the Black Forest (Schwartzwald) of Germany (the dominant Norway spruce trees here seem especially vulnerable) have aroused public opinion to a much greater degree than any amount of scientific study; "forest death" *(waldstirben)* is perceived as a national disaster.

3. **Building materials.** Acid rain also attacks most of the building materials favored by humans, especially the traditional ones from which most of Europe's great historical legacy is constructed, aging and degrading structures well before the close of their normal life span. This is true of wood, metal, and stone; but it applies especially to that most beautiful and characteristic of classic European building materials: marble. Marble is a

Figure 13.1 Percentage of forest affected by rain: selected European countries

Country	Percentage of Forest Affected
Netherlands	55
Germany*	54
Switzerland	50
United Kingdom	49
Czechoslovakia	41
Austria	37
Bulgaria	34
France	28
Spain	28
Luxembourg	26
Norway	26
Finland	25
Hungary	25

Source: Adapted from L. Brown, and others *State of the World 1988*, p. 14.

*Data for former West Germany only.

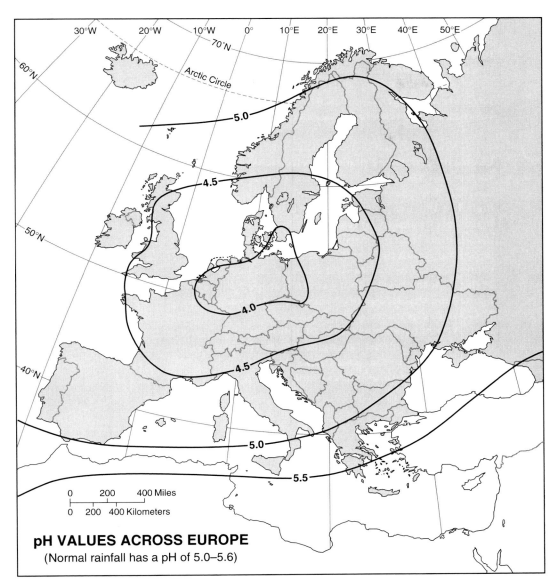

Figure 13.2 Europe: Concentrations of low pH values.

metamorphosed, geologically hardened form of limestone, often showing subtle colors or elaborate veining, which has been favored for costly and dramatic construction projects: the modern value of the famed Carrara marbles of Italy has already been noted. As anyone who has had a first course in chemistry knows, however, sulphuric acid reacts strongly with limestone (calcium carbonate): a drop of acid fizzes on the stone as the solid rock becomes a powdered form of calcium sulphate. At a slightly reduced speed, this simple process is also deteriorating many of the classic buildings and monuments of Europe.

More damage of this sort has been done in the past fifty years than during the preceeding two centuries; more damage in the last two centuries than in the lifetimes of these structures. Some cases are extreme, as revealed by the desperate solutions attempted. The magnificent Parthenon, on its hill in Athens, glory of ancient Greece, has now been encased in a plastic "house," thus protecting it from further ruin, even though its drama and effect are also largely lost. Elsewhere, many other famous urban monuments have been covered, or even removed altogether and replaced with acid-resistant plastic substitutes.

Acid rain is largely a product of large industrial units, particularly coal-fired electricity generating plants. As electricity consumption has soared in recent years, such plants have multiplied in size, production, and the capacity for environmental mischief. The quality of much of the coal burned is low, and its sulphur content often high. For this use, some eastern European countries, notably the former East Germany, rely on their vast reserves of lignite, a low value and very dirty fuel. Adding to the problem is an inconsistent pattern of emission control. The western nations have in general gotten beyond the "taller stack" solution to local fallout, and have installed sophisticated mechanical and chemical devices to remove at least a moderate share of the airborne residue, but eastern Europe has been slow to change; the light brown lignite haze that frequently hangs over cities such as Berlin is the environmental price of German electricity.

Another common source of atmospheric pollution in Europe—as worldwide—is the internal combustion engine. The burning of gasoline and diesel fuel (often low grade) in automobiles and trucks adds not only additional sulphur to the atmosphere, but also various oxides of nitrogen (generally abbreviated "NOx"). Moreover,

Figure 13.3 European industries have until recently been little concerned with environmental quality: L'Hôpital, France in the 1970s.

use of lead as an anti-knock compound results in further air pollution, in this case by a poisonous heavy metal.

Europe entered the age of mass automobile ownership considerably later than did North America: only in the late 1950s did postwar recovery and economic growth begin to put small cars within reach of average families. Similarly, European attention to the environmental problems posed by rapid expansion of the number of vehicles on the road has also lagged behind that of North America, in this case by about a decade. While strict emission controls and unleaded gasoline were introduced in the United States in the middle 1970s, helping to reduce or at least stabilize air pollution levels, Europe in the 1990s is still in the process of setting standards and phasing in controls. Swedish and German automobile producers have taken a lead, especially in their larger, more expensive models (for which the addition of catalyzers and other equipment is proportionately cheaper), since many of these are destined for export to the American market; but manufacturers in other countries—if they have plans at all—have begun making reduced-pollution cars in significant numbers only since about 1992. Expanded membership of the European Union (see Chapter 14), has at last put substantial pressure on most European countries to conform to more demanding environmental standards.

In this context, the political fragmentation of Europe plays a major role. A catalyzer-equipped German car, for example, would have no trouble finding unleaded gasoline in Germany, Switzerland, Scandinavia, or the Benelux nations, where nearly every station has a pump. Beyond this, however, the situation is considerably more approximate. Driving a German car through the peaceful countryside of France as recently as 1988 became a desperate exercise in finding the rare lead-free pump; while in 1990, Czechoslovakia had a grand total of three lead-free outlets nationwide: two in Prague, one in Bratislava. Travel in a nonpolluting automobile to the farther corners of the nation was thus impossible. By 1996, considerable progress has been made in making lead-free gasoline the European norm; but the long lag time in implementing this reform has not helped the continent's environment.

Finally, mention should be made of a practice which has further contributed to darkening the skies of certain European nations: what might be called "pollution importing." In industrialized nations with increasingly rigorous controls on air pollution from manufacturing plants—Germany or the Netherlands, for example—industries faced with obligatory major investments related to cleaning up existing operations or expansion in the 1970s or 1980s were often lured to other countries where standards were more lax. Some of the latter even launched low-key but effective promotional campaigns to build their economic bases at the expense of their environments. Spain was long a notorious case of this practice; as a result, air quality in Madrid (which also suffers from pollution-retaining atmospheric inversions as a result of its location in a basin) is so poor that the capital is generally awarded the dubious prize for having the dirtiest air of any major west European city. With Spanish entry into the European Union in 1986, compliance with the more rigorous rules being developed by the Union has put a stop to this practice; but the environmental damage already done will not easily be reversed.

Freshwater pollution. A second major category of environmental stress in Europe concerns pollution of the continent's lakes and streams. Once again, the dense population, urban concentration, industrial economy, and political multiplicity of Europe have all been major factors in creating the various problems, as well as in making their solutions more difficult and complex. Also, as with air pollution, a long history of ignoring water in any environmental sense, or at least of taking it much for granted, has meant that many situations have escalated to serious levels, which are only now beginning to be widely appreciated.

Since people have been on the European scene for a very long time, easily acquired water has been a critical factor in the location of their settlements and activities. As noted, lakes and streams have traditionally represented domestic water supply, waste disposal, vital transportation routes, and important power sources along which much early industry was sited. Over time, many of the villages and small towns of antiquity and the Middle Ages have grown into large cities, often with sizeable industries, still occupying the familiar locations and placing even greater pressures on the waterways. Each of the motivations for waterside location creates its own potential for environmental stress, and there are others as well.

Agricultural pollution is a widespread, if low-key, problem in a continent where farming has been a basic livelihood for so long. The addition, through runoff, of excessive nutrient material to streams, lakes, and canals—first from animal wastes, later from heavy use of chemical fertilizers—has caused explosive increases in populations of algae and other aquatic plants. This disrupts the ecological balance by reducing the oxygen content of the water (as vegetation decomposes), lowering fish populations, and generally speeding up the aging process of lakes and ponds (eutrophication).

Increased human pressure on lakes and streams as sources of water supply and outlets for waste disposal has long since made it necessary for control and treatment

Figure 13.4 Heavy use of Europe's waterways adds to environmental stress: The Main River near Frankfurt, Germany.

systems to be installed. Water is drinkable and sewage treatment adequate in most if not all corners of the continent (in fact, most of the earliest technical innovations in these areas were European); but there remains a certain residual mistrust of anything flowing from a tap. Europeans are great consumers of bottled mineral water!

Industrial concentrations, especially of traditional water-using industries (agricultural and wood processing, textiles, etc.) were long casual users of lakes and streams as free "natural" dumping sites for waste materials. Most of these abuses have now been corrected, partially as a result of the demise of many such industries; but the scars often remain in the valleys and basins. The important transportation function of Europe's waterways has also played a role in heightening environmental stress. Motorized barges and canal boats, some very large and many very old, ply a dense navigable network. Exhaust fumes, oil leaks and spills, and untreated wastes inevitably enter the waters, and while it is difficult to single out the responsibility of any individual vessel, the overall impact is considerable. Finally, it should be noted that water pollution is a European problem with very long roots. In London, the British Houses of Parliament stand directly on the banks of the Thames River, and reports from the seventeenth century indicate that even then, members were forced to hang disinfectant-soaked sheets at the windows to cut the river's stench sufficiently that business could be conducted!

Coastal pollution. A third major type of environmental stress that profoundly affects Europe as a continent is pollution of its offshore waters. As has been shown, Europe is largely a collection of islands and peninsulas, and the sea has played a vital role in the history and economic development of most countries (of the European total of forty-four nations, only thirteen—including four "mini-states"—are landlocked). Most of the seas that surround Europe, however, are in their turn nearly surrounded by the land, and are particularly vulnerable to environmental perturbations. The Mediterranean (and its extension into the Black Sea) is nearly a closed lake; the Baltic opens only by way of a few narrow channels through Denmark; and the North Sea itself opens to the world ocean only to the northwest and by way of the narrow channel between France and Britain. Only along the Atlantic south of the Brittany peninsula, on the rugged coast of Norway, and on the western exposures of Great Britain and Ireland does Europe face the open sea directly; and even here, embayments and coastal indentations create an environment in which land and water are closely associated. Perhaps only Iceland, an island entirely surrounded by the vast Atlantic, is an exception to the pattern; and it is also of some significance that those limited faces of Europe turned directly to the sea are largely outside the regions of intensive economic development.

Given this maritime orientation, Europe's economic focus over the centuries has consistently been trading and commerce—much of it long-range—which has created great wealth and political power; but it has also caused an unusual concentration of population, industrial activity, and urban growth in the immediate coastal zones surrounding the North, Baltic, Mediterranean, and Black Seas. In recent times, the European need to import vast quantities of industrial raw materials has further exaggerated the significance of coastal location, and important new industrial complexes have sprung up at numerous points accessible to large, ocean-going vessels. In fact, the industrial age has actually moved out to sea with the spread of offshore production of oil and natural gas in the North Sea. Exploration is also being carried out in other waters. The July 1988 explosion and fire that destroyed a North Sea drilling platform, and the 1995 controversy over the proposed destruction by sinking of an outdated platform showed clearly the environmental threat posed by these activities.

Modern tourism has also focused heavily on the coasts, large numbers of both European and foreign vacationers seeking sand as well as sun. This extra addition of people, both seasonally and spatially concentrated, creates substantial problems for water supply and sewage disposal systems, and generates an impressive volume of solid waste, at least part of which finds its way out to sea. Finally, it should be remembered that the major European rivers, many of them heavily polluted, flow inevitably into one or another of the surrounding seas, adding yet another burden to the ocean's natural recycling system.

Oil pollution is a particularly pervasive problem in Europe's offshore waters. Europe is the leading oil-importing continent, and petroleum in both crude and refined forms moves steadily and in heavy volume around its coasts and through its waterways. Given the dense maritime traffic, large tanker sizes, unpredictable weather changes, and tricky problems of navigation, the possibility of accidents spilling large amounts of oil into the sea is a very real one, as the well publicized wrecks of the Torrey Canyon off Land's End, England (1967) and the Amoco Cadiz on the coast of Brittany (1978) have amply demonstrated. Ironically, however, it is the relatively small amounts of oil added steadily to the sea by spillage during transfer and refinery operations, by the tankers themselves, and by the residual oil discharged as ballast tanks are cleaned, that add the greatest proportion of hydrocarbon pollution.

While Europe's seas are being negatively affected by unwelcome human additions, they are also being changed by processes of subtraction. Massive exploitation of coastal wetland environments and large-scale drain-and-fill operations to create more industrial space, condominium complexes, and entire new recreational resorts, are modern evidence of European humanity's continued determination to bring order and profit to the landscape. What Jacques Cousteau has called "mechanical destruction" of the coastal ecosystem is well advanced on all European shorelines. This is a particularly important problem, since the immediate coastal zones, with their abundant energy in the form of light and nutrients, are among the most productive of all natural systems. Fish spawning grounds, migratory bird stopovers, natural pollution abatement, and the multimillion dollar shellfish industry are all lost when wetlands become concrete: small wonder that fresh oysters, once a common and inexpensive delicacy, have virtually vanished from all but the most expensive European menus.

Other forms of pollution. Air and water pollution are clearly the most widespread and pervasive environmental problems facing Europe; but the same factors that make these significant—many people and nations in a small continent; an industrial economy; and a consumption-oriented lifestyle—operate to ensure that Europe displays nearly all of the various other forms of environmental stress found in the world. For example, over the past forty years, Europe has changed from a frugal, instinctively recycling society to a throwaway one, and the volume of solid waste—most of it non-biodegradable—that litters roadsides and picnic sites is sadly impressive, even by American standards. Seeking to tackle this problem, many European nations have recently adopted legislative proposals designed to make recycling mandatory in some impressive ways. In Germany, for example, all packaging must now be recycled at the producer's expense, and even automobiles must, at the end of their useful lives, be recycled into their component parts.

Since many European nations—most notably France—make use of the nuclear option to meet their energy needs, radioactive waste is another potentially serious environmental hazard. In keeping with its status as the world's leading producer of nuclear energy (in per-capita figures), France has taken the lead in controlling and reprocessing nuclear waste, including pioneering some innovative technologies for the long-term neutralization of extremely hazardous materials; and France also contracts to deal with such waste from other European nations.

Nonetheless, questions remain. Until recently, considerable amounts of low-level nuclear waste were simply dumped into the Atlantic off the northwest coast of Spain, while minor accidents and small discharges of radioactive materials have plagued plant operations in many countries. Industrial accidents are an inescapable fact of advanced modern economies; but where highly toxic and psychologically frightening nuclear materials are concerned, the problem escalates into a new dimension: the winds from Chernobyl touched many parts of Europe.

PERCEPTION: ENVIRONMENTAL STRATEGIES: THE GREENING OF EUROPE

As an awareness of the gravity of environmental deterioration has gradually penetrated the European consciousness, the question of an appropriate response has been raised in many countries. Recourse to the legal system, which has played such a vital role in citizen involvement in the United States, is generally more difficult in most European countries; a result of bureaucratic complexities which the public at large finds difficult to overcome. Similarly, the tactics of protest—strikes, marches, confrontation—are often both ineffective and personally risky. What remains is the political system. Most European countries—particularly in the west, but increasingly including Poland, the Czech Republic, and Hungary to the east—have electoral systems which make it relatively easy for minority and special-interest parties to appear on ballots and to have an impact on public opinion and governmental actions. It has been through the political process, therefore, that European environmental interests have found their most effective expression.

The immediate human costs of the April 1986 nuclear disaster at the Chernobyl plant in Ukraine were severe, the long-term health and economic burden immense; but as the resultant nuclear cloud brought problems and near panic to many parts of Europe, one thing became clear: the most significant fallout from Chernobyl was political. No single event has done more to destroy Europe's environmental complacency. A prominently painted sign seen on a barn in remote rural Austria in 1988 sums up the impact: "Chernobyl is everywhere!" When the worst chemical pollution of the Rhine river in history occurred a few months later, the change in European attitudes was complete. Political response to the newly perceived dangers has been strong, both in individual nations and in such international bodies as the European parliament, where environmental or "green" parties have shown surprising strength in recent elections.

Figure 13.5 The threat of nuclear pollution weighs heavily on Europe: Nuclear electricity plant: St. Laurent-des-Eaux, France.

The "greens" are as politically varied as the nations and constituents they represent. The great problem faced by the "greens" in all countries is whether to be a single issue "voice of conscience" party focusing uniquely on the environment and its problems; or to become a more broadly-based political party—with some real chance of sharing power—by adopting policy positions on the wide range of issues with which any government must deal.

The debate over this issue has fragmented "green" strength in many countries. In Germany, for example, the most successful "green" movement in Europe is badly divided between the "fundis," who seek a pure, uncorrupted party dedicated to replacement of the present system by a nonpolluting, socially responsible, economically minimalist government; and the "realos," who insist that the best chance of initiating proenvironmental changes rests in cooperating with and influencing the mainstream political parties. Similar internal conflicts cause dissension in most other national "green" movements. Nonetheless, the "greens" are one of the most vocal and innovative parties on the European political scene today. In Germany, they share power in several state governments and have an increasingly strong voice at national level.

Figure 13.6 Green strength, recent parliamentary elections selected countries

Country	Last Pre-Chernobyl		Post-Chernobyl	
	Year	Green (%)	Year	Green (%)
Belgium	1985	6.2	1991	8.2
France	1986	1.2	1993	7.6
Germany	1983*	5.6	1994	7.3
Great Britain	1983	0.2	1988	1.3
Sweden	1985	1.5	1991	3.4
Switzerland	1983	6.4	1987	8.3

*West Germany only.

ADDITIONAL READING

ALLISON, L., *Environmental Planning: A Political and Philosophical Analysis.* London: Allen and Unwin, 1975.

BROWN, L., et al, *State of the World, 1990.* New York and London: Norton, 1990. (Subsequent editions through 1996.)

CAPRA, F., and C. SPRETNAK, *Green Politics.* New York: Dutton, 1984.

LIEFFERINK, J.D., P.D LOWE, and A.J.P. MOL, eds., *European Integration and Environmental Policy.* New York: Wiley, 1993.

MULLER-ROMMEL, F., ed., *New Politics in Western Europe: The Rise and Success of Green Parties and Alternative Lists.* Boulder CO: Westview, 1989.

VOLGYES, I., ed., *Environmental Deterioration in the Soviet Union and Eastern Europe.* New York: Praeger, 1974.

DISCUSSION QUESTIONS

1. What is "acid rain"? What impact is it having on the European environment? What are some ways in which the future output of this pollutant might be reduced?
2. In many parts of Europe, especially coastal areas, tourism is contributing substantially to environmental degradation. How can this be the case?
3. What factors combine to make Europe arguably the world's most polluted continent?
4. The Rhine is one of Europe's (and the world's) most polluted rivers. What factors are responsible for this? Why is cleaning up the Rhine such a difficult task?
5. "Green" political parties have recently become influential in many European nations. Who are the "greens"? Do their programs have a chance of success?

14
Supra-national Organizations in Europe

Europe is the only continent where international groupings, for military and particularly economic purposes, have been truly successful in recent times. As a result, Europe appears to be moving gradually away from the nation–state ideal toward some concept of a true internationalism; and although the process is a slow and complicated one, an increasing amount of contemporary European history and economic geography can be understood only in the context of international institutions. Moreover, the profound changes that have altered Europe's political map, military configuration, and economic structure since 1989 have resulted in a general rethinking of these institutions, improving the future prospects of some while seemingly dooming others to extinction. Whatever their origins and destinations, European international institutions fall logically into two types: economic and military, both with important political overtones.

INTERNATIONAL ECONOMIC/POLITICAL INSTITUTIONS

The dream of European unity is an old one. Almost before the continent began to segregate into discrete linguistic, cultural, and eventually national units, visionaries of every age began to devise ways to put the whole back together. The Roman Empire at its height came close to achieving this dream, Napoleon I had a substantial piece of the continent under unified control for a few years in the early 1800s, and Hitler subjugated much of Europe in the early 1940s. These efforts at unity, however, were accomplished mainly by military means, and unifying people at the point of a gun is rarely a recipe for long-term success.

The disasters that befell Europe during the period of intense nationalism (three major wars and numerous minor ones in the 75 years between 1870 and 1945) led to a gradual awakening in the European consciousness that there must be some better way to resolve differences. Fortunately, in the bitter aftermath of World War II, a farsighted group of leaders arose in many European nations and began to work diligently for an internationalization of key European institutions and a neutralization of the stress factors that had so often led the continent to war. Guy Mollet and Robert Schumann of France, Konrad Adenauer of West Germany, and Paul-Henri Spaak of Belgium were among this group. Not surprisingly, it was the smaller nations, notably Belgium, the Netherlands, and Luxembourg who, having been twice in fewer than forty years overwhelmed by conflicts between their larger neighbors, pushed hardest for some international accommodation. Bringing France and Germany into some sort of organization where their aggressiveness could be controlled seemed to be a key to survival for these nations.

Additionally, there was another powerful incentive for the development of some sort of larger economic unit within Europe. As noted earlier (see population discussion, Chapter 3), only the largest European countries have had populations in the 50–60 million range (reunited Germany now totals some 80 million), while many have fewer than 10 million citizens. The small internal markets represented by these populations represented a growing handicap as countries were increasingly forced to compete in world markets with much larger nations where sales at home can support substantial research and development investment. No matter how well organized their industrial production, or how clever their technological experts, it is difficult for countries such as Belgium (10 million), Denmark (5 million), or even France (58 million) to compete in such advanced technologies as aerospace, computers, information, or military requirements against such world powers as the United States (250 million), Russia (200 million), or Japan (122 million). While it is only a European joke that "one computer can run all of Luxembourg," building an internationally competitive computer industry obviously depends on the research investments that only a large internal market can provide.

Out of these ideas and fears the European Community (now the European Union) was born. In its original conception, all of the nations of Europe were to be invited to join. However, given the political and economic realities of the 1950s, many were forced to withdraw. The Eastern European nations, which had fallen under the shadow of the Soviet Union, were unable to participate, while Spain and Portugal, still ruled by fascist dictatorships, were considered unacceptable. Switzerland, Sweden, Finland, and eventually Austria were constrained by their traditional or enforced neutrality to decline. The Greek economy was simply too weak to compete, while Great Britain considered that she still had a world role to play that was incompatible with European integration. The result, after all the objections had been accounted for, was that six nations: France, Italy, West Germany, and the Benelux states (Belgium, Netherlands, Luxembourg) formed the original Community (or "Common Market").

Because of their notorious role in raising international tensions and provoking Franco-German clashes (see discussion of resources, Chapter 10), the coal mining and iron and steel industries were the first areas of attention for the new Common Market. In 1952, the European Coal and Steel Community (ECSC) was formed; over some forty-five years, this has very likely been the most successful of all the institutions of the Community. Working quickly to remove tariff restrictions and make critical resources available to all members on equal terms, the ECSC has been instrumental in removing a classic cause of tension and actual warfare in western Europe in recent centuries. All member nations have equal access to the coal reserves of France, Belgium, and Germany, and to the iron ore deposits of Lorraine. In addition, the ECSC has been invaluable in controlling production and cushioning the impact of slowdowns in these critical industries in recent years. Closing inefficient mines, operating retraining schemes for the unemployed, and trying to adjust steel production so that no member suffers more than another have been among the greatest successes of the ECSC.

In 1957, the six original members signed the Treaty of Rome, which established the full European Community (Union-EU). Although designed to have general legislative and judicial functions as a complete community of nations, the greatest successes of the Union have come in the economic sphere. This is logical since, although most nations have strong reservations about surrendering their independent decision-making powers to a larger supranational body, they can find themselves in complete agreement on the desirability of turning a greater profit—precisely the promise of the economic union. The Union has done best in the industrial area, where international tariffs and duties on a wide range of manufactured products have been removed, so that member states quite literally have an internal market of much larger size to contemplate. On the other hand, the area of agriculture has been the principal stumbling block in the path of greater economic integration, and has indeed come close to breaking up the Union on several occasions. Many member states (particularly Germany) have large numbers of farmers on marginal lands who are heavily supported by the government. This is considered unfair in the context of the Union by such agriculturally productive nations as France, the result being a seemingly continuous round of difficult negotiations, threats, and protests, with a view to establishing a viable "Common Agricultural Policy" in terms of which agricultural subsidies are gradually eliminated. Nearly two thirds of the Union's budget is typically spent on agricultural support programs and the weight of subsidies in world markets was dramatically illustrated in clashes between American and EU interests during the protracted negotiations seeking to bring the "Uruguay Round" of the General Agreement on Tariffs and Trade (GATT) to a successful conclusion in 1994.

In addition to economic policy, the EU also strives to promote and expand its political influence, although this is generally a more contentious proposition. An EU judicial arm functions fairly effectively in regulating and controlling legal matters related to the economic aspects of integration, but the most intriguing body is certainly the European Parliament, designed to be the legislative assembly of an eventual United Europe. Parliament was originally made up of members appointed by the parliaments of member countries, and was thus a fairly dull and predictable body, particularly since none of the member nations are really ready to surrender many of their national prerogatives to an international organization. Beginning in 1979, however, the European Parliament was elected by direct suffrage from each member state (subsequent elections were in 1984, 1989, and 1994), and the result has been a much more vocal, visible, and influential body. Members are seated by political persuasion (socialist, conservative, etc.), rather than by country, and since any citizen may run for election, some strange ideas are represented (including those of "greens," regional activists, and even a small group that favors abolition of the entire EU!). Elections to the European Parliament have become an important test of popular opinion in many member countries, and whether or not the Union ever acquires a properly political dimension, the European Parliament seems to be a small step in that direction.

Over the years, other nations have acceded to membership in the Union, testifying to its growing importance in European affairs. In 1973, Great Britain, Ireland, and Denmark joined the Union as a result of national plebescites (Norway turned down membership at the same time), and in 1981, Greece became the tenth member (having been an associate member). In 1986, membership applications from Spain and Portugal were also accepted. Having cleaned up their political acts to make themselves more acceptable to other European democracies, these countries have much to gain from membership. However, a major stumbling block was their largely agricultural economies, which feature products already surplus in the existing Union (especially wine and other Mediterranean products). As of January 1, 1995, Sweden, Finland, and Austria officially became members of the Union, bringing the total to fifteen (Norwegians once again rejected membership in a plebescite). Thus, although progress is slow, the idea of a unified Europe remains strong, with benefits seeming to outweigh disadvantages for most nations. Whether full integration in a political sense can ever be achieved is a moot point; but it seems doubtful on the other hand if Europe will ever return to the "dog-eat-dog" philosophy that prevailed before the Union was established. Evidence of the power of the idea of unity is provided by the lengthening list of prospective members who view affiliation with the Union as an important component of their future success and security. Turkey has long been a hopeful Associate Member, while Malta and Cyprus are other Mediterranean candidates. Farther to the east, Poland, Hungary, the Czech Republic, Slovakia, and Slovenia hope to become members by early in the twenty-first century, while other eastern European nations, including the Baltic states, Ukraine, Belarus, and even Russia retain hopes of future admission. The comparatively weak economies and unsettled political conditions of many of these countries make their eventual adhesion to the Union problematic, but they all view it as a goal to be aggressively pursued.

In the early 1980s, progress toward more comprehensive harmonization and integration of the Community economy seemed to be lagging, and the member nations reached a determination to expand their efforts in this direction. The result was the "Single European Act," which proposed a much more fully unified Europe, in both economic and social sectors, by the end of 1992. Under provisions of the act, free movement of people, commodities, services, and capital among the various member nations will be assured; even a common currency is under consideration.

As a final point of interest regarding the European Union, it should be noted that several European cities have profited considerably from being the site of various EU organisms. Notable among these are Brussels (Belgium), which is the headquarters of the executive and judiciary arms, Strasbourg (France), which will be the eventual permanent site of the European Parliament, The Hague (Netherlands), seat of the European Court of Justice, and Luxembourg City, where a variety of other EU functions are centered. As the EU expands its economic and political activities and ambitions, other cities compete strongly for the right to host the relevant

Supra-national Organizations in Europe

(and financially rewarding) institutions. An eventual European central bank, for example, will probably be awarded to Frankfurt-am-Main, Germany.

As the European Union becomes both broader and deeper, more serious problems begin to appear in the structure of its institutions and in the vision that its members share. Managing a Union of fifteen members—and perhaps twenty or more early in the next century—is a vastly more complex operation than was that called for by the original six-member "Common Market." What weight, for example, should the various nations have in votes on Union policy? Should this be determined by population, wealth, or some more complicated formula? As of now, moreover, the influential Presidency of the Union has rotated among the twelve members every six months. This means that each nation has held this post once in six years. Additional members will clearly lengthen this time: should larger countries such as Germany or France receive more preferential treatment?

Two other problems bedevil the Union and make its ultimate success as a true "United States of Europe" problematical. First, a substantially greater level of economic integration, including a common currency, is proposed, possibly to take effect by 1999; yet some members are either philosophically opposed to such an extension of supranational authority (notably Great Britain), or have economies that are clearly too weak to envision such a timetable (e.g.: Greece, Portugal, and most of the prospective new members). Second, the concept of a closer union implies progress in such areas as foreign policy and defense, and here the European Union seems hopelessly adrift. Apart from an inability to display a coherent strategy for dealing with troubled areas outside of Europe, the Union has also failed to chart a decisive course regarding problems nearer home: for example, the disaster that has befallen the former Yugoslavia.

Nonetheless, the European Union moves on, with a momentum that now seems impossible to reverse. A major long-term conference to review its institutions and its progress toward more complete unity was convened in 1996, and it seems safe to say that the future economic—and perhaps even political—structure of Europe will inevitably pass by way of the EU's Brussels headquarters.

At the time that the European Union was being established, several countries that for various reasons could not see their way clearly to apply for membership, yet had a serious interest in promoting increased trade without the political overtones of the Community, formed the European Free Trade Association (EFTA). Founded in 1960, EFTA grouped Great Britain, Denmark, Norway, Sweden, Switzerland, Portugal, and Austria, with Finland as an associate member. Reasonably successful in stimulating trade and reducing tariff barriers among its members, yet handicapped by the duplication of production on the part of many of its members (note the number of "Alpine" or "Scandinavian" economies), EFTA became a relatively insignificant body when Great Britain and Denmark joined the Community in 1973. With membership in the European Union being awarded to Austria, Finland, and Sweden in 1995, EFTA has effectively become a memory from Europe's past, but one that played a major role in stimulating trade and preparing its members to play on a larger stage. Of all the former members of EFTA, Switzerland has had perhaps the most agonizing decisions to make, given its reliance on export markets to retain Europe's highest living standards and its simultaneous dependence on a nonpolitical image. However, in a 1994 referendum, Swiss citizens again rejected the idea of applying for membership in the European

Figure 14.1 Membership of the European Union (dates of accession shown).

Union, demonstrating once more their determination to remain free of political entanglements and their confidence in an economic system that has rarely failed to adapt successfully to altered conditions.

In eastern Europe, any form of integration was a far more difficult step, since the nations involved had almost no history of interaction with one another. Relationships in this part of Europe have nearly always been directed to east or west, rather than toward neighboring states to the north or south. In the Soviet

Supra-national Organizations in Europe

Figure 14.2 The European Union

I. Known until 1993 as The European Community (or "Common Market"), including:
    A. The European Coal and Steel Community (ECSC)
    B. The European Economic Community (EEC)
    C. The European Atomic Community (EURATOM)
II. Fully organized by 1957 (Treaty of Rome)
III. Original Members:
    A. Belgium   ⎫
    B. Netherlands ⎬ Benelux
    C. Luxembourg ⎭
    D. France
    E. Italy
    F. German Federal Republic (now all of Germany)
IV. Members as of 1973
    A. Denmark
    B. Great Britain
    C. Ireland
    D. (Norway rejected membership.)
V. Members as of 1981
    A. Greece
VI. Members as of 1986
    A. Spain
    B. Portugal
VII. Members as of 1995
    A. Austria
    B. Finland
    C. Sweden
    D. (Norway again rejected membership.)
VIII. Associate Members
    A. Turkey
IX. Approximate Population (full members): 350 million

The "waiting list" of other potential members (whose political and economic structures are not yet up to Union standards) includes Malta, Cyprus, Hungary, Poland, the Czech Republic, Slovakia, and—farther off—Russia and Ukraine. As of 1996 it seems that there will be little further enlargement of the EU for several years, as there are substantial problems of integration among the existing members to be addressed (a common currency and further economic convergence are major issues, as are defense policy, international relations, and the degree to which each member nation is willing to abandon its often cherished political and economic independence).

Union, which forced its client states in eastern Europe to decline membership in the original European Community, that body was seen for many years as "the last gasp of capitalism." Faced ultimately with the striking success of the Community in many economic areas, however, the Soviets finally decided to emulate the western Europeans by reviving the Council for Mutual Economic Aid (CMEA or COMECON) in 1960 (it had been relatively inactive since its founding in 1949). Grouping the Soviet Union and its satellite states in eastern Europe (adding some non-European states and accepting the defection of Yugoslavia and Albania), COMECON had a modest history of success once the Soviets were convinced that real partnership did not lie in all member states coordinating their economies to the requirements of the Soviet Union.

While a great deal less impressive than the advances made in the Community nations of western Europe, COMECON was successful in a limited way in promoting a much greater volume of trade among eastern European nations, and in building links of communication and interaction among them which the long sweep of history had never before provided, and which have proven useful in the post-Soviet era.

At the end of 1989, the spectacular upheavals in Eastern Europe, in which many of the prevailing political and economic structures were swept away or at least substantially modified, also marked the end of COMECON. An international grouping based on the presumption of centrally planned economies seemed increasingly irrelevant, and most of the emerging governments of these nations have been forced to reassess their options in the area of international economic cooperation. COMECON was formally disbanded in February 1991, and its former members are now exploring various forms of regional association, as well as considering the advantages of membership in the European Union.

DEFENSIVE GROUPINGS

In addition to these largely economic groupings, Europe in recent times has also witnessed a variety of other initiatives toward interdependence. Militarily, the ideological division of Europe following World War II led to the formation of two opposed camps: the North Atlantic Treaty Organization (NATO), sponsored by the United States, on one hand; the Warsaw Pact nations, sponsored by the Soviet Union, on the other. Concerned almost exclusively with the strategic and tactical aspects of military preparedness from either side of the ideological line that divided Europe, these organisms had little interest in the broader sphere of economic growth or international cooperation. In this area as well, the events of 1989 and those subsequent have required considerable re-evaluation on all sides. Yet European security, against outside forces as well as internal threats, is still a real concern that all nations will seek to address. NATO, even including united Germany, is a possible defensive umbrella; but many countries have mixed feelings about the formal presence of the United States in Europe's military establishment, while the dark days of the "cold war" have given it a negative image in the eyes of Russia and other former Warsaw Pact members. NATO may indeed expand the concept of collective security to the east; but its evident inability to deal effectively with such an internal European threat as the continuing crisis in Bosnia and other Yugoslav states certainly weakens its credibility. NATO is attempting to expand its influence in eastern Europe and the former Soviet Union by organizing a loosely affiliated group known as "Partners in Peace." This groups twelve of the fifteen new nations formed from the Soviet Union, as well as most of the eastern European nations and some western countries (Sweden, Finland) which have never felt comfortable joining NATO itself.

OTHER INTERNATIONAL BODIES

Much attention is currently being given to the Organization for Security and Cooperation in Europe (OSCE), a thirty-five-member body grouping most European states in addition to the Soviet Union and the North American nations. Originally created as a forum for human rights discussions, the OSCE thus far lacks the kinds of permanent structures and expanded mandate that would enable it to serve as a true defense-coordinating body; but it enjoys a positive image that might enable it eventually to replace both NATO and the Warsaw Pact as the guarantor of Europe's peace.

There is a variety of other international bodies that reflect some vision of a united Europe, and which, although they have until now had little impact on European events, are suddenly being cast into the spotlight by the dramatic changes taking place on the continent. It seems, in general, to be a time to review the potential contributions of many such bodies. The Western European Union (WEU) is one of these bodies; the Council of Europe, another. The Council, whose twenty-three members include all of the European Union and former EFTA nations, plus their associate members and even two of Europe's "mini-states" (San Marino, Liechtenstein), is postwar Europe's oldest international organization (1949). Founded to encourage greater European unity and cooperation, multiparty democracy, and human rights, the Council is another body whose time has clearly come, and it has been steadily increasing its influence with the Union, with eastern Europe, and in the larger world as well.

Figure 14.3 The European Union is Europe's most important international organization: EU headquarters, Brussels, Belgium. Photo courtesy of the European Commission.

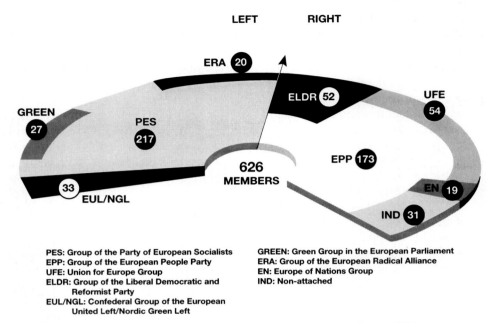

Figure 14.4 The European Parliament: Representation by political group, 1996.
Source: *Europe,* publication of the European Commission.

Finally, mention should be made of the Organization for Economic Cooperation and Development (OECD), sometimes referred to as the "rich man's club." The OECD is a group of twenty-four nations (primarily European) which is particularly notable for providing accurate and useful statistics about the world economy. Based in Paris, the organization studies trends in employment, education, taxation, and welfare in a unique global perspective. Several European nonmembers are in line to join, although the OECD clearly has no economic or political ambitions.

The European Union, with its strong emphasis on economic betterment, is at this point the organization most clearly pointed to the European future. Nonetheless, the dramatic changes in Europe's political and economic alignments fostered by events of the late 1980s and early 1990s will clearly make internationalism, mutual security, and cooperation more compelling ideas. New groupings of nations for these purposes, or the revival and expansion of existing organizations are thus likely trends.

ADDITIONAL READING

BLACKSELL, M., *Post-war Europe: A Political Geography.* Boulder CO: Westview, 1978.

CARNEY, J., R. HUDSON, J. LEWIS, eds., *Regions in Crisis: New Perspectives in European Regional Theory.* New York: St. Martin's, 1980.

DEMKO, G., ed., *Regional Development: Problems and Policies in Eastern and Western Europe.* New York: St. Martin's, 1984.

NORTH ATLANTIC TREATY ORGANIZATION, *NATO Handbook.* Brussels: 1995.

PARKER, G., *The Countries of Community Europe: A Geographical Survey of Contemporary Issues.* New York: St. Martin's. 1979.

PINDER, D., *Regional Development and Policy: Theory and Practice in the European Community.* London: George Allen and Unwin, 1983.

WILLIAMS, A., *European Community: The Contradictions of Integration.* London: Blackwell, 1991.

DISCUSSION QUESTIONS

1. What is the membership of the European Union (EU)? What are its goals? Does it have political as well as economic ambitions?
2. What was COMECON? With the collapse of communist regimes in eastern Europe, what kinds of international organizations will these countries seek to join?
3. The thirty-five-nation Organization for Security and Cooperation in Europe (OSCE) may well come to play a major role in European security. Why should this group, rather than the North Atlantic Treaty Organization (NATO), become increasingly popular?
4. The European Parliament has become a useful forum for the expression of minority views outside the mainstream of European politics. How is the Parliament chosen? What do you see as its future role in a more closely integrated Europe?
5. Why was the European Free Trade Association (EFTA) formed? Why did it fail to survive into the 1990s?

PART FOUR
National and Regional Profiles

15
Northern Europe

The nations of northern Europe have a distinctiveness in their physical landscapes, cultural and historical patterns, and levels of social and economic development that make this region a well-defined and generally cohesive part of the European mosaic. Building on what would appear to be an unpromising resource base, the countries of northern Europe have achieved remarkable levels of development; they are among the world's wealthiest and most successful by nearly every standard of measurement. Sweden, Denmark, and Norway, in addition to the islands of Iceland, the Faeroes, and Greenland, comprise the Scandinavian component of northern Europe, while Finland is generally considered separately because of its linguistic and cultural uniqueness. The Baltic states of Estonia, Latvia, and Lithuania, enjoying their freedom again following the collapse of the Soviet Union, complete the list of northern European nations.

PHYSICAL FEATURES

Physically, much of northern Finland and Sweden are part of the ancient Fenno–Scandian shield, which also extends over the Kola Peninsula of Russia. These are among the oldest rocks exposed on earth, and erosion processes over millions of years have worn them smooth. The Scandinavian mountains divide Norway from Sweden, with their core in southern Norway and the peninsula divided such that most of the gentler slopes are on the Swedish side. The highest peaks of this range are over 2,400 meters (the Glitterfinden rises 8,104 feet), and the mountainous interior of Norway—the *fjeld*—is the most remote and least populated part of the European continent. Glacial erosion has substantially modified this entire section (see discussion of physical geography, Chapter 1), with the Norwegian fjords being a particularly distinctive, spectacular feature.

Central and southern Finland, with its hundreds of lakes, is a classic example of a glaciated landscape, exactly analogous to northern Minnesota in North America. Erosion features to the north and deposition features (eskers, kames, etc.) to the south are textbook evidence of recent glaciation, and have created a difficult environment for the Finnish people. Farther south, in Denmark, southern Sweden (the *Skåne*), and the valleys of southern Norway, glacial evidence is in the form of outwash features, and the underlying limestone formations have produced some areas of rich soil. Denmark relies on this resource to make it one of the more successful agricultural nations of Europe, while Sweden and Norway have very few other areas of significant agricultural production. The Baltic states have also been strongly affected by glaciation, particularly including depositional features.

The extensions of northern Europe in the Atlantic are Iceland, the Faeroes, and Greenland. Iceland, one of the world's geologic "hot spots," occupies the junction of two major seams in the earth's crust. It is marked by constant volcanic activity, with new craters frequently forming (and often being submerged in the Atlantic). Not surprisingly, geothermal energy, created by converting water to steam in the hot rocks of the earth's interior, accounts for a substantial portion of Iceland's heating and power needs.

Greenland, a huge island midway between Europe and North America, has recently gained its cultural and economic independence from Denmark (which remains responsible for defense and foreign affairs). It is, with the exception of Antarctica, the world's largest ice cap (ice in the center of the island may be nearly two miles thick!), and under these conditions human settlement is confined to a few coastal locations. Greenland is sometimes considered part of North America because it is somewhat closer to Canada than to any part of Europe. However, because of its long association with Denmark and Scandinavian culture, it is probably more correct to think of it as European. As is Greenland, the remote Faeroe Islands are a largely self-governing dependency of Denmark.

CLIMATIC PATTERNS

Climatically, northern Europe largely reflects its latitudinal location, and climate poses severe problems for conventional human activities in most areas. Only the moderating influence of the North Atlantic Drift brings relatively mild winter temperatures to higher latitudes (see discussion of weather and climate, Chapter 2). The blocking effect of the Scandinavian mountains, however, since they run at right angles to prevailing westerly winds, means that much of interior Norway, Sweden, Finland, and the Baltic States suffer very harsh winter conditions. As the climatic map indicates, climates with mild winters, abundant precipitation, and cool summers extend northward along the Atlantic coast to latitudes beyond the Arctic Circle. Ice-free ports are found at Narvik in Norway, and even around the North Cape to Murmansk in Russia. Even the south coast of Iceland, washed by the Gulf Stream (or its extension into the North Atlantic), enjoys mild winters, although true summers are virtually nonexistent.

Away from the coast, climates quickly become continental, with long, cold winters and only brief summers. It is little wonder that citizens of Sweden and Finland are among the world's great "sun worshipers"! Population and economic activity are concentrated in the southern regions of Scandinavia and Finland, as much because of climatic advantage as any other factor. Denmark, with a large area of Marine West

Northern Europe

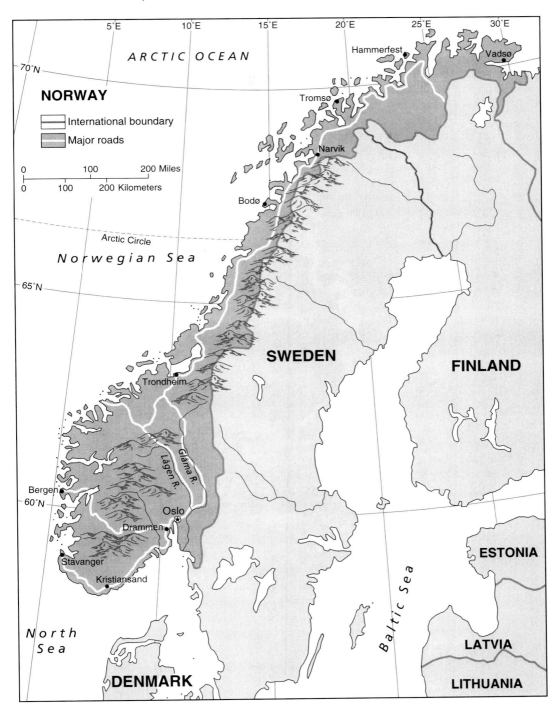

Figure 15.1 Norway.

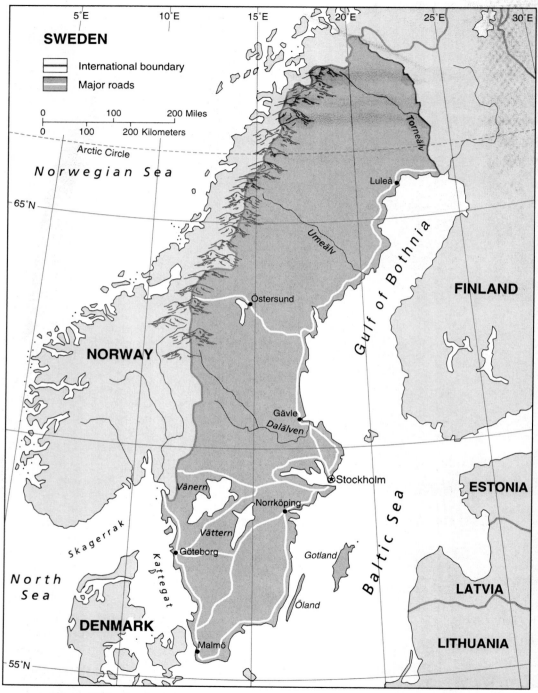

Figure 15.2 Sweden.

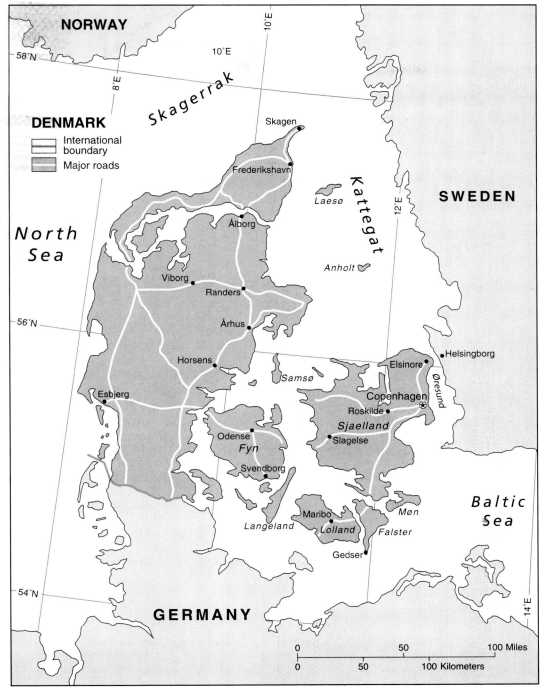

Figure 15.3 Denmark and its strategic location at the entrance to the Baltic.

Figure 15.4 Finland.

Northern Europe

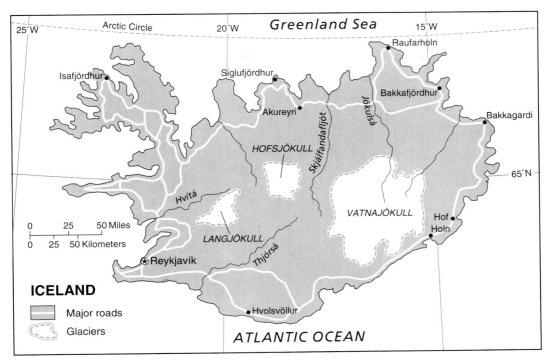

Figure 15.5 Iceland.

Coast climate, is more typical of western than northern Europe; and the rich agriculture of that country and the southern tip of Sweden (where even wheat can be grown) owes much to the warmer summers that occur there.

Greenland, dominated by its ice cap and located at very high latitudes, has essentially a climate which makes it impossible for vegetation to survive. Only along the coasts, where tundra conditions may prevail, is there any break in the icy desert. The Gulf Stream, which warms even the southern coast of Iceland, is here too far south to have an impact.

Climate, always marginal in terms of basic human activities, has played a key role in the history of northern Europe. For example, it is widely believed that deteriorating climatic conditions—longer winters and shorter growing seasons—during the eighth and ninth centuries constricted the agricultural base of the lands that are today Norway and Sweden to the point where desperate populations were forced to take to their boats and embark on a period of pillage and conquest throughout western Europe that left cultural traits clearly evident today. The small areas of flat, cultivable land at the ends of the Norwegian fjords are known in local language as "viks," and it was from these overcrowded lands that the "vikings" set sail.

HISTORICAL BACKGROUND

Culturally, the peoples of Sweden, Norway, Denmark, Iceland, and Greenland share a great deal of history, and have similar ethnic, linguistic, and religious backgrounds, while the Finns and Baltic nationalities are quite different at many points. The entire

Figure 15.6 The Baltic nations.

region was colonized by groups arriving from the east: the Finns settled Finland about 2000 years ago, while the Danes arrived in the seventh and eighth centuries and soon became masters of the rest of the region. Danish and Norwegian "vikings" ravaged most parts of Europe that were accessible by water; Iceland was settled in 874, Greenland in 983, and North America was probably reached by 1000; a Viking chief was

Figure 15.7 Northern Europe: Basic data.

Country	Area 1000 km.²/mi.²	Population million, 1996	GNP $/capita	Capital	Population million, est.
Denmark	43.1/16.6	5.2	28,110	Copenhagen	1.40
Estonia	44.9/17.4	1.5	2,820	Tallinn	.48
Finland	337.1/130.2	5.1	18,850	Helsinki	.50
Iceland	102.8/39.7	.3	24,590	Reykjavik	.10
Latvia	63.5/24.6	2.5	2,290	Riga	.90
Lithuania	64.9/25.2	3.7	1,350	Vilnius	.60
Norway	387.0/150.0	4.3	26,480	Oslo	.47
Sweden	450.0/173.7	8.9	23,630	Stockholm	1.50

Source: U.S. Department of State, Population Reference Bureau.

named Duke of Normandy in 911. The Danes succeeeded in temporarily conquering England in 1042, and from 1375 to 1523 were masters of all the Scandinavian kingdoms.

By the seventeenth century, however, Sweden was becoming the most powerful of the northern states, and played a major role in European affairs until the disastrous wars of King Charles XII (culminating in his defeat by the Russians in 1709) forced a rethinking of the future of this sparsely inhabited region. Norway, controlled first by Denmark, then by Sweden, did not become independent until 1905, while Greenland achieved a measure of local self-government from Denmark in 1979, and the Faeroe Islands now enjoy the same status.

Finland's recent history has been somewhat different as a result of its proximity to an expansionist Russia. Controlled by Sweden from the ninth century until 1808, when Russia defeated the Swedes after a century of warfare, the devastated country then became part of the Russian empire. Only with the Russian revolution of 1917 did Finland become independent, and this cherished status was barely preserved in fighting with the Soviets in 1918, 1939–40, and 1941–44. Forced to cede some of her most vital territory to the Soviet Union and to sign "treaties of friendship" with the Russians, Finland nonetheless clings fiercely to its democratic institutions, cultural distinctiveness, and links to the rest of Europe. With the restored independence of the Baltic republics of the former Soviet Union, Finland (whose language is very similar to that of Estonia) is now playing an enhanced role in the political and economic geography of the north.

ECONOMIC PATTERNS

Economically, the relatively sparse populations of these northern lands (which altogether group only about 31 million inhabitants) have been able to achieve a modern level of success that allows them to enjoy one of the world's highest overall standards of living, although the Baltic states are still recovering from their enforced membership in the Soviet Union. Agriculturally, production is severely limited across much of the north by the severe climate of most areas away from the Atlantic coast, and by the rugged landscapes and thin soils that are also dominant. Thus, only 3% of Norway is considered arable land, a figure that rises merely to 8% for Finland and 10%

for Sweden. In the more southerly areas, however, notably Denmark and the southernmost region of Sweden (*Skåne*), agriculture is not only possible, but is often extremely productive. Until World War II, in fact, Denmark was a nation that earned over half its total income from agricultural exports, and even today the Danes rely on their agricultural sector more than any other of Europe's "rich" countries.

Wheat, the favored food grain, is grown where possible, with potatoes and other grains (barley, oats, rye) more important in the less favored areas. As a reflection of dietary change across Europe, fresh fruits and vegetables are important seasonal products, while animal husbandry is widespread in all parts of the region. Poultry and cattle have been replacing pigs in the favored areas (such as Denmark), while hardy sheep, goats, and even reindeer provide some slender economic base for the remote interior districts. As noted (Chapter 9), fishing is also a crucial activity for this region, in terms of both food supply and income. A vital component of the economies of Norway and Iceland, fishing is also significant in Denmark, but is less of a factor in Sweden and especially Finland, both of which lack coastline on the open Atlantic.

PERCEPTION: THE CHANGING ROLE OF DANISH AGRICULTURE

The economic and social success of many smaller European countries in modern times is often puzzling, given their limited populations and often deficient resource base. The keys to this paradox are the locational advantages enjoyed by these nations, their ability to recognize and exploit whatever opportunities geography or history offers them, and the flexibility to alter their strategies and philosophies in response to changing circumstances. Denmark, with its important agricultural vocation, is an outstanding example of this phenomenon.

Denmark's only assets are a land base of reasonable quality, an industrious, productive population, and a strategic location between the Baltic Sea and the open Atlantic. Copenhagen, the capital and one of the most important cities of the north, profits greatly from controlling the flow of maritime commerce into and out of the Baltic. In addition to commercial activity, agriculture has long been a vital component of the economy. Adaptability has played an important part in this success.

Denmark began to be a major exporter of agricultural products during the eighteenth century, cultivating especially the markets in Britain, where the industrial revolution and rapidly increasing urban populations created a strong demand for food, especially for wheat, which Denmark produced in quantity, despite marginal climatic conditions. To facilitate the British trade, a new port city, Esbjerg, was built on the west coast of Jutland in 1864, closer to Britain than any existing port, and was linked by rail to the crop growing regions.

This happy relationship promoted Danish prosperity through the middle of the nineteenth century, when the more favored wheat-growing lands of North America and Russia came into world competition. Their enormous productivity and low prices quickly drove Danish producers out of the market, and disaster loomed. However, Danish agriculture was quickly able to recycle itself into the new product in demand: beef.

Figure 15.8 Denmark's capital, Copenhagen, controls shipping to and from the Baltic Sea.

As prosperity increased in Britain, the quality of diet also improved, and demand for meat increased more rapidly than local farmers were able to fulfill. Denmark again proved equal to the task; the country was populated with beef cattle as another opportunity presented itself. This prosperity, too, proved illusory, for in the early years of the twentieth century perfection of the refrigerated ship opened new low-cost production areas to the European market. Australia (linked to Britain by colonial traditions), and Argentina quickly began to undercut the price of Danish beef, and the country again faced difficult times.

Once again, the Danes adapted. Agricultural production moved into new areas of specialized output featuring high quality processed goods that could command a premium price on world markets. Danish hams, bacon, cheeses, and eggs quickly acquired a reputation that enabled them to be sold at competitive prices across Europe and in North America. Uncompromising quality, aggressive marketing, and a keen sense of consumer tastes have kept Danish agriculture competitive into the twenty-first century.

Today, Denmark's agriculture provides about a third of the nation's export income, down from nearly 75% in 1955 but still one of the highest ratios in "rich" Europe, as is the 7% of the population still engaged in agriculture. Moreover, the lessons of the agricultural past—adaptability and the vision to find new niches in world markets—have been successfully translated into Denmark's rapidly growing industrial and service activities.

Resource distribution is also uneven in northern Europe. Energy sources are naturally headed by the vast North Sea oil and gas fields, which have become a major support of the Norwegian economy (Norway is second only to Great Britain among European oil-producing nations). Very little coal was ever mined; this came mainly from the Norwegian Arctic island of Svalbard. The swift, steady streams that are found in most parts of Norway, Sweden, and Finland are ideal sites for the production of hydroelectricity, and this energy is not only used to power most industry in these countries, but is also profitably sold to energy-hungry nations farther south. Iceland, as a volcanic hot spot, makes good use of geothermal energy, so that only Denmark is a true "have-not" country in terms of energy requirements. The Danes, as a result, are leaders in the development of unconventional energy sources; a new generation of windmills is one example (see Chapter 10).

As far as metallic minerals are concerned, iron ore is the most important product of the north. Mining has gone on for many centuries in central Sweden (the Bergslagen district), and was the basis of the historic iron-making tradition of that country. Later, Europe's largest iron ore reserves were discovered in Sweden's far north, at Kiruna-Gällivare, and open-pit production from that field makes Sweden Europe's largest iron-ore producer today. Norway also has a modest amount of iron ore. Sweden is clearly the most mineralized of the northern states, as small amounts of copper, lead, zinc, other industrial minerals, and even gold are produced. Finland was forced to cede her richest mineral asset, the nickel mining district around Petsamo, to the Soviet Union as part of the peace settlement following World War II.

Figure 15.9 Norway traditionally faces outward to the Atlantic: Bergen.

Although Norwegian oil, Swedish iron ore, and hydroelectricity are important sources of income for the north, it has been the remarkable level of industrial development of these nations that has created their considerable modern prosperity. Again considering their small populations, and the substantial poverty that prevailed in northern Europe well into the twentieth century, these achievements are even more impressive. The leading industrial power, Sweden (8.5 million population), has for example a range of successful high-technology engineering industries (e.g., aircraft and two makes of automobiles) that would be the envy of many much larger nations.

In dozens of small cities and towns, widely scattered in a belt from Stockholm, the capital, to the major port city of Göteborg, Swedish industry (much of it powered by hydroelectricity) produces for export a variety of goods whose high quality assures them worldwide markets.

Denmark is the other major industrial economy of the north, relying on its traditional ability to identify market possibilities and move quickly into them to overcome its resource deficiencies. Denmark's proximity to the major nations of Europe (especially Germany and Great Britain), the vital role of Copenhagen as a trans-shipment port controlling commercial movement into and out of the Baltic (and the consequent range of port-associated industries), and its membership in the European Union have all been advantages to this small nation.

Industry in Norway is highlighted by a surprising amount of heavy metallurgy (iron and steel, aluminum and zinc refining), which takes advantage of cheap power and fine natural harbors for the movement of materials and products. Finland's industry is based heavily on exploitation of the country's vast forest

Figure 15.10 Göteborg is Sweden's gateway to the open sea.

resources, which are not only exported as raw materials, but are also processed into more valuable products. Numerous pulp and paper mills are a good example of this process.

THE BALTIC STATES

Because of their special role in Europe—rejoining the life of the continent after half a century within the confinement of the Soviet Union—the Baltic states certainly deserve a special mention within the context of northern Europe. Tucked along the eastern end of the Baltic Sea south of the great Russian city of St. Petersburg, these nations: Estonia, Latvia, and Lithuania, share highly developed and distinctive European cultures, economic bases similar to those of the other northern countries, and strong senses of historical significance and national identity focused especially on their most recent period of independence: 1918–1940. Throughout the centuries, Latvians and Estonians have usually been politically dominated by various Baltic powers (including Sweden, Germany, and Russia), but Lithuania was in itself a major force in the region for many years, combining with Poland to defeat and contain Russian expansion in the fourteenth and fifteenth centuries.

As a result of a cool, humid climate and marginal soil conditions, the Baltic states have resembled their north German and Polish neighbors in evolving a rye–potato cultivation base; but this environment is ideally suited for raising livestock—especially cattle—and this is a major economic support of the region. Mineral resources are nearly nonexistent, but diversified light industry has developed, especially around the three capitals: Tallinn (Estonia), Riga (Latvia), and Vilnius (Lithuania). In addition, the location of these states on the Baltic has led to the growth of commercial and port-related activities, as well as to increases in cruise ship tourism.

The Baltic states are closely linked to Europe by strong ties of language, culture, religion, and history. Their period of isolation within the Soviet Union was bitterly resented, and it is in no way surprising that they were the first of the Soviet republics to opt for total independence. Estonia shares linguistic ties with Finland (and with the adjacent Russian province of Karelia), and in all of the states western religions are dominant (Roman Catholicism in Lithuania, Protestantism in Estonia and Latvia).

Despite their enthusiasm for independence and the excitement of rejoining the European family of nations, the Baltic states face some daunting problems. Isolation on the northeastern fringe of Europe, economies based on very limited resources, and the looming presence of a Russia whose own future is uncertain make the outlook less than totally optimistic. Substantial Russian minorities (many Russians were settled in these states to take industrial jobs, and military officers often chose to retire in the Baltics) have their own linguistic and cultural demands, which place stress on the political framework. Nonetheless, support from Finland and the Scandinavian nations is considerable, and the Baltic nations should survive and gradually become integrated into the fabric of European life. Already there are proposals for these states ultimately to join the European Union.

Figure 15.11 National pride is now evident in the Baltic nations: Tallinn, Estonia. Photo courtesy of Christopher J. McDonald.

PERCEPTION: BRIDGING THE GAPS

Communication has always been a problem in the north. The Baltic, with its various arms and embayments, fragments the region considerably; the fjords make travel in Norway difficult, while the Scandinavian mountains effectively divide Norway from Sweden. Overland connections between Finland and Scandinavia, and between the Baltic states and the west have been virtually nonexistent, with the result that most interaction across the region has been by water. Denmark has been particularly impacted by this problem. A nation of islands (except for the Jutland peninsula, which forms its only land boundary—with Germany), Denmark suffers from an isolation that substantially raises its costs of integration into the Baltic economy. Any tourist or trucker who has waited for hours to catch a ferry across the Øresund between Denmark and Sweden needs no reminding of the problem.

Now, this is changing. Denmark has for some time been linking together its various islands by an impressive system of bridges, and construction is now planned on a bridge connection to Sweden. The 10-mile (16-kilometer) link between Copenhagen and Malmö (first proposed over a century ago), is expected to be completed about the year 2000, at a cost of some $2.2 billion. This will create an international urban area of some 2.3 million people, and one of the most impressive economic regions in Europe. Not surprisingly, not all Danes and Swedes are in favor of the project, seeing in their existing isolation a certain guarantee of cultural "purity."

Across the European continent, similar projects of linkage are in the works, or at least in the dreams of many governments. The Channel between France and England has at last been traversed by a tunnel (see Chapter 16), while a serious project has been proposed to link Spain with Morocco across the narrowest part of the Mediterranean Sea. Italy is also contemplating a bridge/tunnel connection between the mainland and Sicily. As the European Union embraces more members and becomes more economically and politically pervasive, and as the advantages of direct overland connections across Europe become more evident, additional projects will no doubt be considered.

ADDITIONAL READING

BRONDSTED, J., *The Vikings* (tr. K. Skov). London: Penguin, 1960.

JOHN, B.S., *Scandinavia: A New Geography.* London and New York: Longman, 1984.

MALMSTROM, V., *Norden: Crossroads of Destiny and Progress.* Princeton, NJ: Van Nostrand, 1965.

MEAD, W.R., *An Economic Geography of the Scandinavian States and Finland.* London: University Press, 1964.

RAUCH, G. VON. *The Baltic States: The Years of Independence; Estonia, Latvia, Lithuania, 1917-1940.* Trans. G. Onn. Berkeley: University of California Press, 1974.

SOMME, A., ed., *A Geography of Norden.* Oslo: J.W. Cappelens, 1960.

DISCUSSION QUESTIONS

1. Finland is generally considered part of northern Europe, but not part of Scandinavia. Why is this?

2. What advantages of site location does Copenhagen have? How has its site influenced the city's economic development?

3. Many Norwegians consider the wealth of North Sea oil to be a very mixed blessing. What negative impacts could oil money be having on Norway?

4. Do you feel that Greenland should be considered part of Europe or part of North America? What are the arguments on each side of this question?

5. Of the northern European countries, only Norway has refused to join the European Union. Why are the Norwegians so insistent on their political and economic independence?

6. The Baltic states were leaders in gaining independence from a collapsing Soviet Union. Why were they so keen on this? What problems do these small countries face as they attempt to distance themselves from Russia and become an integrated part of northern Europe?

16
The British Isles

The British Isles—Great Britain and Ireland plus numerous smaller islands—constitute one of Europe's most distinctive and familiar regions. The very fact of their insularity sets them apart from the remainder of the continent, and the cultural and economic impacts attributable to the English language and the vanished grandeur of Empire give this region an historic and even modern significance out of all proportion to its population or resources. Politically, the United Kingdom consists of England, Scotland, Wales, and Northern Ireland (Ulster), while the Republic of Ireland occupies a majority of the smaller island. The Shetland, Orkney, and Hebrides island groups, as well as the Isle of Man, are important additions to Britain.

PHYSICAL FEATURES

Physically, the central fact in this region is the 22-mile- (36-kilometer) wide separation of Britain from France by the English Channel (*La Manche*, to the French). Although not a particularly formidable barrier by the standards of late twentieth century aircraft and hovercraft technology, the Channel has been a substantial physical deterrent to continental influences (it is worth remembering that Britain has not been successfully invaded since the arrival of William the Conqueror in 1066 AD), and the isolation has both permitted distinctive economic and cultural development and led to the persistent British psychological feeling of being somehow apart from (and, in some periods, implicitly superior to) continental Europe. This attitude is well reflected in the classic British weather forecast: "Fog in Channel; continent isolated."

PERCEPTION: THE CHANNEL TUNNEL

In 1994, after seven years of technologically impressive construction and cost overruns that finally ran the cost of the project to some $15 billion, a dream of many centuries was realized when a tunnel was completed under the Channel (roughly from Calais, France to Dover, England). There are actually three tunnels: one for rail traffic in each direction plus a service tunnel; and cars and trucks can now move directly (by train) between Britain and the continent. As rail linkages become more sophisticated (particularly on the British side), Europe will be stitched together much more closely in terms both of the import and export of manufactured goods and of tourism.

Despite the numerous economic advantages of such a link, there are deep traditional misgivings on the part of many British at the prospect of having a physical overland connection to the rest of Europe. Napoleon, it is remembered, actually began construction on an earlier version of the "chunnel" with a view to facilitating an invasion of Britain!

Between the large island and Ireland, the Irish Sea and St. George's Channel form physically at least as important a division, but have rarely been able to prevent the heavy hand of English power from dominating. Only since 1922 has most of Ireland been able to exist as a separate state in Britain's shadow.

PHYSICAL FEATURES

The Scottish Highlands and Southern Uplands, most of Wales, and the Lake District of England are areas of rough, hilly country, with elevations rising to 4406 feet (1343 meters) on Ben Nevis in Scotland. The younger Pennines run north to south for some 120 miles (190 kilometers), forming a backbone for England from Scotland to the Midlands. Only in southern and eastern England—extensions of the North Sea Plain—and in the Scottish lowlands are there substantial extents of flat to gently rolling land.

As is true of most European countries, the history of Britain has been written to the pattern of her river valleys. Thus the Thames, Severn, Mersey, Tyne, and others, although small and short by world standards, became the location of the ports and industrial activity which have accounted for so much of the nation's strength.

In Ireland, the most significant natural region is the basin of the Shannon River, which occupies much of the central part of the island. Often swampy and poorly drained, this area has played much less of a role in Ireland's economic life than have the basins of the British rivers in that nation. To north and south, ranges of low hills—extensions of similar features in Scotland and Wales—are the dominant landforms.

CLIMATIC PATTERNS

Climatically, both islands are almost entirely in the Marine West Coast type area. Cool summers, mild winters, high humidity, and frequent rain, drizzle, and fog are the normal condition. The landscape is thus usually green and lush ("The Emerald Isle" is not just a casual nickname for Ireland); but clear days with bright sunshine are rare, and are treasured accordingly.

The British Isles

Figure 16.1 The United Kingdom.

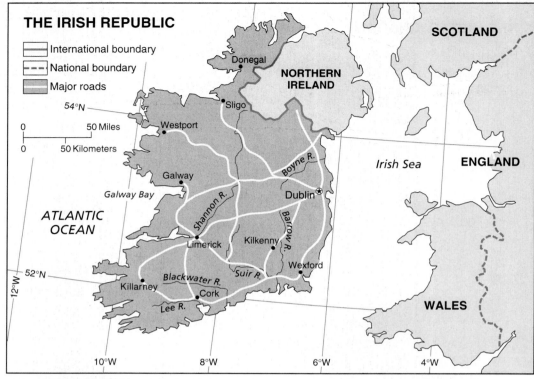

Figure 16.2 The Irish Republic.

At higher elevations near the coast, as in Scotland, North Wales, or parts of the Pennines, the strong, nearly constant westerly winds cause excessive evaporation and have a mechanical erosion effect on vegetation. The result is a type of tundra climate in which normal forest vegetation is replaced by such characteristic plants as heather and gorse.

Important variations in the region's microclimates include the warmer, dryer summer conditions found in East Anglia and other areas of southeastern England; and the extreme southwest of the country (Cornwall and the Scilly Isles), where an almost totally marine environment creates an area of such mild winters that it has become a favored retirement center, a region of year-round horticulture (fresh flowers from the Scillys arrive in urban markets regularly), and a landscape dotted with palm trees

Figure 16.3 The British Isles: Basic data.

Country	Area 1000 km²/mi²	Population million, 1996	GNP $/capita	Capital	Population million, est.
Ireland	70.3/27.1	3.6	13,630	Dublin	1.0
United Kingdom	244.1/94.3	58.8	18,410		
England				London	6.8 (metro.)
Scotland				Edinburgh	0.63
Wales				Cardiff	0.63
Northern Ireland				Belfast	0.69

Source: U.S. Department of State, Population Reference Bureau.

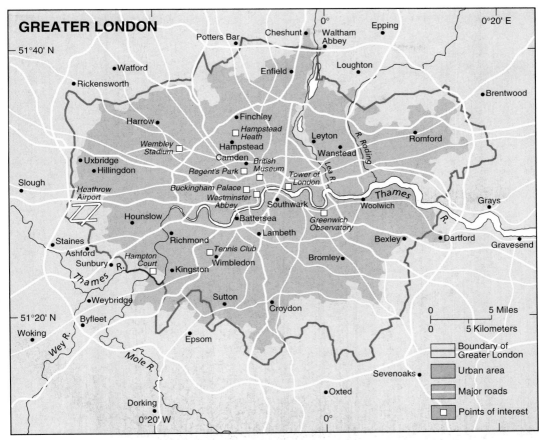

Figure 16.4 Greater London: One of Europe's largest metropolitan areas.

and other exotic flora. Snowfall is predictably abundant in Scotland and North Wales, where a modest winter sports industry has developed. Elsewhere, snow is rare and does not persist for long: a curiosity rather than a fact of winter life. Violent weather is also rare, although strong winter storms in the Atlantic or the North Sea may frequently combine with high tides to threaten coastal communities.

HISTORICAL DEVELOPMENT

Historically and culturally, the pattern of British geography until the final conquest from the continent in 1066 AD was a successive series of waves of invasion and influence from the east. Various European groups, either adventuring or themselves fleeing from other pressures, made their way across the North Sea or the Channel. The earliest group still recognizable today—the Celts—were thus pushed continuously westward, and so occupy today the western fringes of the British Isles (and thus of Europe). Ireland, Wales, and Highland Scotland are today areas of Celtic speech and tradition, which are also remembered in Cornwall and on the Isle of Man, as well as in Brittany, western France, whence Celtic groups returned from England in the fifth to seventh centuries.

Figure 16.5 The rugged Scottish highlands are home to an independent-minded Celtic folk: Near Aviemore.

Angles and Saxons brought the basic Germanic elements of the modern English language, and were in their turn pressured by Viking raiders and a long period of Danish settlement. Earlier, the Romans had for a few centuries controlled all of the larger island up to the Scottish Lowlands; they left numerous cultural, architectural, and linguistic reminders. Finally, from 1066, the Norman conquest gradually unified the nation in its own image.

Ireland, the seat of monastic learning in Europe during the Middle Ages, fostered strongly independent traditions in literature and the arts, but was unable politically to resist the numbers and economic strength of its larger neighbor, and thus passed through many centuries of troubled and contentious English occupation and ownership until 1922, when the modern Irish Republic was created. In fact, with the persistent "troubles" that continue to plague Northern Ireland (see discussion of religions, Chapter 5), it is no exaggeration to say that the "Irish question" is still one of the most perplexing issues on the British political agenda.

ECONOMIC ACTIVITY

Economically, the British Isles (especially Great Britain) are particularly associated with modern industrial development and the importance of international trade. However, it should be remembered that the evolution of a productive, profitable agriculture not only provided both food for a growing population and raw materials for industry, but also generated the capital that was eventually cycled into industrial growth.

Figure 16.6 The Welsh coast presents a rugged front to the Atlantic: Near Fishguard.

Modern agriculture still sees grain crops as a prominent feature: wheat where it does well under the sunnier, dryer climate of the southeast; rye, barley, and oats in the cool, wet areas of the north and Scotland. Many traditional root crops, as well as cabbages of all kinds, are still important, while the white potato, ideally suited to the lighter soils and high humidity, has been a popular crop since the 1700s. Its importance in the history of nineteenth century Ireland has already been noted (see discussion of migrations, Chapter 4). Fruit crops of excellent quality are also widely grown, and British farming is overall one of the most productive in Europe.

It has been the animal economy, however, that has particularly characterized British agriculture, and this remains perhaps its most distinctive feature today. Cattle, grazing on the lush pastures produced by a wet climate, transform this energy into forms more suitable to man's use than nearly any crop that might be grown: beef and dairy products are traditional staples of diet. Even more closely associated with the region, however, is that classic multipurpose animal: the sheep. Able to thrive on the most meagre of pastureland, sheep have long been the main economic support of vast tracts of marginal land in Wales, Scotland, and the Pennines, as well as other regions. However, the great value of the wool produced to the general prosperity of Britain, particularly with the onset of the industrial revolution in the 1700s (woolen textile manufacture was a key industry in this vast technological upheaval), meant that sheep were often accorded a favored place in regions of good quality farmland. The demand for wool was such that virtually the entire Highland region of Scotland was forcibly cleared of tenant farmers in the 1700s to make way for sheep; and the coat-of-arms of more than one prosperous little city in such English regions as Yorkshire or East Anglia prominently features a sheep; these places literally "rode to riches on the back of a sheep"!

If agriculture and animal husbandry laid the foundation for a British economy that would become the wonder of the world, however, it was industrial production, the export of goods and ideas, and the perfection of the colonial system that made Britain the "workshop of the world" and for some two centuries the world's wealthiest nation. Even today, despite the loss of colonial possessions, competition from larger nations richer in basic resources, and a decline in traditional "red brick" industries (such as iron and steel manufacture), the economy is still among the world's most successful, and still depends heavily on manufacturing and the export trade.

The distribution of industry in Britain follows a logical pattern: original location in areas of basic resources, followed by a marked shift in the later twentieth century to regions where technological concentration, human skills, and proximity to the markets of Europe are paramount. Thus, many traditional manufacturing regions were based on coal mining, iron and steel manufacturing, and related "heavy" industries (shipbuilding, railway equipment construction, etc.).

Located naturally on major coal fields, such areas as the Scottish Lowlands (Glasgow), South Wales (Cardiff, Swansea), and the Northeast coast or Tyneside (Newcastle) find themselves poorly situated to compete now that the advantages of being near coal have lessened and the heavy industries have declined in importance. These regions are accordingly among the most economically depressed parts of the modern United Kingdom, along with Northern Ireland, where political isolation from the Irish Republic, remoteness from continental Europe, and the decline of shipbuilding have raised unemployment to desperate levels, thereby exacerbating the explosive religious issue (see discussion of religions, Chapter 5).

Other major manufacturing regions of Britain form a semicircle around the western, southern, and eastern sides of the Pennines. This was the home of the industrial revolution and, although coal seams on the Pennine flanks were very important in establishing manufacturing sites, these areas are better located in twentieth century terms, and have been generally able to diversify into modern engineering industries (automobiles, for example), and thus to maintain their position. To the west, Lancashire (Manchester, Liverpool) specialized in cotton textiles utilizing imported cotton, and has now diversified into a wide range of export-oriented manufactures. Across the Pennines, what was formerly known as the West Riding of Yorkshire (Leeds, Sheffield) focused on woolen textiles and fine cutlery, while to the south, the West Midlands (Birmingham, Coventry) became the automobile capital of the nation, as well as producer of a great range of other engineering goods.

The greatest of all concentrations of British manufacturing is found in London, its metropolitan area, and increasingly throughout the rapidly growing southeast. The large population of this region has long made it a logical location for the manufacture of light consumer goods, while the variety of advanced economic activities and international linkages focused on London make it a favored site for production of all sorts of high technology/modern lifestyle items (scientific instruments and fashion clothing, to name but two). So important is location in or near London that a major share of all new industry in Britain in recent years can be found in the metropolitan area or in attractive smaller cities within the capital's orbit (Reading, Oxford, Cambridge, etc.). With completion of the Channel Tunnel (see Perception box above), this region will continue to prosper based on improved connections with the continent.

As Britain moves toward a new century, shifting economic patterns remain foremost among national concerns. The exploitation of North Sea oil and gas has provided a welcome—some would say providential—windfall to the country in the senses of providing employment both directly and indirectly (Aberdeen, Scotland is a classic "boom town"), of reducing dependence on imported oil, and of generating a profit at a time when few other sectors of the economy have been doing as well. The oil, however, will probably largely be gone by about 2025, and other structures must be in place by that time.

As is the case in nearly all advanced economies, employment and the generation of wealth is shifting from industry to the services sector (health care, education, financial services, food services, etc.). Britain's economic future probably depends largely on the extent to which the nation can successfully promote these sectors as manufacturing continues to stagnate, and to which it can continue to build on the traditional strengths of export orientation, extensive worldwide connections, and one of the few truly international currencies. It also seems likely that Britain will inevitably move closer to her European neighbors and partners within the European Union, no matter how distasteful this may seem to that sizeable segment of the population that still recalls nostalgically an era when Europe—and in fact most of the world—danced to the British tune.

Figure 16.7 London reflects the glories of British history: Tower Bridge.

The Republic of Ireland, long supported by an agricultural and pastoral economy and suffering from the economic domination of its larger neighbor, has recently been working diligently to improve its manufacturing base, mainly by encouraging foreign investment in export-oriented production. Although this strategy has been moderately successful, and Irish membership in the European Union has been particularly helpful, the relative isolation of Ireland from Europe and the lack of an urban structure (Irish cities are very small, reflecting the country's strong rural traditions) are difficult problems to overcome. In the sad traditions of the 1850s, Irish emigration continues to be heavy today, and includes a surprisingly large flow of illegal migrants to the United States, where a few months of steady work can generate enough earnings to go a long way back home.

Tourism is a major money-maker for Ireland, and indeed for all of this region, and is heavily promoted. The common English language and the "roots" syndrome are important factors in attracting visitors from the United States and Canada, and it has been estimated that a favorable rate of currency exchange prompted some 4 million North Americans to visit Britain and Ireland in 1985 alone.

ADDITIONAL READING

BARR, J., *Derelict Britain.* Harmondsworth: Penguin, 1969.
GIBB, R., ed., *The Channel Tunnel, a Geographical Perspective.* New York: Wiley, 1994.
GOUDIE, A., *Discovering Landscape in England and Wales.* London: Routledge, 1992.
GRAHAM, B.J., ed., *Historical Geography of Ireland.* New York: Academic Press, 1993.
HOSKINS, W.G., *The Making of the English Landscape.* London: Hodder and Stoughton, 1955.
JOHNSON, J., *The Human Geography of Ireland.* New York: Wiley, 1994.
JOHNSON, R.J., and J.C. DOORNKAMP, eds., *The Changing Geography of the United Kingdom.* London and New York: Methuen, 1976.
MITCHELL, J., ed., *Great Britain: Geographical Essays.* Cambridge: University Press, 1967.
STAMP, L.D., *The Land of Britain: Its Use and Misuse,* 3rd. ed., London: Longman, 1962.

DISCUSSION QUESTIONS

1. England has not been successfully invaded since 1066 AD. In what other ways has the "English Channel" influenced development of the British nation?
2. Both Scotland and Wales have strong nationalistic movements. On what factors is this political unrest based?
3. Britain is one of only two European countries (with Norway) that profit directly from North Sea oil. What has this windfall meant to the British economy?
4. Emigration from Ireland continues to be an important flow at the present time. Why is the Irish economy unable to provide adequate employment for the nation?
5. A century ago, Britain was clearly the world's most powerful and influential nation. To what extent does the country continue to play this role today? Why?

17
France

With the obvious exception of Russia, and also excluding Ukraine (see Figure I.3), France is Europe's largest country. Its 212,913 square miles (543,998 square kilometers) shelter a population no larger than other much smaller countries (Italy, Germany, Great Britain), resulting in a relatively low density of population. France is, in fact almost exactly comparable to Ukraine in both size and population. France is one of Europe's most varied nations from nearly every point of view, with a wide range of natural environments, cultural backgrounds, and economic activities that give great diversity to the country.

The French are fond of referring to their nation as "the hexagon," and a glance at the map reveals that it does in effect have six distinctive faces. To the northwest, west, and south, France faces the major seas surrounding Europe: the North Sea (by way of the English Channel, or *La Manche*), the Atlantic, and the Mediterranean. To the northeast, east, and southwest, the nation borders many of those continental neighbors which have played such a prominent role in its own history: Belgium, Luxembourg, and Germany to the north and east, Switzerland and Italy across the Jura and Alps, and Spain beyond the Pyrenees. France has thus been open to the influence of ideas, immigrants, and invasions from many directions, and has also exported its own products, philosophies, and culture to much of the world.

PHYSICAL FEATURES

Physically, France is composed of several distinctive natural units which are often quite complex in their geologic history. The northern part of the country, particularly the region of the Île-de-France, is dominated by the Paris Basin, a vast surface which seems quite flat in some places, and which has been compared to a series of saucers of decreasing size stacked one inside the other. This results in gently sloping surfaces toward the interior and steep escarpments on the outer edges. The city of Paris sits at the center of this basin, and the outward facing limestone ridges have historically provided strong defensive positions looking to the east, and are seen again in the famous chalk cliffs along the Channel to the west. Northward, the Paris Basin blends into the plains of Flanders and so to the flat lands of the North Sea Plain; while rougher hill country to the northeast (Ardennes, Vosges) is a continuation of a larger zone extending beyond the Rhine through central Germany.

To the west, much older rocks (mainly granites and schists) form the hill country of the Brittany and Normandy peninsulas, but to the southwest, the Paris Basin is connected by a low divide to the basin of Aquitaine, the other large region of generally flat land in the country. The south-central part of France is dominated by a large area of mainly volcanic uplands known as the Massif Central. With peaks rising into the 4,000 to 6,000 feet (1,200 to 1,800 meters) range, this area (the regions of Auvergne and Limousin) is one of difficult access and weak economy, and remains one of the poorest parts of France.

Some of Europe's most impressive mountain barriers border France to the southeast and southwest. The Jura divides France from Switzerland, while the Alps begin at the Mediterranean (the Maritime Alps) and extend northward before turning east to enter Switzerland and northern Italy. Although rugged and complex, the French Alps were heavily glaciated, especially to the north, and as a result have broad valleys providing relatively easy access deep into the mountains. The highest point in Europe (again discounting Russia) is Mont Blanc: 15,771 feet (4,808 meters), which stands near the common boundary with Switzerland and Italy. Not so high, but to many more impressive, are the Pyrenees, a single block-faulted piece of the earth's crust dividing France from Spain with the steep face on the French side. Although the highest French point in this range—the Pic de Vignemale—10,820 feet (3,298 meters) is not notably high, the Pyrenees were never severely glaciated (being farther south and nearer the sea), and the absence of glacial valleys makes crossing them and moving around inside them a much more difficult proposition than is the case in the Alps. Between the Pyrenees and the mouth of the Rhône river, a relatively large area of flat to gently rolling coastal plain is found. This is the region of Languedoc, and is another of the rich agricultural areas of the country.

Most of France is drained by five major river systems. The Seine rises in the east, in Burgundy, flows through Paris, which it connects to the sea, and empties into the Channel at Le Havre. Among its major tributaries are the Marne, Oise, and Eure. The Loire, which has the largest drainage basin of any entirely French river, has headwaters in the Massif Central, and is joined by the Cher, Sarthe, and Vienne before it flows into the Atlantic below Nantes. Much of northeastern France borders the Rhine, or its major tributary, the Moselle; while the southeast is drained by the Rhône, which enters France from Switzerland on its way to the Mediterranean just west of Marseille. Finally, the southwest is the drainage basin of the Garonne, which flows westward to the Atlantic downstream from Bordeaux through a wide estuary known as the Gironde.

CLIMATIC PATTERNS

Climatically, France displays as much variety as her size and physical diversity might suggest. The Marine West Coast climate, with its moderate temperatures and abundant rainfall, covers most of the northern and western regions, including the Paris Basin. Farther

France

Figure 17.1 France.

Figure 17.2 France and Monaco: Basic data.

Country	Area 1000 km²/mi²	Population million, 1995	GNP $/capita	Capital	Population million, est.
France	551.7/220.7	58.4	23,470	Paris	10.7 (metro.)
Monaco	1.5/.58	0.028	*	Monaco-Ville	†

*Not calculated.

†City and country are co-terminous.

Source: U.S. Department of State, Population Reference Bureau

east, away from the moderating influences of the Atlantic, seasonal temperature differences are more marked, while at higher elevations of the Jura and Alps, abundant snowfall is recorded. To the south, along the Mediterranean coast, the usual dry-summer climate is found and temperatures are mild year-round, particularly where mountain ranges (such as the Maritime Alps) shelter the coast from winter storms to the north. The Pyrenees also have a highland climate, although they receive much less snowfall than the Alps.

In the finer detail of this broad pattern, a vast range of microclimates may be found in France, dependent on local variations in latitude, elevation, exposure angles, and other factors. These seemingly minor variations are the basis of much local agricultural specialization, and are also one of the causes of the strongly defined regional character which marks so many parts of France. The traditional school of French geography found its unique character in detailed study of the numerous *pays,* or distinctive local districts—often with important nuances of climatic change—which make up the mosaic of the country.

Figure 17.3 The variety of the French landscape is remarkable: Glacier and church at Argentieres in the French Alps.

CULTURAL AND HISTORICAL FACTORS

As might be expected, the French population shows a great deal of cultural diversity, reflecting a long and complex historical development and the influence of outside ideas from many directions. One indication of this is the language complexity of the nation (see discussion of languages, Chapter 5) in which, in addition to the dominant French, such important minority tongues as Breton, Alsacian, Provençal, Corsican, and Basque are also spoken, and have recently been successful in gaining more official recognition. Foreigners have long made up an important part of the nation's population, with Spanish, Italian, and Polish groups prominent; while the retreat from colonialism and the massive labor immigration of the 1960s and 1970s has added some 5 million immigrants and their families to the population. Portuguese, Algerian, Moroccan, and Vietnamese groups are the most important of these.

If the French population is remarkably diverse, however, the most important theme throughout much of the nation's history has been the steady concentration of political authority and economic power in a centralized government and at a single place: Paris. Of all the national capitals of a continent in which centralization has generally been the rule, Paris has come to concentrate all facets of French life to a remarkable degree that is unsurpassed elsewhere. In the process, the city grew to be a classic example of a "primate city," concentrating as much as 20% of the national population and exceeding the second most populous city (usually Lyon) by a ratio of ten to one. Paris has thus come to symbolize France to the rest of the world, while many more remote regions have remained backwaters outside the mainstream of national development, sending their resources and their people to an all-consuming capital.

In recent years, however, the increasingly unwieldy bureaucracy of Paris, the frustrating difficulty of getting anything done in the country without seeking approval from the central government, and the growing cultural awareness and economic strength of many regions have led to the beginnings of a badly needed decentralization of authority and decision-making. To the traditional structure of small political units tied closely to Paris has been added a group of 22 regional units, many of them duplicating the names and general sizes of the old provinces of France before the revolution. These regions, with sizes and populations large enough to be meaningful, have been granted a surprisingly amount of autonomy in local affairs, including discretionary budgets and many institutions free from the influence of the capital. An increasing number of decisions are being made at local levels, and while Paris remains by far the most important city in all respects, its dominance is much less than in the past, and other centers, such as Strasbourg, Grenoble, Marseille, Lyon, Rennes, Bordeaux and Rouen are becoming more important in their own right.

ECONOMIC ACTIVITY

Economically, France also demonstrates a high degree of diversity, featuring strong agricultural traditions and a manufacturing industry that blends the luxury products for which the country has long been famous with some of the most modern technology found anywhere. The importance of agriculture is reflected in the fact that over 8% of the French labor force is still employed in farming, a relatively large figure among the industrialized nations of Europe. France is also one of the few European nations to be

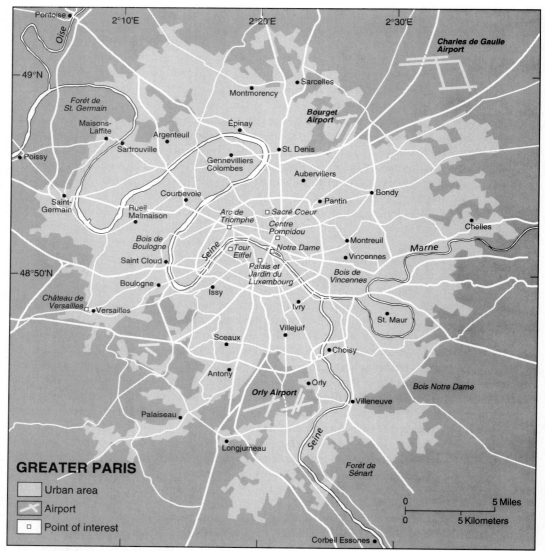

Figure 17.4 Greater Paris.

generally self-sufficient in basic agricultural production, and to earn more revenue from agricultural sales abroad than is spent on agricultural imports. As in many other European nations, however, agriculture is a heavily subsidized activity—it is seen by most French as a necessary part of the national character—and this brings French policy into conflict with countries that combine lower subsidies with calls for more free trade in agricultural commodities (the United States, Canada, and Australia are among these).

The most important agricultural region of the country is the Paris basin, the traditional "breadbasket" of France. Here wheat, the favored food grain of Europe, has long been the basic crop, and France customarily produces over 5% of the world total, enabling it to rank as high as fifth among all nations (see discussion of European

Figure 17.5 Paris, France's centralizing capital, continues to innovate: New entrance to the Louvre museum.

agriculture, Chapter 9). Wheat is grown in many other districts of the country as well, and total production is nearly double that of the next leading European nations (Great Britain and Italy). In recent years, cultivation of hybrid varieties of maize, introduced from the United States and used largely for animal fodder, has assumed an increasing role in the agricultural landscape, even displacing wheat in some areas.

Other critical regions of agricultural production are in the north, where sugar beets are grown, and in the south, where the warmth and sunshine of Aquitaine, Languedoc, and the Rhône valley make them the leading regions for the cultivation of fruits and vegetables, and the major centers of wine production. Europe produces over two thirds of the world's wine, and France and Italy each contribute about 20% of the European total.

Despite this impressive picture of agricultural prosperity, however, there are many large areas of France where slope conditions, poor soils, and climatic problems make farming a marginal occupation at best. The drift of farm populations away from such regions continues (the Central Massif is a good example), and in the absence of other economic supports, depopulation becomes a severe problem.

France's economic development was stimulated during the industrial revolution by the discovery of important coal reserves in the north. This power and metallurgical resource, and the nineteenth century utilization of the iron ore reserves of Lorraine, permitted the country to become a major iron and steel producer, and to develop most of the related types of heavy industry. Today, both coal and steel have declined in relative economic importance (no European country except those of the east still makes steel utilizing the coal-fired blast furnace), and the remaining steel industry, as elsewhere in Europe, has largely shifted to coastal locations where imported iron ore can be easily utilized (notably at Dunkerque on the Channel coast and Fos on the Mediterranean near Marseille.

Many traditional industries have fallen on hard times in the late twentieth century, in France as elsewhere, and many of the coal and steel towns of the north are today among the most depressed areas of the country. At the same time, however, the

Figure 17.6 France has Europe's richest agricultural economy: Grain fields in Burgundy.

traditional fascination of the French for the "high technology" of any era is seen today in a willingness to move rapidly into promising new industrial areas and generally to recycle the nation's industrial establishment. Such technologies as supersonic aircraft (the Concorde is a joint French–British venture), high-speed trains (the *Train à Grande Vitesse* (the "TGV") links Paris with Lyon in only two hours), engineering projects such as the Channel Tunnel, and the nuclear power industry, in which France is the world leader, have given new dimensions to French industry, and have redrawn the industrial map to favor "sun belt" and other attractive locations such as Grenoble, the Rhône valley, and the Mediterranean coast.

PERCEPTION: THE FRENCH NUCLEAR POWER INDUSTRY

As the twentieth century ends, it is reasonable to say that France has made a greater commitment to the generation of electric power using nuclear fuels, and thus to the role of the nuclear option in its overall energy package, than any other nation. Russia and several other successor states of the Soviet Union, pursuing their nuclear aims despite the Chernobyl disaster, perhaps constitute a competitor; but the United States has virtually abandoned the idea of any further expansion of nuclear generating capacity, while most other European countries are scaling back or abandoning the nuclear approach to energy creation.

Not so in France. The nation currently produces nearly 80% of its electricity from fifty-five nuclear plants at twenty-five different sites. By the end of the century, when the three additional plants now under construction have joined the network, this figure may rise to nearly 90%, which would represent as much as a third of the total energy package.

The French government, which despite political fluctuations remains heavily technocratic, has long concluded that there is no reasonable medium-term alternative to nuclear energy if France is to meet her major objectives in this area: maintenance of high living standards, a reasonable rate of industrial growth, and relative independence of action in the international arena. With almost no petroleum, only modest supplies of natural gas, and depleted reserves of generally inferior and expensively acquired coal, France trails nearly all major European nations in supplies of traditional fuels, and has few prospects of significantly increasing production. Most of the country's overall energy needs are met by petroleum imports from an increasingly unstable Middle East and North Africa.

Since the decision made during the Middle East crisis and oil blockade of 1973–74 to proceed at an accelerated rate toward energy self-sufficiency through nuclear power, the French have moved with steady determination. In fact, the massive project of power station construction has been the single largest capital investment program in the country over a twenty-year period. Construction of the world's largest breeder reactor (producing new fuel for other reactors, as well as energy) at *Creys-Malville,* the world's most sophisticated nuclear fuel reprocessing center at Cap de la Hague, and a range of technologically advanced means of coping with nuclear waste, the French feel that they have closed the nuclear cycle and created at least a partially satisfactory solution to their energy shortages until the next generation of energy options (e.g., fusion, solar) becomes a reality.

The question of nuclear energy in France is clearly a complex one which touches on numerous aspects of the nation's political as well as economic life. Moreover, the French experience has substantial implications for other nations contemplating a future of scarce and/or costly energy. Among the critical issues are:

1. the problematic future availability of nuclear fuel supplies
2. the clouded question—particularly after Chernobyl—of environmental protection
3. the long-term problem of finding acceptable nuclear waste disposal procedures (despite French confidence in their current technology)
4. the international menace of weapons proliferation based on the reprocessing of nuclear fuel exports
5. the true cost-effectiveness of nuclear energy compared to other options, especially if estimates of reactor life prove to be exaggerated
6. the limitations of electricity as a primary energy source
7. the inevitable technical problems of such a massive network (the breeder reactor at *Creys-Malville*—built at enormous cost—has never overcome its problems to become fully operational)
8. the potentially profitable role of French nuclear technology in those parts of the world (notably Asia) where nuclear electricity is still a growth industry
9. the enormous, inevitable costs of shutting down the plants at the end of their useful lives
10. the evolution of French politics as the "nuclear question" becomes an increasingly important part of the national debate

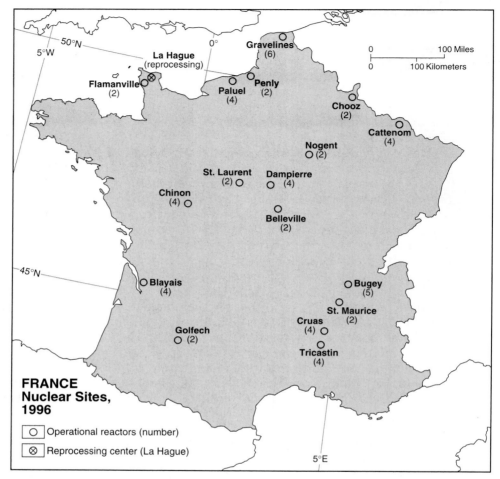

Figure 17.7 France: Nuclear electricity plants, 1996.

If the French have guessed correctly, the nation may be deriving substantial advantages from a cost-competitive, independently controlled energy supply well into the twenty-first century. If they are wrong, they will be saddled with an economic and environmental white elephant of monumental proportions.

Finally, mention should also be made of another sort of industry with which France is closely identified in the world's imagination: the production of luxury goods of all kinds. The high fashion industry still revolves mainly around Paris, and the production of perfumes, furs, gourmet foods, fine wines, and similar expensive items not only earns money for the nation, but also conveys a positive, cherished image. The identification of these industries with France dates from the time of Louis XIV when Colbert, the king's minister and partisan of mercantilism, established the silk industry at Lyon to eliminate the outflow of currency needed to pay for such luxury imports.

France

Figure 17.8 France produces some 80 percent of its electricity in nuclear plants: Cows and reactors co-exist uneasily at Cattenom.

PERCEPTION: EUROPEAN "MINI-STATES"

It has already been made clear that Europe is a finely divided continent in terms of both space and population, and that the political, cultural, and economic interactions of small but usually fiercely independent nations have led to both the successes and the tragedies of European history. Beyond the multiplicity of small nations which continue to play a significant role in the world, however, close observation of the European map reveals a scattering of even smaller nations which often measure their size in a handful of square miles (or less), their populations in a few thousand individuals.

Leftovers from the past, survivors usually of some economic necessity or political compromise, these countries are nonetheless proud of their heritage, their distinctiveness, and their cultures, and resist strongly any efforts to absorb them into some larger entity. Although scarcely significant on a world scale (with the exception of the worldwide religious influence associated with the Vatican), these "mini-states" (as they are frequently called) have for us that fascination which stems from survival against seemingly impossible odds: the underdog somehow beating the system. In the already rich socioeconomic mosaic of Europe, these small states provide an instructive lesson in adaptability as well as a model which many of the continent's restless ethnic or linguistic minorities would like to emulate.

Some of Europe's smaller nations have had sufficient size and population, and have profited from some scarce resource or other economic advantage, to play at least a modest role on the continental stage. Luxembourg, with 999 square miles (2,586 square kilometers), 360 thousand people, and a rich iron ore deposit (see Benelux regional discussion, Chapter 18), is a classic case. More commonly, however, the true "mini-states" survive on a much more tenuous basis. The first of these appealing if unlikely nations—Monaco—is a very small neighbor of France.

MONACO

Certainly the most glamorous of all the "mini-states" is Monaco. Totally surrounded by France except for its spectacular facade on the Mediterranean, ruled by a prince of one of the great hereditary European families (the Grimaldis), and associated with the most dramatic circuit on the Grand Prix automobile racing tour and

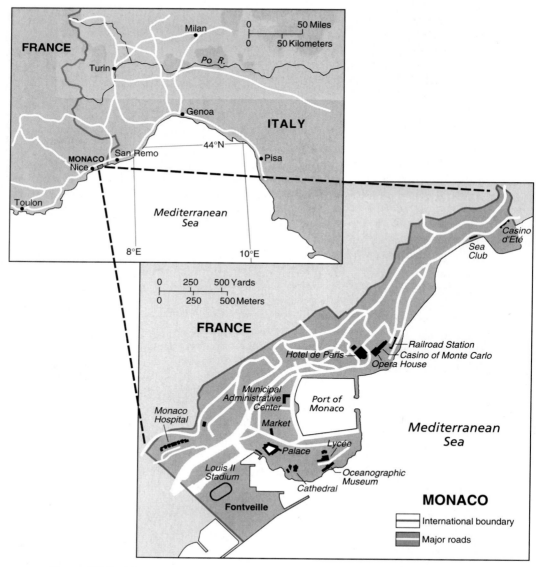

Figure 17.9 Monaco: A mini-state on the Mediterranean coast.

Figure 17.10 Monaco is Europe's most famous and glamorous mini-state: The Mediterranean port.

unquestionably the world's most snobbish gambling casino (in the Monte Carlo district), Monaco seems the classic fairy-tale kingdom. Beyond the glitter of the casino and the shared rush of tourism along the Riviera, Monaco also profits as a tax haven and a center of international finance. A permanent population of about 27,000 occupies just over half a square mile in a country which has been an independent principality for three hundred years.

ADDITIONAL READING

BEAUJEU-GARNIER, J., *France (The World's Landscapes Series)*. London: Longmans, 1975.

BRAUDEL, F., *The Identity of France; v.1: History and Environment*, tr. S. Reynolds. New York: Harper & Row, 1988.

"Five tiny secrets of success," *The Economist*, August 27, 1988, pp. 39–41.

HOUSE, J.W., *France: An Applied Geography*. London: Methuen, 1978.

HUDSON, G.L., *Monaco*. World Bibliographical Series, vol. 120. Oxford: Clio Press, 1991.

KEATING, M., and P. HAINSWORTH, *Decentralization and Change in Contemporary France*. Aldershot: Gower, 1986.

PINCHEMEL, PH., *France: A Geographic, Social, and Economic Survey*, tr. D. Elkins. Cambridge: University Press, 1986.

TUPPEN, J., *France, Studies in Industrial Geography Series*. Boulder CO: Westview, 1980.

DISCUSSION QUESTIONS

1. France is Europe's leading agricultural nation by most criteria. What advantages does the country have to give it this economic benefit?
2. France is often referred to by its citizens as the "hexagon." Why? Describe the country's various frontiers.
3. As most European countries, France is not culturally homogeneous, but contains numerous distinct nationalities. Who are some of these people? What is their position in France? What are their aspirations?
4. Paris rivals London as modern Europe's largest city. Why has it grown so large? What problems does the city face because of its size?
5. Despite serious environmental concerns, France is deeply committed to the nuclear energy option. Why is this? What other energy choices does the nation have?
6. Europe's "mini-states" are especially interesting because of their continuing survival. How have they accomplished this? What difficulties might they encounter in a rapidly changing Europe?

18
Regional Profiles: Benelux

At the very heart of western Europe, tucked between the North Sea and the major rivers of the area and wedged (often uncomfortably) between their powerful neighbors—France to the south, Germany to the east—the three Benelux states (Belgium, the Netherlands, Luxembourg) epitomize in many ways both the problems and the successes of Europe. So small in size as to be better suited perhaps to the scale of the Middle Ages than that of the twenty-first century, these three nations, in various individual ways and in combination, have overcome often staggering adversity not only to become economically strong, but also to serve as the catalyst for the movement toward a more united Europe. Employing the combination of an almost perfect location with reference to Europe's political and economic centers of power, the good luck of a surprisingly diverse and timely assortment of energy and other mineral resources, and a great deal of human initiative and stubborn hard work, the Benelux countries have built economies that are as successful and societies that are as advanced as any in the world, for which reason their influence both inside and outside Europe remains remarkably strong.

PHYSICAL FEATURES

Physically, the Benelux states generally divide into two types of terrain: the flat lands (often below sea level) of the Rhine delta and the North Sea Plain, which dominate in the Netherlands and northwestern Belgium; and the hills of Hercynian age in the interior (primarily the rough, wooded area of the Ardennes), which are seen in Luxembourg, southeastern Belgium and the southern Dutch province of Limburg. The Netherlands, especially its most productive provinces (north and south Holland and

Zeeland), are quite literally the product of the mighty Rhine, which has changed its course many times and built a vast, marshy delta as it flows into the North Sea. As early humans moved into the area, the only lands for settlement were a fairly unpromising interior covered with glacial gravels and a series of sand bars along the coast; all the rest was water. "God made the world, but the Dutch made Holland," runs the rather smug proverb; but in fact, ever since the fifteenth century, the Dutch have been using state-of-the-art water engineering technology to reclaim land from the clutches of the sea.

Actually, some 40% of the modern territory of the Netherlands is reclaimed land that lies below sea level, including the nation's economic heartland where nearly 80% of its fifteen million people live. The famous Dutch windmills (national monuments today) were used primarily to pump water back into the sea from inside the early dikes, creating "polders," or very welcome pieces of new land. As steam engines and other new technology replaced wind power, projects became more ambitious. The *Harlemmermeer* (site of the world-famous flower bulb industry) was drained in the nineteenth century, and the twentieth century has seen the culmination of Dutch water engineering with completion of two impressive projects.

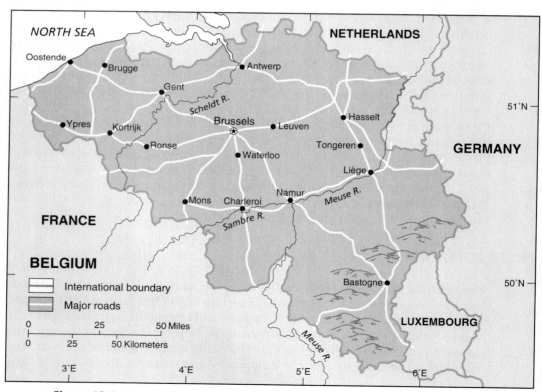

Figure 18.1 Belgium.

Figure 18.2 The Netherlands.

In 1932, a national dream was realized when the Zuider Zee was dammed by a dike more than twenty miles long (usually called the North Sea dike), and the resultant lake (the *Ijsselmeer*) has seen four large polders carved out of it while still retaining abundant fresh water and recreational space. Later, disastrous 1953 floods in the south of the country, where the Rhine has its modern outlet, stimulated work on the mammoth, recently completed "Delta project," by which the area is protected from flooding and brought into the economic mainstream of the country through a series of dams and other devices (such as elaborate storm-surge barriers) that close off the old, unused arms of the Rhine and connect previously isolated islands to the mainland.

These massive projects have awakened certain social and environmental protests in the Netherlands, based on the alteration of traditional ways of life and the disruption of sensitive ecosystems, but it is difficult not to admire the level of Dutch dedication to

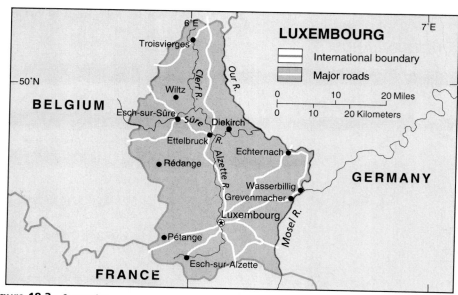

Figure 18.3 Luxembourg.

taming the sea and creating an environment as much as possible in tune with human requirements. At the same time, the Dutch have shown a remarkable ability to adapt to changing conditions (almost always a key to the success of various European countries). Thus, while the early polders (including the first three major ones in the *Ijsselmeer*) were designed to increase the amount of farmland, it is now recognized that the Netherlands can never really hope to feed its dense population, no matter how efficiently its limited land base is cultivated (even the open spaces at the vast Schipol international airport near Amsterdam are farmed!). Instead, more recent new land (the fourth *Ijsselmeer* polder and land relaimed by the Delta project) is being put to urban–industrial–recreational use, while the increasing value of fresh water means that the remainder of the *Ijsselmeer* will almost certainly never be polderized.

The remainder of the Benelux lowlands—the flat provinces of the northeastern Netherlands and the coastal areas of Belgium—have had less dramatic geologic histories. This part of the Netherlands, as the north German region it adjoins, is comparatively poor agricultural land, and has been sparsely populated. In Belgium, conditions are better, the rich plains of Flanders being an extension of northern France, and prosperity here—even through a troubled history—has been greater.

Figure 18.4 Benelux States: Basic data.

Country	Area 1000 km²/mi²	Population million, 1995	GNP $/capita	Capital	Population million, est.
Belgium	30.5/11.8	10.2	22,920	Brussels	1.0
Luxembourg	.26/.10	.4	39,850	Luxembourg City	.076
Netherlands	41.5/16.5	15.5	21,970	Amsterdam	.72
				The Hague	.45

Source: U.S. Department of State, Population Reference Bureau.

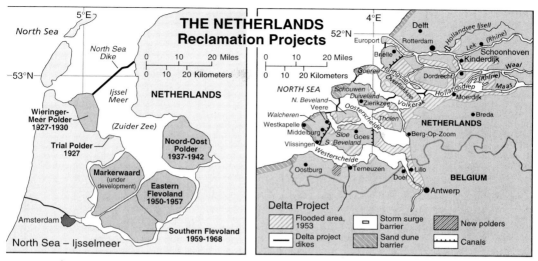

Figure 18.5 Major Dutch reclamation projects.

In the interior, the forested hills of the Ardennes and pieces of other Hercynian formations create a moody, difficult, and less promising landscape, with the valleys of streams tributary to the Rhine deeply incised into it. Ironically, it is likely that Luxembourg, the Dutch province of Limburg, and eastern Belgium would have remained among Europe's poorest regions were it not for the impressive quantities of critical industrial raw materials (coal, iron ore) that were found there. The varied topography, dense forests, wildlife, swift streams, and proximity to the dense populations of the lowlands make this region an important area for recreation and tourism.

CLIMATE

Climatically, the entire Benelux region clearly falls under the Marine West Coast regime, with typically mild winters, cool summers, and abundant precipitation at all seasons. While snowfall and freezing conditions are rare along the coast, the higher elevations—over 1,000 feet (305 meters)—of the interior result in a considerably harsher winter season. Strong winter storms off the North Sea are a serious threat to the low-lying coastal areas, especially when they coincide with high tides.

HISTORICAL AND CULTURAL FACTORS

The Benelux states display the cultural diversity and often troubled history that might be imagined, given their prime location at the heart of Europe. The Dutch struggled against Flanders, the powerful trading cartel known as the Hanseatic League, and the Spanish throne (it was a Hapsburg possession), before becoming independent in 1578. Many historians have concluded that it was the intelligence, sense of strategy, tenacity, and sheer stubbornness developed by the Dutch in their clashes with both other humans and nature that led them out into the world in the seventeenth and

Figure 18.6 Much of Luxembourg and eastern Belgium lies in the Ardennes uplands: Vianden, Luxembourg.

eighteenth centuries to challenge English dominance of the seas, build a large colonial empire, and make a fortune from the long-distance trade in spices and other highly-prized commodities.

Belgium, based on the three small, often wealthy provinces of Flanders, Hainaut, and Brabant, has had an even more checkered history. Following domination by the French under Napoleon, the territory became part of the Netherlands in 1815, and only achieved independent status in 1839.

Luxembourg has perhaps the strangest history of the three, tracing its origins back to the division of Europe following the death of Clovis in the sixth century. Its independent survival has been a result of mutual jealousies among the larger neighboring powers, skillful maneuvering by successive generations of leadership, and the development of a surprisingly powerful economic base. Controlled (though hardly subjugated) by various large powers over the centuries, modern Luxembourg became fully independent in 1867.

Belgium and Luxembourg formed a customs union in 1922, and all three of the Benelux states have been linked in an economic union since 1948. They are thus the forerunner of the modern European Union (see discussion of international institutions, Chapter 14), and have also been vitally concerned with international security.

With over 1,000 persons per square mile, the Netherlands is—with the exception only of the Mediterranean island nation of Malta—Europe's most densely populated country. The obvious need to organize limited space in the most efficient way has made the Dutch world leaders in social, economic, and urban planning. When Rotterdam was rebuilt after being totally destroyed during World War II, the *Lijnbaan* was laid out as

Figure 18.7   Belgium's historic prosperity is based on its foreign trade: Bruges/Brugge.

one of the world's first pedestrian malls. A culturally and linguistically homogenous people, the Dutch are reminded of their colonial past by the sizable Indonesian minority which adds variety and a certain amount of stress to the urban scene.

Cultural stress, however, is much more evident in Belgium, a country created in the nineteenth century out of quite disparate pieces, where tensions focused on language differences are among the most severe in Europe. The conflict is between the Flemish community of the north and west and the French-speaking Walloons of the south and east. The two groups differ culturally and politically at many points, but it is language that has become the battleground between them, and as the economic balance of the nation has shifted from the Walloon to the Flemish region, matters have been further complicated (see discussion of languages, Chapter 5). Street rioting has occurred, governments have fallen over the language issue, and the duplication of institutions insisted upon by both sides (political parties, newspapers, educational systems, etc.) seems a wasteful strain on the nation's resources. The capital, Brussels, with its many important European associations (North Atlantic Treaty Organization headquarters, many of the European Union's agencies) is too busy a city to dwell on language differences; and there are some hopeful signs that a distinctive Belgian language may slowly be evolving in the metropolitan area.

In little Luxembourg, a country of only 999 square miles—about the size of two average counties in the United States—both French and German are widely spoken, and there is a distinctive dialect that combines the two. In 1985, release of the first feature film made in "Letzeburgesch" was a national occasion! Luxembourg's cultural scene is also enlivened by a large immigrant community, mainly workers from Portugal.

ECONOMIC ACTIVITIES

The economies of the Benelux states are consistently among the strongest in Europe, reflecting clearly the cardinal rules for success: maximizing whatever advantages are offered by location or the natural environment, and adapting quickly to changing economic conditions. Controlling the outlet of the Rhine, and thus the movement of goods to and from a large portion of the continent, has for example been a constant source of wealth to the Dutch, and the port of Rotterdam (with its twenty-mile downstream industrialized extension known as "Europort") is Europe's largest and busiest.

The Dutch have also had the unlikely good fortune, however, to have discovered important and timely energy reserves. The Campine (or Kempen) coal field that extends from Belgium to Germany also passes through the narrow "tail" of the Netherlands in the province of Limburg. This windfall enabled the Dutch to build a heavy industrial base, notably around Eindhoven, and the later discovery of a modest oil field on the coast at Scheveningen (a geological preview of the great North Sea fields, in which the Dutch unfortunately do not share) was also important. Best of all, however, has been the discovery in the northern province of Groningen of the largest onshore natural gas field in Europe. Not only has this provided an abundant source of energy for the nation, but the profits from its sale to energy-hungry neighbors have also paid for one of Europe's most elaborate social welfare systems.

Finally, the Dutch have shown great ability to change with the times, moving rapidly from an agricultural to an industrial economy (witness the altered philosophy of how best to use the newly reclaimed polders), and from a colonial power to a self-contained nation living on its trading and exports. Even the nature of Dutch industry has shown steady change, with coal mining, steelmaking, and shipbuilding giving way to electronics, communications and transportation equipment, and other sophisticated engineering industries.

Belgium has been more closely identified with and more strongly committed to the heavy industrial sector, and as this type of manufacturing has become increasingly obsolete, the country has had somewhat more difficulty in adjusting to modern conditions. The rich coal deposits of the Borinage field (which extends into northern France), and the nation's central location in Europe, made rapid expansion of the industrial revolution and its technology a natural step. The first modern blast furnace outside Britain was set up in Belgium in the 1820s, and such cities as Liège and Charleroi became important industrial centers.

Figure 18.8 Rotterdam, Europe's leading port, is at the heart of the Dutch economy.

The decline of coal and steel has brought hard times to this region, however, and the center of economic gravity has shifted to the well-situated port of Antwerp (not as large as Rotterdam, but still one of Europe's most important ports), to areas of growth near the Dutch border, and to the large and wealthy market represented by Brussels, in many ways the most important continental urban center north of Paris. Generating new employment and rebuilding the economic base of the "black country" is a difficult but important national priority.

Finally, the economy of Luxembourg, strong enough to ensure the small nation's survival, has also been successful in changing to accommodate altered circumstances. A modest agricultural economy was revolutionized with the discovery that the large iron ore deposits of French Lorraine also extend slightly into the southern part of Luxembourg. On a world scale, these ores are barely significant, but to a country of fewer than 400,000 inhabitants they become a major resource indeed! Exploitation of these deposits, often by open-pit mining, enabled the Luxembourgers to establish one of Europe's largest integrated iron and steel mills, located literally a few hundred yards from the French frontier.

For many years, iron ore and steel were practically the sole supports of Luxembourg's economy; but, as elsewhere, this industry is now generally stagnant or in decline. In a diverse economy, slowdown in one sector can be compensated by growth in others, but in one-dimensional Luxembourg this might have produced a real crisis. Fortunately, the nation has been able to move strongly into quite different activities, and thus to maintain its high living standards and nearly zero unemployment rate. First, the international banking and related financial services industry has been expanding rapidly, to the point where Luxembourg rivals more famous Switzerland in certain areas. This is now the single most important support of the economy. Second, Luxembourg City has fought strongly to become one of the "European" cities associated with the financially lucrative agencies of the European Union. While Brussels and Strasbourg (France) have received the lion's share of this economic benefit, Luxembourg City retains a very useful share of the pie, and seems likely to maintain its valued position as a truly "European" city.

Location, human energy, timely resources, and adaptability. In Benelux, more even than elsewhere, these are the keys to economic success. These countries have been more responsible than any others for building the secure foundation of a peaceful, united Europe, and it seems just that they should be able to profit from this evolution.

ADDITIONAL READINGS

LAMBERT, A., *Making of the Dutch Landscape: An Historical Geography of the Netherlands,* 2nd. ed. Orlando FL: Academic Press, 1985.

LAWRENCE, G., *Randstad Holland. (Problem Regions of Europe).* Oxford: University Press, 1973.

PINDER, D., *The Netherlands. (Studies in Industrial Geography).* Boulder, CO: Westview, 1976.

RILEY, R.C., *Belgium. (Studies in Industrial Geography).* Boulder, CO: Westview, 1976.

RILEY, R.C., and G.J. ASHWORTH, *Benelux: An Economic Geography of Belgium, the Netherlands, and Luxembourg.* London: Chatto and Windus, 1975.

DISCUSSION QUESTIONS

1. With only 360,000 people in an area of 999 square miles, Luxembourg is one of Europe's smallest independent countries. How have the Luxembourgers been able to maintain their independence?

2. Some 40% of the territory of the Netherlands lies below sea level. What are some of the major projects that have enabled the Dutch to push back the sea to this extent? How has the philosophy of reclamation changed over time?

3. Brussels is perhaps Europe's most "European" city. What does this mean? Can the city's increasing international importance help solve the nation's linguistic and cultural divisions?

4. Rotterdam is by all measurements Europe's most important port. Why? What locational advantages does the city have?

5. The Benelux countries have been in the forefront of the European movement toward further economic and political integration. Why have these nations been quick to take the lead in this area?

19
Germany

No European country has been more profoundly affected by the wave of political, social, and economic change that has swept the continent since 1989 than Germany. A divided nation since the end of World War II, when the descent of the "iron curtain" split Europe into two opposed ideologies with physical barriers separating them, Germany has again recovered its national identity by means of an astonishingly swift process. The reunified country, facing substantial problems of reintegration, is at once a source of satisfaction and concern, the object of congratulations and nagging doubts, on the part of nearly all other European nations, as well as among the Germans themselves.

In less than a century and a half, Germany has evolved from a large collection of minor kingdoms and principalities, each jealously guarding its prerogatives (traveling across Germany in the early 1800s was a customs nightmare!), to a role as the fountainhead of European intellectual achievement, to a unified state of awesome military capabilities, to the European bully of three increasingly disastrous wars from 1870 to 1945, to a divided nation the western part of whose economic recovery and cultural renaissance over the past half-century has been widely hailed as a "miracle." At the time of the events of 1989, the Federal Republic of Germany—commonly known in the United States as "West Germany"—was Europe's most populous and economically most powerful country. The country has been in many ways the economic motor that powers the European Union, has shown itself deeply committed to keeping the peace under the umbrella of NATO, and has been slowly reclaiming its intellectual credibility in the great traditions of the eighteenth and nineteenth centuries. With the restoration of a united Germany, new challenges, new opportunities, and perhaps new uncertainties have clearly been added to the European equation.

PHYSICAL CHARACTERISTICS

Physically, Germany is a perfect cross-section of Europe from the Baltic Sea to the Alps. In the north, the North Sea Plain stretches flat and featureless for some 120 miles (190 kilometers) south of the Baltic. Here, the plain is covered with sands and gravels of glacial outwash, agriculture is poor, forest cover is common, and only the great north German port cities—Hamburg, Bremen, Lübeck, Rostock—hint at the prosperity of the lands farther south.

Central Germany is dominated by the Hercynian-age hills that are such a feature of middle Europe from France eastward. Between the North Sea Plain and the Hercynian zone, however, lies the most important region of all Germany: the Börde. The Börde, which extends east to west across Germany—from the Ruhr in the west to the Polish and Czech borders in the east, is crucial for three reasons. Initially, it defined the major early routeway of invasion, trade, and communication from east to west in Europe. Important market cities such as Dresden and Leipzig sprang up, many of which cherish their ancient traditions and privileges yet today, and commercial prosperity became general.

Second, winds blowing off the vegetation-free mud flats revealed by the retreating glaciers swirled and eddied against the hills, depositing a thick layer of fine particles (organic material, clay, silt) known here and to geomorphologists everywhere as loess. The addition of this fertile top-dressing to the soils of the Börde have made this one of the richest agricultural zones of Germany, and even of Europe.

Finally, the sedimentary deposits beneath the North Sea Plain contain a major coal seam, and this, bent upward by the uplift of the Hercynian hills, comes to the surface along the same contact zone, producing the richest coal fields and the most important concentration of traditional heavy industry in all of Europe.

South of the vital Börde, the central portions of Germany are largely within the Hercynian hill zone. Wooded, rough, but of low elevation, sparsely inhabited except where the natural environment offers some advantage of site, central Germany is a region of marginal agriculture but attractive landscapes that draw large numbers of tourists. The Schwartzwald, or "Black Forest," Taunus, Thuringerwald, and Erzgebirge are among the best known of these features.

Exceptions to this rule are found in the Rhine valley and those of its major tributaries, where the flat land of flood plains and water access to the world not only permit better agriculture, but also support highly developed industrial concentrations. The major urban center of Frankfurt-am-Main is a classic example of this locational advantage.

Continuing south, a transect through Germany next encounters an Alpine foothill region of about 1,000 feet (305 meters) elevation (higher to the south), but more gently rolling than the Hercynian hills. Much of this zone is in the state of Bavaria, and the great south German regional capital, Munich, is typical of the many prosperous cities of this region. The former territory of East Germany did not extend this far south.

Finally, in the extreme south, Germany finds its share of the Alpine environment along its borders with Switzerland and especially Austria. Summer pasture for dairy cattle, heavy snowfalls and active winter sports, and spectacular glacial scenery are as much a reality here as in any other of the Alpine countries.

Germany 281

Figure 19.1 Reunited Germany.

Figure 19.2 Germany: Basic data.

Country	Area 1000 km²/mi²	Population million, 1995	GNP $/capita	Capital	Population million, est.
West Germany	248.2/95.8	63.2	18,530	Bonn	0.3
East Germany	108.2/41.6	16.3	*	East Berlin	1.14
Germany†	356.4/137.4	81.7	25,580	Bonn‡	0.3

Note: Although Germany is now reunified, data for the two parts of the nation are presented as being of historical interest.

*East Germany did not calculate accurate GNP data.
†The reunified country as of 1996.
‡As of 1996. Reunified Berlin (3.4 million) will eventually become the national capital.
Source: U.S. Department of State, Population Reference Bureau

CLIMATE

The climate of Germany is nearly as varied as its topography. At an intermediate location between the Marine West Coast and Continental climatic zones, Germany thus experiences both major European climates, while the large percent of national territory in slope ensures that a great diversity of microclimates will also be found. The maritime influence is strongly felt across the expanse of the North Sea Plain and in the western parts of the country, along the borders with France and Benelux: summers are cool and often rainy, winters gloomy and wet, but without long periods of sub-freezing temperatures. Farther from the coast, however, in the eastern regions, and especially at the somewhat higher elevations of the Hercynian hill belts, winters are colder and snowier, making the climate Continental: January in Berlin averages 6°F colder than in the western city of Cologne. The higher parts

Figure 19.3 The Rhine valley is Germany's economic and cultural heartland: St. Goarshausen.

Germany

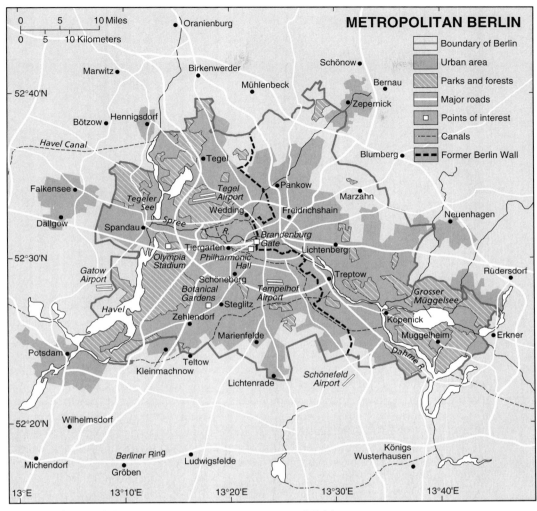

Figure 19.4 Berlin, a city reunited after 45 years of division.

of the Alpine zone have of course a more severe Highland climate; but just as in Switzerland or Austria, elevation differences and exposure angles create a wide range of microclimates.

HISTORICAL AND CULTURAL FACTORS

As part of the successive waves of invasion from the east, Germanic tribes put great pressure on earlier Celtic arrivals and caused the Romans to fortify the Rhineland and establish the "Watch on the Rhine." In fact, the constant, expensive struggle to repel German invaders was among the factors that led to the eventual collapse of the Roman Empire. At the same time, however, the Germans began to feel pressure from Slavic groups advancing westward by the fourth century, who pushed them south and west, beginning a struggle for the "Germanic" lands which has echoes down to the present time.

Figure 19.5 Timbered houses characterize the architecture in Germany's Black Forest region: Schiltach.

Austria, Prussia, and the Rhineland emerged as the poles of German cultural and economic development; but the region as a whole was fragmented into an incredible patchwork of duchies, principalities, kingdoms, bishoprics, and other minor political entities, many, but not all under the loose organization of the Holy Roman Empire. A modern tourist traveling down the Rhine is struck by the large number of well-positioned, fortified castles. Most of these played a role in the political games of the German "mini-states," and all of them had as a primary purpose the collection of taxes and tolls from passersby. Moving any sort of merchandise across Germany was for many centuries a bureaucratic commercial nightmare. At the same time, the German states became the focal point of European music, art, science, philosophy, and university life.

The unification of the German states, with the notable exception of Austria, was only accomplished in 1871, under the powerful hand of Otto von Bismarck. From a harmless collection of small, attractively romantic nations, Germany suddenly became a unified, dangerous, and belligerent player on the European stage; and as its economy grew to match its political strength, it became the most powerful and most feared of Europe's states. The German military machine was largely responsible for devastating much of Europe during three major wars in some seventy years, culminating in the human disaster of World War II in which, its government fallen into the hands of the Nazis, Germany was viewed by most of the world as an absolutely evil nation.

Following the war, Germany was divided into occupation zones by the Allied powers; but as the "cold war" between the Soviet Union and its former allies intensified and it became clear that the Soviet sector would not be allowed to rejoin the other zones on any reasonable terms, France, Great Britain, and the United States merged their occupation areas and restored sovereignty to the Federal Republic of Germany.

The country, which had Europe's largest population even before reunification (although it is currently losing numbers as a result of very low birth rates), consisted of eleven states, or *Länder*, each with extensive local decision-making powers: a far cry from pre-war Germany, when all decisions flowed from Berlin. With reunification, an additional five *Länder*, which had been abolished by the East German regime, have been added to the nation. Berlin itself, until recently still technically under four-power occupation, was well within East Germany, so that even though its three western zones formed one of the states of the Federal Republic, it was unrealistic to make it the capital of the new nation. This honor went to Bonn, a small city on the Rhine near Cologne, which was chosen because of its central location, its relative insignificance (symbolizing to many Germans the idea that Berlin remained the "true" capital of an eventually reunited nation), and its personal associations with the great postwar German leader, Konrad Adenauer.

PERCEPTION: WHAT CAPITAL FOR A REUNITED GERMANY?

With the reunification of Germany a fact as of October 1990, the question of selecting a capital for the nation has inevitably been raised. Just as inevitably, the choice of Berlin had already been made, leaving the former West German capital Bonn to return to being a "small town in Germany" on the banks of the Rhine. No third choice was apparently even contemplated. Yet the choice was perhaps not so obvious, and there are substantial implications in the outcome.

Berlin was of course the nation's capital from the time of unification until the debacle of World War II, and East Berlin, the Soviet occupation sector, became capital of the now vanished German Democratic Republic. The city thus has a number of unpleasant associations with modern German history, discounting its role as one of Europe's most sparkling cultural centers both before World War I and during the interwar period. It is for many non-Germans the city of Bismarck and the Kaiser, the city of Hitler, and more recently the city of a discredited communist regime. These ghosts are difficult to exorcise. Berlin is also the capital of Prussia, traditional stronghold of German militarism, and the message sent by its selection as seat of government of the "new" Germany is not an altogether soothing one.

At a different level, selection of Berlin to again be Germany's capital clearly shifts the balance of political interest in Europe to the east. The primary goal of Soviet foreign policy in Europe since World War II had been to maintain a "buffer zone" of friendly states against attack from the west—specifically German "revanchism." With an economically powerful Germany again united on the border with Poland—itself a rapidly westernizing nation—many Russians will feel less secure, and this may influence decisions on defense and foreign policy.

And what of poor Bonn? A textbook example of the growth and benefits that becoming a capital can confer on even a modest city—population has more than doubled over the past twenty years—Bonn will have difficulty articulating a new identity. As embassies, government offices, and all of their attendant support systems drift to Berlin, Bonn will become a footnote to history: a pleasant small city which will again be best known for its associations with Ludwig von Beethoven.

But perhaps not just yet. The German finance ministry has made clear to the government that the enormous cost of transferring the national capital to Berlin is an expenditure that the nation—given the immense costs related to unification and other financial pressures of the 1990s—would be ill-advised to assume. For the moment, Bonn will continue to enjoy its role as (by far) the smallest capital of a major European nation, although by early in the next century the transfer of government to Berlin should be completed.

ECONOMIC ACTIVITY

West Germany has been by far the strongest economic power in modern Europe; a remarkable fact, considering the physical devastation and financial ruin brought on in this century by two savage wars, with the great depression sandwiched between. In fact, the restoration of Germany to the front rank of the world's industrialized nations in little more than twenty years following the close of World War II has justly been termed an "economic miracle."

The essence of this economic success is primarily industrial, rather than agricultural. German farming, in fact, because of the poor soils, hilly topography, and short growing seasons which characterize much of the natural environment, is often marginal, despite the unquestioned skills of German cultivators; and is heavily subsidized by the government, largely for political reasons. Only along the Börde, where loessal deposits have enriched the soil; in the middle and upper Rhine valley, where the vine has a favored place; and in the rolling pre-Alpine landscapes of Bavaria, is agriculture truly productive and competitive in a European context.

East German farming, which shares most of the physiographic and climatic patterns found in the west, except for the southern lands, has been further handicapped by the pattern of collectivization enforced as part of the communist command economy. The inefficiencies of very large farm sizes, the price and resource distortions that seem inevitable without some form of market control, and the indifference of many farmers who see little incentive in the system have meant that East German farmers have rarely produced at levels which might be visualized by their skills and the inherent quality of the land. Integration and restructuring of agriculture in the two parts of reunited Germany will not be the least of the country's problems.

The mineral wealth of Germany, which permitted the rapid expansion of the nation's industry, is dominated by one product: coal. Along the Börde, where the sediments of the North Sea Plain were folded upward by the Hercynian uplift, Europe's major coal seams (in terms both of quality and quantity) are found. The Ruhr is Europe's leading producing region, while the related Silesian fields to the east now support the industrial development of Poland, the Czech Republic, and Slovakia. Coal is also found in the Saar basin, where it has been one of the bases of bitter contention between France and Germany until the recent success of the European Coal and Steel Community (part of the European Union) in denationalizing these key products (see discussion of resources, Chapter 10). In former East Germany, vast deposits of lignite (or "brown coal") have been utilized to generate electricity, but the combustion of this low-quality fuel with virtually no emission controls has been a major contributing factor to the poor air quality over much of eastern Europe. Nuclear energy is also significant in Germany, but the increasing strength of "green" political parties and the

Germany

frankly hazardous condition of the East German plants (which will lead to their early closing) means that this option will decline in importance.

The other important raw material of German industry has been the country's large deposits of various mineral salts, which have given rise to the powerful chemical industry. Beyond this, however, resources are scarce, and have been largely exploited to the point of exhaustion. Iron ore deposits in the hills near the Ruhr were soon depleted as the steel industry expanded rapidly; and Germany produces only about 4% of her oil needs. With reunification, the Soviet oil that was provided to East Germany at discount prices has disappeared; in fact, Russia has eliminated its subsidized oil shipments to all its former East European client states. The foreign purchase of these commodities, along with natural gas (Germany is the world's leading importer) and most other mineral resources, represents a substantial drain on national revenues, and can only be compensated by the large export earnings generated by the manufacturing sector.

Fortunately, German manufacturing is more than equal to the task, combining efficient organization and high quality to produce a wide range of goods that are much in demand. Engineering industries are featured, and German technology remains the equal of any in the world. The country is Europe's leading producer of a great variety of products, including automobiles, plastics, artificial fibers, and many chemicals and pharmaceuticals.

East German industry, grouped into large, unwieldly *kombinats*, was unproductive by western standards, although the country produced the best quality and most highly prized manufactured goods in the eastern bloc. Placing East German industry on a market, rather than a command, basis, and integrating it with that of the west, is proving to be a difficult, time-consuming task. A special agency, known as the *Treuhand,* was established to privatize much of the East German economy, and has been generally successful despite numerous difficulties. Unemployment has risen sharply in the east, however, as inefficient industrial "dinosaurs" have been forced to close; while wage levels are still

Figure 19.6 German prosperity is evident in the marketplace: Baden-Baden.

below those of comparable jobs in the west. East German labor, however, is generally as skilled, educated, and adaptable as its western counterpart, and the transition could well produce another "economic miracle." Encouragingly, migration from east to west in a Germany without internal frontiers has slowed to a modest level after an initial period following reunification when it appeared that all East Germans would move west.

German industry, traditionally heavily concentrated in the Ruhr, is today more spatially diversified. The Ruhr, however, remains a remarkable region of industrial might, grouping within a rough square some fifty miles (eighty kilometers) on a side about 10% of the country's total population, and forming the most important and most impressive single industrial concentration in all of Europe (see industrial discussion, Chapter 11).

In addition to the Ruhr, modern German industry is concentrated at several other points. Of particular interest is the lowland area where the Rhine is joined by a major tributary, the Main. Focused on the great commercial and transport center of Frankfurt, several cities make a particular specialty of high-technology engineering industries. Farther south, such centers as Stuttgart (home of Mercedes-Benz) and Munich, the large regional capital of Bavaria, are also industrialized; while to the north Hamburg and Bremen are important commercial and trading centers with industries related to their port functions.

In the eastern sector, the attractions of the Börde made it the leading industrial area as well. Such cities as Leipzig, Dresden, Halle, and Chemnitz have a strong tradition of industrial quality and may be expected to figure prominently in future expansion. A good geographic example is the city of Gotha, famed for the production of high quality wall maps (as used in classrooms). At the close of World War II, when the division of Germany became a fact, many eastern companies were able to move at least part of their operations to the west. The "reunification" of these firms will be another feature of the post-reunion industrial landscape: the famous optical firm of Zeiss, traditionally headquartered in the eastern city of Jena, is a classic case.

ADDITIONAL READING

BURTENSHAW, D., *Economic Geography of West Germany.* London: Macmillan, 1974.

DICKINSON, R.E., *Germany: A General and Regional Geography.* New York: Dutton, 1973.

MORETON, E., *Germany Between East and West.* New York: Columbia, 1987.

TURNER, H.A., *The Two Germanys since 1945.* New Haven, CT: Yale University Press, 1987.

WILD, M.T., *West Germany: A Geography of Its People.* Totowa, NJ: Barnes and Noble, 1980.

DISCUSSION QUESTIONS

1. Would Bonn or Berlin be the better capital of a reunited Germany? What are the arguments in favor of each city?
2. West and East Germany have a combined population of nearly eighty-two million, by far the largest in Europe (excluding Russia). Does this weight of numbers pose problems for the increasing economic and political integration of Europe?
3. The West German economy has been the strongest in Europe and one of the world's most powerful in recent years. Will the reintegration of East Germany change this? What problems of assimilation might be expected as the two Germanys reunite?
4. The Börde is Germany's most important region in terms of population, urbanization, economic concentration, and numerous other factors. What makes this region so important?
5. What factors have combined to make the Ruhr Germany's (and Europe's) most important concentration of heavy industry? How is the region coping with changes in economic realities?

20
The Alpine States

Although the various ranges of the Alps are a prominent part of the physical geography of several European countries, there are two nations contained almost entirely within these mountains, nations for which the highland environment is the most important factor in their cultural and economic patterns. These countries are Switzerland and Austria (tiny Liechtenstein, between the two, has a similar setting).

The Alpine environment, which imposes such strict limits on many conventional economic practices, offers generally the same challenges and opportunities to both nations, yet there are very substantial differences between them. Switzerland is that rarity in Europe: a self-contained country that has studiously practiced neutrality during the continent's seemingly endless upheavals, while Austria has played a very large, involved role indeed, and has known the extremes of triumph and defeat to a much greater degree than has its neighbor. These differences in philosophy and historical development have left their marks on the modern nations, which have adjusted to change in often divergent ways despite the similarity of their natural environment.

THE PHYSICAL SETTING

The physical geography of Switzerland divides the country logically into five generally parallel regions with a northeast–southwest trend. Along the northwest margin, the Jura mountains are the common boundary with France. Folded mountains of the "Appalachian" type, with long, parallel valleys and ridges, the Jura is difficult to cross against the grain, so, despite the fact that its peaks rise only to about 5,000 feet (1,520 meters), is a region of surprising isolation. Moreover, as the first substantial mountain range confronting storms moving off the North Sea, the Jura receives unusually heavy amounts of precipitation, including winter snowfall.

Southeast of the Jura is the most important region of the nation: the Central Plateau. Although by no means at low elevation—it averages nearly 2,000 feet (610 meters)—this gently rolling region, some 150 miles (240 kilometers) long and 25 to 40 miles (40 to 70 kilometers) wide, is the focal point of Switzerland's population, urban development, agriculture, and industry.

Beyond this plateau come the major ranges of the Swiss Alps, still trending northeast to southwest. These are usually considered in two sections, divided by the glaciated valley occupied by the headwaters of both the Rhine and the Rhône that provides the major route of access into and through the area. Northwest of this line, the northern Alps, with their glaciated valleys opening out onto the plateau, offer perhaps the classic image of Switzerland: mountains, lakes, glaciers, but at the same time a distinct human presence and a reasonable level of accessibility. It was in these valleys at the heart of the country that fiercely independent tribes banded together for defense to form the first Swiss confederation in 1291 (one of these groups, occupying the modern state, or *canton* of Schwyz, gave both a name and a flag to the nation as a whole). South of the Rhine–Rhône corridor, the Interior Alps are more isolated, more rugged, and less civilized. With the exception of a few fashionable winter resorts, a gradually vanishing pastoral economy, and concentrations of activity along the major pass routes to Italy, the mountains here reign supreme, with several peaks exceeding 13,000 feet (3,960 meters).

Finally, south of the Alps, the Swiss canton of Ticino, although small, is in many ways a distinctive region. An extension of the Po valley in Italy, the Ticino has that rare commodity in Switzerland: flat land; and its climatic, cultural, and economic patterns also set it apart from the other regions.

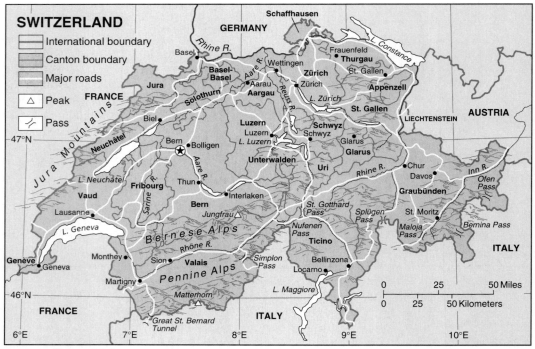

Figure 20.1 Switzerland

The Alpine States

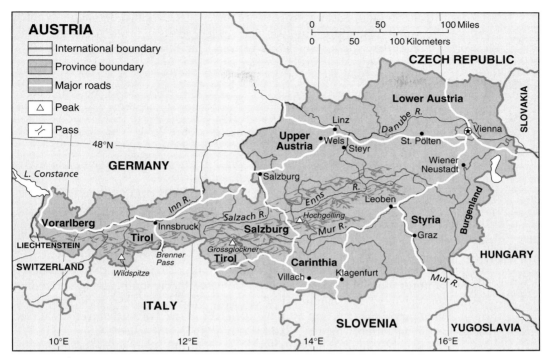

Figure 20.2 Austria

The major chains of the Alps continue into Austria, and constitute some 70% of that nation as well, including nearly all of the west and south. The orientation of the mountains here becomes more nearly east to west, and control of the few good north–south passes (such as the famous Brenner) has long been crucial. South of the Inn valley (an important tributary of the Danube) is found one of the most inaccessible and rugged of all the Alpine regions: the Tauern. With some peaks exceeding 12,000 feet (3,660 meters), this remote area is little used except for mining and resort activities.

The heart of modern Austria lies to the east, along the broad valleys of the Danube and its major tributaries. In a curving line from Salzburg in the northwest to Graz in the south are found most of the population and economic activity of the nation; these focus especially on the great capital, Vienna. Finally, to the southeast of Vienna, Austria includes a small slice of the great, flat east European plain known as the Alföld. This region, the Austrian province of Burgenland, with its broad landscapes and

Figure 20.3 Alpine states: Basic data.

Country	Area 1000 km²/mi²	Population million, 1996	GNP $/capita	Capital	Population million, est.
Austria	83.8/32.4	8.1	24,950	Vienna	1.6
Liechtenstein	.16/.062	.03	22,300	Vaduz	.0049
Switzerland	41.3/15.9	7.1	37,180	Bern	.26

Source: U.S. Department of State, Population Reference Bureau.

"eastern" feeling, is the exact opposite of the majestic Alps which dominate so much of the country, and reminds the observer of the rich diversity of physical environments, as well as human groups, over which the Austrian empire once ruled.

CLIMATIC PATTERNS

Climatically, the Alpine regions of both Switzerland and Austria are characterized by the relatively harsh conditions imposed by elevation. Winters are long and fairly cold, and the abundant moisture provided by the constant flow of maritime air from the Atlantic produces heavy snowfall. Summers are cool and moist; bright, sunny days are much appreciated by residents and visitors alike.

The climatic pattern of the Alps, however, is far from being simplistic. Differences in elevation, slope, and exposure angles produce an elaborate pattern of microclimates: 7,000 feet is not 10,000 feet, the north face is not the south face, and a valley running north to south is not one running east to west. Faced with this complexity, systems of climatic classification generally compromise by using only the designation "highland;" but this generality actually obscures a great, often surprising, climatic variety. For example, the north shore of Lake Geneva *(Lac Leman),* facing south, receives abundant warmth and sunshine. The vine flourishes, producing some excellent wine, palm trees are grown in public gardens, and even the banana will survive in sheltered areas (see discussion of vegetation, Chapter 2). South of the Alps, in the low elevations of the Ticino, the climate resembles that of the Po valley, with cool, damp winters and warmer summers; crops grow here that can be produced nowhere else in Switzerland.

The "highland" climate complex naturally dominates much of Austria as well, and produces the same difficulties for man. Since Austria is farther from the Atlantic, however, the maritime influence fades, and the lower areas in the east are likely to have a continental climate, with colder winters but warmer summers than regions farther to the west at the same latitude. In the Burgenland, stations with a humid subtropical climate can be found—a preview of conditions across much of southeastern Europe. Moderately severe winters characterize this region, while summers tend to be hot and humid.

CULTURAL AND HISTORICAL CHARACTERISTICS

Switzerland and Austria display much greater diversity in their cultural and historical development than in their physical characteristics. Switzerland is absolutely unique among European states in its form of government: a confederation, dating back as far as the thirteenth century, in which each canton (state) has considerable authority over most of its own affairs, while the central government plays a distinctly secondary role. The federal capital, Berne, is only a small city, unlike such strongly centralizing urban concentrations as Paris or London; and in reality only Zurich, among all the well-known Swiss cities, can stretch its suburbs to encompass a population of close to one million. If you wish to become a citizen of Switzerland (as many people would), you must first satisfy the requirements of your village or city and canton; only then can you be accepted at national level.

This strong feeling of local initiative has marked the history and conduct of the Swiss nation in many ways. Banding together for defense against Austria in 1291, the four original cantons (who were joined by most of the important Germanic cantons by 1353) took great care to specify the limits of their agreement, and thus to ensure local democracy at the most basic level. Throughout the subsequent history of the nation, during the period when Swiss mercenaries were the most valued soldiers of Europe during the bitter conflicts and divisions of the wars of religion (which were strongly felt in Switzerland), and during the turbulent Napoleonic era, when the French-speaking cantons were added to the confederation, the traditions of bitterly defended independence and a growing neutrality in Europe's countless conflicts became deeply engrained in Swiss philosophy, and have in fact served the nation remarkably well.

Austria's case has been quite otherwise. From humble beginnings on the Danube, the country emerged under the strong guidance of the Hapsburg monarchy to become one of the most powerful of all European states, controlling an empire extending over a major portion of southeastern Europe. By the turn of the twentieth century, the Austro-Hungarian Empire (Hungary had been granted equal status in 1867) was clearly one of the "major" powers, and its capital, Vienna, was the only serious rival to London and Paris in size, prestige, art, music, technology, and all the other characteristics prized by contemporary Europeans.

In the debacle of World War I, however, defeated Austria lost both its monarchy and its empire, and was reduced to its present modest size. Even before the country could adjust to these new conditions, it was annexed by Hitler's Germany and plunged (again on the losing side) into World War II. A period of Allied occupation followed, and it was not until 1955 that the Soviet Union agreed to a reunified, neutral Austria. Thus, while Switzerland's studied non-involvement in European quarrels goes back several centuries, Austria's less rigorous neutrality covers only a modest thirty-five years. Through it all, Vienna has remained one of Europe's most impressive cities. Grouping over a quarter of the national population, much more than the average of most capitals, Vienna—capital of a vanished empire—is sometimes referred to as "a head without a body."

Except for small Italian and Hungarian minorities, German-speaking, mainly Catholic Austria is a culturally uniform country today. This is totally different than the case of Switzerland, where three major languages and one obscure one have official status, and adherents to the Protestant and Catholic varieties of Christianity are nearly equal in number. The Swiss nation, in fact, is held together precisely by its insistence on local decisions and its respect for individual opinions. Unlike Belgium, where linguistic divisions threaten the stability of the nation, or Northern Ireland, where religious strife has completely polarized a people, Switzerland has defused these issues by allowing each canton to have an "official" language and religion (not obligatory for all citizens, of course), and requiring the nation to respect these choices (see discussion of languages, Chapter 5).

ECONOMIC ACTIVITIES

Economically, the two essentially Alpine nations again vary somewhat, although the constraints posed by their natural environments are much the same. In the highland regions of both countries, common problems have led to common solutions, and

Figure 20.4 Geneva, Switzerland is Europe's most international city.

forestry, scattered mining, and year-round tourism (with an emphasis on winter sports) are the main supports of a sparse population. Highland agriculture has been traditionally based on dairy cattle, which are able to make use of precious energy resources (the brief summer pastures) to produce food and marginal income for human groups. To the great uneasiness of the Austrian and especially Swiss governments, to whom maintenance of the economic vitality of the mountains is an article of faith and national self-sufficiency, the Alpine animal economy is fading, as cash earnings are rarely adequate for modern lifestyles. Thus, in Switzerland, the practice of "Alpage" is promoted, whereby urban residents unemployed during the summer are encouraged to "board" cattle from local farmers in highland pasture areas, turning the milk to cheese, perhaps feeding the remaining skim milk to hogs, sharing profits with the farmers when the cattle are returned in the fall, and generally keeping the mountains alive (see discussion of Alpine economy, Chapter 8).

Both countries have long traded on their control of the vital communications and transport routes between northern Europe and the Mediterranean. The famous St. Bernard dogs, taking their name from the pass where they rescued stranded travelers, were symbolic of the lucrative income to be made by helping early travelers across the formidable Alps. The inns and hospices established to promote this flow of wealth were in fact the original basis of today's highly developed Swiss tourism industry. Later, development of electrified railways, using the abundant hydroelectric power of the Alps, reinforced the transit function, again to the particular benefit of Switzerland.

Agriculture in both countries, restricted by the natural environment, is inadequate to meet requirements, but is efficient and strongly promoted in the name of national self-sufficiency. The Central Plateau region of Switzerland produces a reasonable volume of basic European food crops, plus a substantial amount of seasonal vegetables and fruits, while the Ticino is a vital source of such "southern" products as tobacco, off-season vegetables, and cut flowers. In Austria, the Danube valley and the rolling landscapes of the states of Lower Austria and Carinthia provide basic commodities, while the rich fields of the Burgenland are the source of most fruits, vegetables, and "exotic" crops.

Industry in Switzerland is highly diversified, and is based on the widespread use of hydroelectric energy, on a long tradition of technological innovation and manufacturing skill, and on a location at the heart of Europe's major axis of economic growth. In the absence of basic industrial raw materials, Switzerland has long focused on such highly profitable industries as top-quality watch and clock making, the manufacture of heavy electrical equipment and pharmaceuticals, and export-quality agricultural products (chocolates, cheeses). The industrial landscape of the nation is thus divided among numerous cities, some of them quite small (Winterthur, St. Gallen), while only Basel, the major Swiss outlet to the world via the Rhine, has a predictable industrial base associated with the processing of imported raw materials.

Austrian industry has a somewhat different focus based on the nation's historical development. As a major European power until 1918, Austria participated at a fairly late date in the industrial revolution, and still produces basic iron and steel at Wiener Neustadt, south of Vienna. Graz is also a center of heavy industry, and the Austrian industrial plant is in general more concentrated than that of Switzerland. Technologically advanced industries have recently been heavily promoted, and Austria is making good use of its traditionally strong ties with eastern Europe to increase trade and economically "colonize" this evolving region.

Figure 20.5   Highland agriculture is still a vital link in the Alpine economy: Gosau, Austria.

Additionally, a number of other important economic supports of the Alpine states should be noted. First of all, tourism is vitally important to both Switzerland and Austria, although the Swiss have been stressing it much longer than the Austrians. The Alps offer the possibility of nearly year-round recreational activity, especially given the spectacular rise in the popularity of winter sports since about 1960, while the strong cultural and historical appeals of such cities as Salzburg and Vienna, Zurich and Luzern are major attractions for two nations which rely heavily on tourism to balance their books. Switzerland has the more highly developed tourism infrastructure (hotels, roads, guides, activities) based on several centuries of experience in providing for the needs of travelers; but Austria is not far behind, and has the advantages of lower costs and an often less "spoiled" environment than Switzerland.

Additionally, both nations trade heavily on their neutral political position in Europe and the world to attract international agencies and corporations, peace conferences and negotiations, and other activities for which a neutral venue is vital. Geneva, Switzerland, is the classic example of such a place, being or having been the headquarters of the post-World War I League of Nations, several modern United Nations agencies, the International Red Cross, the International Labor Organization, and numerous other agencies, as well as the locale of countless international conventions.

Figure 20.6 The Rhine is Switzerland's connection to the world: the port of Basel.

The political neutrality, absolute security, and generous hospitality offered by Geneva when antagonists decide to discuss their differences make it a preferred venue, and of course guarantee to the Swiss a steady and substantial source of income. Although a more recent player in the game, Austria has recently been using its lately-acquired neutrality to rival Switzerland in a modest way as an international meeting place. The Organization of Petroleum Exporting Countries (O.P.E.C.) has its headquarters in Vienna, as does the International Atomic Energy Agency. In addition, as noted, Austria uses her longstanding relationship with eastern Europe to promote both dialogue and trade between east and west. It was no coincidence that many of the refugees fleeing unsettled conditions in the east at various times over the past forty-five years came first to Austria. The concept of neutrality still differs, however, between the two nations. In 1994, Austrians voted to join the European Union, thus broadening its political as well as economic horizons. Swiss voters, on the other hand, have consistently rejected similar opportunities. Because of its political character, Switzerland has never joined the United Nations, although it is an active participant in most specialized UN agencies.

Finally, the special case of Switzerland as a financial haven should be mentioned. The country makes an enormous profit out of insurance, investments, and the advantages of a strong, stable economy; and the income earned by its secretive and closely protected banking industry is a key factor in balancing the nation's accounts. Although Swiss banks are notoriously tight-fisted in matters of interest paid on accounts, the rigidly enforced laws on banking secrecy (details of accounts can only be released if active criminal prosecutions are underway), and the jealously guarded reputation of Switzerland as a stable, secure country where funds will be—above all—safe, have made the nation a magnet for external investment. This in turn solidifies the Swiss position as the world's richest developed country.

LIECHTENSTEIN

Tucked between Austria and Switzerland in the heart of the Alps lies another of Europe's "mini-states." Tiny Liechtenstein consists of a narrow strip of the Rhine river flood plain and a larger piece composed of rugged mountains: a total of sixty-two square miles (160 square kilometers). Another leftover of history, independent since 1866, the country owes its survival largely to rivalries between its two larger neighbors. It is still ruled over by an ancient line: the Count of Liechtenstein can survey virtually his entire domain from the castle perched dramatically above the small capital, Vaduz.. The country's 28,000 people profit from being linked by a customs union to Switzerland. Although tourism and tax-advantaged financial dealing pay most of the bills, Liechtenstein's economy is perhaps best known for its issues of colorful postage stamps, much prized by collectors; while a modest industrial base has also been developed.

In September 1990, Liechtenstein had the distinction of becoming the 160th member of the United Nations. It also became the smallest member in both size (lowering the record of St. Vincent and the Grenadines in the Caribbean: 150 square miles/387 square kilometers) and population (edging out the 36,000 of another Caribbean member: St. Kitts and Nevis).

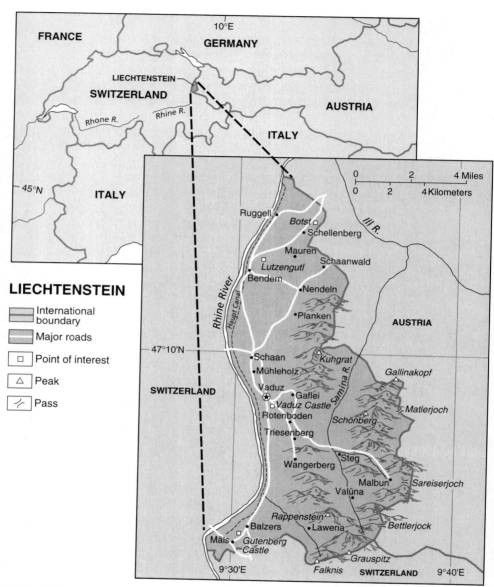

Figure 20.7 Liechtenstein: Alpine mini-state.

The "mini-states" were never invited to join the League of Nations following World War I, and have not until now bothered to apply to the UN. Liechtenstein's successful application has prompted the others to take similar action, especially as the UN appears to be emerging as a more influential force in the world, and since the political evolution of many small nationalities within larger countries may eventually lead to a substantially increased total number of nations. Thus, Monaco, Andorra, and San Marino are among the new members which have raised total UN membership to 184 since 1992. Liechtenstein's unquestioned wealth is also a factor; the tiny nation is proud to be participating in the new European bank for the reconstruction of eastern Europe.

Figure 20.8 Liechtenstein is on the upper Rhine between Switzerland and Austria: Castle and factory at Balzers.

ADDITIONAL READING

JENKINS, J.R., *Jura Separatism in Switzerland.* Oxford and New York: Clarendon, 1986.
LICHTENBERGER, E., *The Eastern Alps.* Oxford: University Press, 1975.
MAYER, K.B., *The Population of Switzerland.* New York: Columbia, 1952.
SULLY, M.A., *A Contemporary History of Austria.* London and New York: Routledge, 1990.
WHITE, P.E., *The Social Impact of Tourism on Host Communities: A Study of Language Change in Switzerland.* Oxford: Oxford University School of Geography, 1974.

DISCUSSION QUESTIONS

1. Switzerland and Austria have many physical similarities, yet have had vastly different histories. Why have the centuries treated the two countries so differently?
2. Why is Switzerland's industrial structure divided among a large number of cities, rather than concentrated, as in the German Ruhr? Why has Basel become the most rapidly growing industrial center of the nation?
3. Vienna is sometimes called "a head without a body." Why? Why is Vienna about twice as large in population as the largest Swiss city (Zurich), even though Austria has only about 20% more people than Switzerland?
4. In 1976, a new canton (or state)—Jura—was created by popular initiative in Switzerland. What does this reveal about the nature of the nation's government?
5. Control of crucial trans-Alpine transportation routes has always been a key factor in the Swiss economy. Trace the evolution of this control, noting how the Swiss have consistently been able to keep up with changing circumstances.
6. Liechtenstein (as well as the other "mini-states" of Andorra, Monaco, and San Marino) has recently joined the United Nations. Is it appropriate for such small nations to join this world organization? What might they hope to contribute?

21
Italy, Greece, and the Mediterranean Islands

ITALY

To the world at large, Italy typifies the pleasures and problems of the Mediterranean region. The landscape of whitewashed farmhouses with red-tiled roofs, tall, stately, dark green cedar trees, neatly tended vineyards and olive groves, and perhaps in the distance a glimpse of the "wine-dark sea" is an unforgettable image. Yet this is only one of many Italian images, which vary from the snow-capped peaks of some of Europe's highest mountains to the flat grasslands of the Po delta; from such bustling, thoroughly European cities as Milan through the incomparable beauty of Florence, Venice, and other reminders of the Middle Ages and the Renaissance, to the drowsy poverty of countless small villages that seem to have remained unchanged for literally hundreds of years.

The most populous nation of the European south, Italy, from Roman times onward, has generally played a dominant role in the Mediterranean, while at the same time often managing to be a significant nation in the European scheme of things. Today, much of Italy has moved successfully into the mainstream of European affairs, yet parts of the country remain deeply rooted in the philosophies of the past. Italy probably shows a greater differentiation between rich and poor regions than does any other European nation, and the reconciliation of these regional differences has long been—usually with indifferent success—a major policy goal of all Italian governments since the modern country was unified in the 1860s.

Physical Characteristics

Physically, northern Italy is dominated by two features: the Alpine mountain border and the valley of the Po River. Italy has a largely mountainous border with all of her continental neighbors: France, Switzerland, Austria, and Slovenia. This border has, over the centuries, tended to insulate the country from events to the north (although both France and Austria have frequently been invaders), and to focus Italian attention on the Mediterranean. The Maritime Alps, forming the border with France, run north from the sea, turning eastward into the major ranges of the Alps where Italy joins both France and Switzerland. As on the north side of the Alps, Pleistocene glaciation has carved deep valleys to the south, and many of these are filled with naturally dammed glacial lakes. Lakes Como, Garda, Maggiore, and others are justly famed for their natural beauty, and are a popular tourism destination for both Italians and foreigners.

Farther east, the border with Austria is also rugged, the Dolomite Alps (named for the particularly resistant form of limestone of which they are composed) having been carved by ice into spectacular jagged peaks. Italy and Austria struggled inconclusively across this difficult terrain during most of World War I. Finally, the border with Slovenia is also generally mountainous, although somewhat less imposing. A narrow coastal plain, on which the city of Trieste is located, provided relatively easy access to southeastern Europe. The steep slopes, harsh climate (although this is more moderate than on the north face of the Alps), and isolation of the mountain districts give rise to most of the same cultural and economic patterns found elsewhere in Europe's mountains (see discussion of Alpine economy, Chapter 8).

South of the Alps lies the richest and most productive region of the Italian peninsula: the Po valley. Long considered by historians and political geographers as the "heartland" of the Italian nation, this area is today the most "European" part of Italy, and its prosperous agriculture and industry are the major support of the national economy. A glance at the map makes it clear that the Po plain is simply an extension of the Adriatic Sea which happens to be slightly above sea level. Just as the Adriatic is a very shallow sea, so the Po valley has very little elevation: Torino (Turin), some 200 miles (322 kilometers) from the sea, is only about 500 feet (152 meters) above sea level.

The extreme flatness of the plain has led to an historic range of problems involving flooding, poor drainage, and waterlogging, as the Po, which drains the entire south face of the Alps plus the north slopes of the Apennines, moves slowly to the sea. Because of the unpredictability of the river, few major cities are actually located along its banks (in contrast to most other European rivers), and water control and management projects have long had high priority. The river itself has actually been displaced toward the south side of its valley by an accumulation of sand, gravel, and other materials washed down from the Alps. Where the bedrock surface intersects these loosely consolidated deposits, many natural springs occur. These springs, or "fontanelli," as a reliable water supply, have been an important and unusual locating factor for many towns and cities in the area.

The remainder of the Italian peninsula is formed largely by the Apennine mountains. Beginning at the Mediterranean border with France, this range trends diagonally across the peninsula at the southern border of the Po plain, then swings southward as far as the "toe" of the Italian "boot." The Mediterranean Italian islands of Sicily

Figure 21.1 Italy.

Figure 21.2 Italy, San Marino, Vatican City: Basic data.

Country	Area 1000 km²/mi²	Population million, 1995	GNP $/capita	Capital	Population million, est.
Italy	301.2/116.3	57.3	19,270	Rome	2.8
San Marino	.062/.235	.03	8,520*	San Marino	.0045
Vatican City State (Holy See)	.00044/.00017	.001	†	‡	‡

*Per-capita income estimate, rather than GNP.
†No data calculated.
‡City and country are co-terminous.
Source: U.S. Department of State, Population Reference Bureau

and Sardinia are also mainly mountainous. The highest points of the Apennines are to the east of Rome, where elevations exceed 9,000 feet (2,740 meters): the highest peak is Monte Corno (2,912 meters), but the range gives a rugged character to much of the southern peninsula, and has been a major factor in the region's weak economic development. At least seven active volcanos have been identified in southern Italy, including the famous Mt. Etna on Sicily, along with Mt. Vesuvius and several others, either on the mainland near Naples or on smaller islands near Sicily, such as Lipari or Vulcano (from which the generic term "volcano" is derived).

Apart from the Po plain, high quality land for agricultural uses is at a premium in Italy. The gently rolling hillsides of Tuscany, north of Rome and west of the main sweep of the Apennines; and the region of Apulia, at the "heel" of the "boot," are reasonably productive. Elsewhere, only the small flood plains of the rivers provide good farm land.

Figure 21.3 The Italian lake district is famed for its natural beauty: Lake Como.

Climatic Patterns

Climatically, Italy is usually considered to be a "classic" Mediterranean country; but this designation is somewhat deceptive. Certainly the central and southern parts of the country are typically Mediterranean, with the warm summers, mild winters, and summer drought so characteristic of that climate. In fact, the prolonged summer drought (an average of about three months) has long been another difficult problem for these regions. The north, however, is another matter. The Alpine fringe naturally shares with other Alpine regions the pattern of short summers (although these are warmer than to the north), severe winters, and abundant precipitation, including substantial snowfall in places.

The vital Po valley is more distinctive. Surrounded by a "horseshoe" of mountains, the Po plain finds itself sheltered at once from the severe winter storms of the north and the worst effects of the Mediterranean summer. Climate here is more often of the humid subtropical type, featuring cool, wet winters and comparatively hot summers with high humidity and enough summer rainfall to support many types of conventional agriculture.

Historical And Cultural Patterns

Italian history has been an intricate and troubled one, and has left many reminders in the modern landscape. Political unity was unknown on the peninsula from the collapse of the Roman Empire some 1500 years ago to the modern reunification of the many Italian states in the 1860s. Diversity did not necessarily mean poverty, however, and the powerful city states that emerged during the Middle Ages and the Renaissance (Florence, Pisa, Siena, Milan, Venice, Genoa, and many others) became fabulously wealthy by controlling trade between the eastern Mediterranean (and Asia) and the developing regions of northern and western Europe. The city-state of Genoa, for example, actually became the banker for the English monarchy. Lavish patronage of the arts and sciences during this period supported such genius as the work of Leonardo da Vinci, and paid for the priceless treasures of art and architecture for which Italy is still so noted today.

The gradual eclipse of these small states, however, made the Italian peninsula and all its assets fair game for the larger, better organized nations that dominated Europe in the eighteenth and nineteenth centuries. Austria, France and Savoy fought and plotted to control such rich areas as the Po plain regions of Lombardy and Piedmont, while the holdings of the Roman Catholic church (the Papal States) were milked for revenues to support many non-Italian causes, and vast areas of the region fell into the hands of absentee landlords who cared little about development of the nation. Only with reunification—less than 140 years ago—did "Italy" appear on the maps of Europe, and the subsequent period has been filled with the same world wars (and a long period of fascist dictatorship) that have afflicted the rest of the continent. One result of this has been that modern Italy still continues to display great differences in levels of regional development, as measured by nearly every economic criterion.

The Italians are proud of their rich ancient heritage, and of their modern European nation. They feel that their history gives them a particularly important role to play in North Africa and the Middle East, and that their artistic and scientific achievements are the equal of any in Europe. Their classical history, in fact, may almost be said to have "created" Europe, at least in the eyes of many Italians.

Figure 21.4   Italy's rich artistic and archaeological heritage attracts millions of visitors annually: Pisa.

Economic Activity

The Italian economy is based on a blend of successful agriculture and modern industry, and while it is a poorer country than many of those in northern and central Europe, it has been steadily closing the gap and is a world leader in many significant technologies. The north, traditionally the richest part of the nation, remains so today, while the south, more isolated from the main currents of European development and disfavored in many ways, remains one of the continent's poorest regions despite major efforts to stimulate economic growth.

Agriculturally, Italy is the leading European country after France. Rice is widely grown on the easily flooded lands of the Po valley, the vine is practically universal except on the peaks of the Alps and Apennines, and olives flourish in most regions as well. Italy is thus the leading European producer of rice, wine (narrowly edging France), and olives (and their oil); it is also a major contributor to the production of wheat, maize, and sugar beets (see Figure 9.1). Production of fruits,

Italy, Greece, and the Mediterranean Islands

vegetables, and fresh flowers is another mainstay of the economy, particularly as these are often shipped to the north at premium prices during the winter months; and the nation's animal herd—well-balanced among cattle, hogs, and sheep—is also one of the largest on the continent.

In terms of energy and mineral resources, Italy has had the misfortune to be one of Europe's most impoverished countries. Only modest amounts of low-quality coal were discovered, mainly on the island of Sardinia, and the small pockets of oil and natural gas found in the sediments of the Po valley were scarcely more impressive. Hydroelectric energy is an important factor in the north, where the abundant potential of the swift Alpine streams can be harnessed, but farther south the dry Mediterranean summers largely rule out this possibility. As a result, Italy has been forced to become a major energy importer.

Other minerals have also been in short supply. Italy does not figure among the principal European producers (scarcely an abundant continent) of iron ore, bauxite, or indeed any major industrial mineral. The modest original reserves of such materials were largely used to create the splendors of the "golden age" of the great Renaissance cities. Only sulphur, traditionally mined on Sicily, and the famous marbles of Carrara, known the world over as fine building stone, can be said to be genuinely profitable minerals in the modern Italian economy.

The lack of an adequate mineral base and the fragmentation of the country until the 1860s into a variety of ministates and colonial holdings, few of which received any attention as far as industry was concerned, meant that Italy was very slow to industrialize. As late as the 1920s, most of the Italian industrial labor force was still employed in such low-technology activities as textile manufacturing and food processing: much the same pattern shown by many third-world nations today.

A heavy industrial base began to take form, however, stimulated by the demands of two world wars and the late Italian drive for a colonial empire; and when in the 1950s Italy's industrial economy began to rebound from wartime destruction, its growth was spectacular. Government influence has played a major role in determining the location and direction of industrial growth, while modern Italian industry has not only been quick to adopt (and often improve upon) the latest technology, but has also added elements of style and good design to its products and has learned to market them aggressively, thus becoming one of Europe's more successful manufacturing economies.

The key region for Italian industry is, predictably, in the north; specifically in an area bounded by Milan, Turin, and Genoa, and known generally as the "industrial triangle." The dynamics of this region, its proximity to the significant sections of economic Europe beyond the Alps, and its comparatively long history (for Italy) of prosperity and industrial growth, make it one of Europe's primary industrial zones. At one pole of the triangle, Milan, the great rival of Rome among all Italian cities, concentrates the financial, cultural, and even political decisions of the nation to a remarkable degree. Farther to the west, on a classic water-power site at the extreme end of the Po valley, Turin is a powerful center of engineering industries, including notably the immense Fiat automobile empire. The third point of the triangle is represented by Genoa, the requisite port city for any large-scale European industrial region, which also boasts a steel plant and all of the import-processing activities so commonly found in or near Europe's major ports. A virtually ideal region for additional industrial location, this triangle continues to expand and to consolidate its role as Italy's premier economic region. Many of the smaller cities within the triangle are in fact among the fastest growing in the country.

Industrial development in the central part of Italy has also been impressive, although less important than that of the north. Probably no European city has gone through more spectacular changes of fortune than Rome, which was the center of the world's greatest empire 2,000 years ago, then a malarial swamp of fewer than twenty thousand people by the end of the Middle Ages, and since reunification the rapidly expanding capital of the new nation. As a result of its size and influence—the "capital city" effect—Rome has attracted a large assortment of industries producing consumer goods and technologically sophisticated products.

Many other of the classical cities of the Renaissance have also acquired a modest industrial base; but these have often been handicapped by locations (away from the coast, not on navigable streams, etc.) that reduce their attractiveness as manufacturing sites. Many thus rely on artistic and artisanal productions of very high quality, which have a worldwide reputation, and which also enhance the tourism vocation of these centers.

PERCEPTION: THE MEZZOGIORNO AND ITS PROBLEMS

The south of Italy—roughly the regions south of Rome, plus the islands of Sicily and Sardinia—has been for centuries one of the most chronically depressed parts of Europe. The area is commonly referred to as the Mezzogiorno, which translates briefly as "midday," and is often more romantically rendered "land of the noonday sun." There is little romantic about many parts of the south, however, and the region epitomizes the difficulties of the Mediterranean zone and the poverty endemic to so much of the basin. The problems of the Mezzogiorno may be briefly summarized as follows:

1. **Climate.** The Mediterranean climate needs no introduction. Its dry summers pose problems for humans across the region. In the lower latitudes of southern Italy, however, as in Greece and North Africa, the drought bites more deeply, the summer temperatures are higher, and without irrigation human economic options become more restricted.

2. **Population pressures.** Population can only be evaluated against the resource base which supports it. Family sizes have traditionally been large in southern Italy, however, and the limited land base has rarely been able to provide reliable sustenance, let alone abundance, for the numbers involved.

3. **Environmental degradation.** Faced with the pressure of numbers, human groups in the south have deforested most of the landscape for fuel and building materials (this pattern was noted even in ancient Roman times), and have increased the numbers of sheep and goats in an effort to augment income. Deforestation plus overgrazing plus steep slopes have contributed to erosion on a large scale, further diminishing the fragile land base.

4. **Absentee ownership.** Until reunification in the 1860s, the south was largely controlled by absentee proprietors. Large landowners had little interest (there were of course exceptions) in the quality of life of their tenants or in the future of their holdings, so long as the rents were paid. The inevitable result was a further decline in land quality.

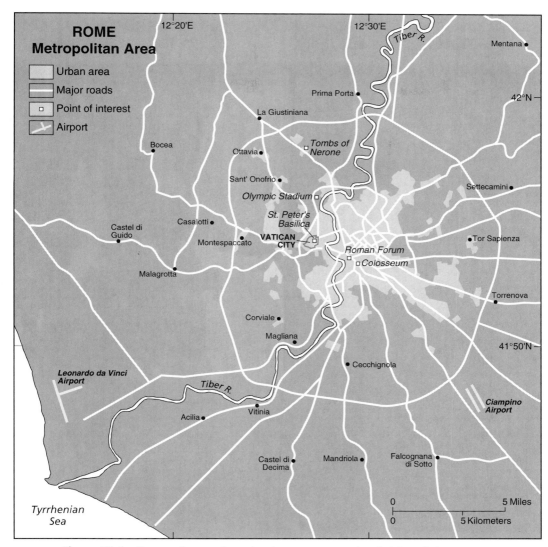

Figure 21.5 Rome rediscovers its ancient importance as modern Italy's capital.

5. **Lack of resources.** As has been noted, the south has never enjoyed an abundant natural resource which might have stimulated local development. In particular, it has been a deficit region in terms of energy supplies, especially after the trees were removed. A special problem posed by the Mediterranean climate is the absence of perennial streams; thus the possibilities of hydroelectric energy—the salvation of many alpine regions—are drastically reduced.

6. **Political realities.** Without an important tax base, the south has always lacked the political influence at Rome that might be expected to ameliorate its condition. Moreover, southern politics are invariably conservative (communist influence in Italy is a phenomenon of the industrial north); and a docile electorate has rarely demanded better treatment.

Figure 21.6 The Mediterranean environment poses special problems for southern Italy: Landscape on Sardinia.

7. **Social problems.** As a culminating problem, the south is also home to the Mafia (largely Sicilian) and other organizations associated with international crime and local terrorism. Originally self-help associations designed to help neighbors cope with the difficulties of life in the region, these groups have evolved into powerful clans more interested in private profit through criminal activity than in welfare. Beyond the bad reputation brought to the south by these associations, their influence within the national justice system produces a never-ending drain on scarce financial and personnel resources.

Faced with this staggering array of problems, the Italian nation has struggled since unity to find a solution that might enable the south to close the gap with the rest of the country (most economic indicators are at levels little more than half those of the north). Among the events and suggestions are the following:

1. **Emigration.** This has been the region of massive Italian emigration, first to the New World around the turn of the twentieth century, later to the post-World War II jobs in northern Europe and, eventually, northern Italy. Emigration, if a necessity, is however always a sad solution to the pressures of population brought on by economic inadequacy.

2. **Land reclamation and land reform.** Breaking up large estates was a popular program, but the poor quality of the land base is not improved by replacing a few large farms with many small ones; in fact, this may even have increased erosion. Abandonment of land, coupled with large reforestation programs, is an

attractive option for rebuilding the south; but turning the whole region into a "national park" poses one crucial question: what to do with the entire population during the century or so until moderate vegetation cover can be restored?

3. **Investment programs.** The south has been the recipient region for a massive program of investment from various sources. Since the 1950s, the Italian Cassa per il Mezzogiorno (Fund for the South) has invested large sums, primarily in infrastructure (roads, water supply systems, etc.). Internationally, the Mezzogiorno is by far the principal beneficiary of the regional development fund established by the European Union. Without an economic motor to stimulate growth and attract investment, however, this improvement has had a limited impact.

4. **Forced investment.** The modern Italian government has enjoyed a substantial degree of control over the basic economic structure of the country, especially before the current trend toward "privatization" became notable. This has enabled investment decisions to be made that will address social, as well as economic, goals. Thus, for many years, 40% of all new investment in industries controlled by the government was required by law to be in the south. This meant, for example, that the state-controlled steel industry, seeking to expand its well-located plant at Genoa, was forced also to open a new steel mill at Taranto, in the south. Similarly, oil refining and petrochemical industries have been established by decree near Bari, on the Adriatic coast.

This investment under legal duress has had mixed results. On one hand, some projects (notably the complex at Bari) have been reasonably successful. On the other, industrial decision-makers faced with this imperative have often opted to site only assembly plants in the south, retaining technological innovation, research and development, and other "cutting edge" operations in the better-located north. The result has been numerous new facilities which actually employ few southerners and have little lasting positive impact on the local economy: Italians have called these "cathedrals in the desert."

SAN MARINO AND THE VATICAN

Embedded within the Italian nation are two more of Europe's "mini-states": San Marino and the Vatican City (see Perception, Chapter 17).

Claiming independence from as early as the fourth century, tiny San Marino may in fact be Europe's oldest country. Clinging to 24 square miles (62 square kilometers) at the top of Mt. Titano, (2,424 feet/739 meters) near the Adriatic coast of Italy, the 23,000 inhabitants of this small republic supplement income from marginal farming with revenues provided by tourism, postage stamps, and a few light manufactures.

For the Sanmarinese, the small size of their country has more often been seen as an advantage than a handicap. Offered more land by Napoleon early in the nineteenth century, they refused, calculating—probably correctly—that more territory and a larger role in regional affairs would only make them more vulnerable to takeover by more powerful outside forces. Following the lead of Liechtenstein, San Marino has also recently (1992) become a member of the United Nations.

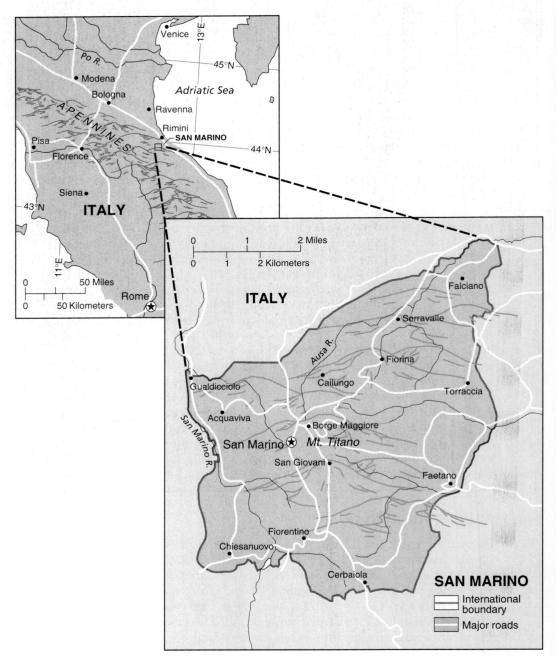

Figure 21.7 The mini-state of San Marino.

Europe's (and the world's) smallest independent political entity is the Vatican City in Rome, seat of the Roman Catholic church. (The Sovereign Military Order of Malta, which occupies a single-walled building on three acres in the center of Rome, is sometimes considered an independent state, although its 9,600 "citizens" actually retain other nationalities and most live elsewhere).

Reflecting the great temporal holdings (as well as worldwide spiritual authority) of the church, the 108.7 acres of the modern political unit (see Figure 21.5) were established in 1929 after a long period of difficult relations between the church and the modern Italian nation. The Pope remains head of state, and the population of about 1,000 is largely concerned with church business; tourism is also a signficant economic support.

ADDITIONAL READING: ITALY

BETHEMONT, J., and J. PELLETIER, *Italy: A Geographical Introduction,* tr. E. Kofman, ed. R. King. London and New York: Longman, 1983.

COLE, J.P., *Italy; An Introductory Geography.* New York: Praeger, 1966.

KING, R., *The Industrial Geography of Italy.* New York: St. Martin's, 1985.

LOUGHLIN, J., ed., *Southern European Studies Guide.* New Providence, NJ.: K.G. Saur, 1993.

SMITH, C.D., *Western Mediterranean Europe: A Historical Geography of Italy, Spain, and Southern France since the Neolithic.* New York: Academic Press, 1979.

WILLIAMS, A., ed. *Southern Europe Transformed: Political and Economic Change in Greece, Italy, Portugal, and Spain.* London and New York: Harper and Row, 1984.

DISCUSSION QUESTIONS

1. What is the Mezzogiorno? Why does its economic performance lag so far behind the rest of Italy? What are "cathedrals in the desert"?
2. What are the poles of the "industrial triangle"? Why is this region so well suited to substantial industrial development?
3. Ancient Rome (population probably at least one million) declined to a small town of some 17,000 by the fourteenth century. Since being made capital of a reunified Italy in the 1860s, it has become a city of about three million. What does this tell us about the role of political influence on the growth of cities?
4. Damage to Florence during the Arno flood of 1962 revealed to the world the vast amount of art and antiquities that is badly cared for in Italy. How much of the national budget should the country devote to preservation and restoration efforts?
5. The Po is a slow, meandering river much given to flooding and of little value for navigation. Why? How far does the river fall over its course from Turin to the Adriatic?

GREECE AND THE MEDITERRANEAN ISLANDS

At the eastern end of the Mediterranean, where Europe blends into the Middle East, one important country of great historical significance and two small island nations complete to the southeast the spectrum of Europe's major political units. As noted in the Introduction, the small part of Turkey that extends west of the straits between the Mediterranean and Black Seas is sometimes considered geographically "European," but is more logically included with its larger "Asian" portion.

These nations summarize in many ways the character of the Mediterranean, in terms both of physical characteristics and of cultural and historical complexity. Moreover, their location on the outer fringes of Europe, far from the economic heartland of the continent, has meant that their development has been slow and difficult. Often drawn as much toward Asia or Africa as toward Europe, they find themselves caught in political turmoil and search for their own European identity as well as for economic success.

GREECE

The major country is of course Greece, the cradle of western civilization, which has survived long periods of decline and foreign control to emerge today as a western-oriented nation fully conscious of both its heritage and its possibilities. Greece is primarily a mountainous nation, the ranges of eastern Europe, notably the Pindos range, extending southward as fingers across the mainland, ending in the Peloponnesian peninsula. Snowcapped Mt. Olympus, traditional home of the gods, rises 9,550 feet (2,911 meters) near the coast of the Aegean Sea.

It is also a country of islands, hundreds of them (some very famous, such as Crete, Rhodes, Mykonos, or Lesbos) being scattered through the Ionian and Aegean Seas, arms of the Mediterranean which separate Greece from Italy to the west and Turkey to the east. Many of these islands are within sight of the Turkish shore, reflecting the long history of Greek colonization on what is now the Turkish mainland. Flat land is rare, the plain of Thessaly and the valley of the Vardar river, both in the northwestern part of the country, being the only major areas of plain.

The climate is characteristically Mediterranean. Moving eastward and southward through the basin, high pressure from North Africa lingers longer and the Atlantic storms are farther away; this results in a summer drought of fully four months. Temperatures are also higher than in most other European parts of the basin, with summer maxima frequently exceeding 90 °F (33 °C) at Athens and other sea level stations. In the highlands, temperatures are cooler and seasonal precipitation more abundant.

Greece is particularly concerned about the possible implications of the "greenhouse effect" (see climate discussion, Chapter 2). If average temperatures are indeed becoming warmer, the already marginal rainfall over much of the country may further diminish, while increased evaporation will leave even less water for agriculture and other uses. Irrigation of crops, using stored groundwater that may not easily be replaced, has been increasing in recent years. In Athens, higher temperatures and some of Europe's most polluted air have resulted in a dramatic increase in heat stress-related deaths; entrepreneurs have found a ready market for small containers of bottled oxygen for individual use!

Despite the glories of its ancient culture, modern Greece has always been forced to struggle to retain (or regain) its independence and to achieve even a modest level of prosperity. Emigration has been a familiar necessity ("Greece's most important export," runs the saying, "is Greeks"), and a subsistence agriculture has been supplemented by remittances from Greeks living and working abroad.

Improvements in agricultural production, especially of cereals and the off-season fruits and vegetables in demand in northern and western Europe, have been a major boost for the economy. Moreover, Greece has made use of the relatively low cost of its labor to achieve a modest level of industrialization, and the country's merchant marine is one of the world's largest. Some minerals, particularly bauxite, are exploited.

Figure 21.8 Greece.

Membership in the European Union, which Greece attained in 1981, should in theory provide an important developmental stimulus; but Greece has become tangled in political and economic disagreements with its Union partners, as well as with its traditional enemy: Turkey (a member, as Greece, of NATO). Following the collapse of Yugoslavia and the formation of several independent successor states, Greece has also quarreled with the new nation of Macedonia on its northern frontier, suspecting attempts by that country to expand into the Greek portion of the Macedonian culture area. As various 1997–99 deadlines for greater integration of the Union nations, as well as for increased intra-Union economic competition among the member states, approach, the Greek economy and many of its political institutions look uncomfortably weak and vulnerable.

As throughout most of the Mediterranean, tourism is a vital support of the Greek economy. Farther from the center of gravity of the tourist-providing nations than is Spain or Italy, Greece nonetheless receives an impressive flow of visitors each year and, since Greeks traveling abroad are much less numerous, has one of Europe's highest net incomes from tourism. The long season of warmth and sunshine is of

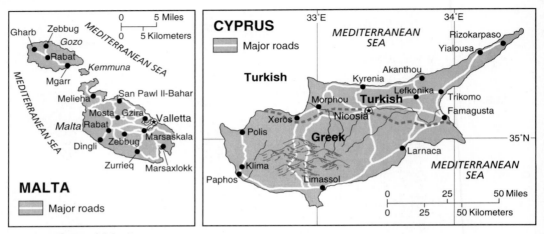

Figure 21.9 Malta and Cyprus.

course a major attraction, and excursions or cruises to the many islands are a popular tourism activity. Beyond this, however, the visible glories of ancient Greek civilizations, especially in architecture and sculpture, have been magnets to visitors for centuries, and continue to exercise their spell today.

MALTA AND CYPRUS

Among the many islands of the eastern Mediterranean, two have become independent nations after having formed part of the far-flung British empire: Malta and Cyprus. Both seek to play a key role as links among Europe, North Africa, and the Middle East in this troubled part of the world, but both are beset by a variety of economic and political problems. To the west, Malta occupies a strategic location south of Sicily and north of Africa. Ruled successively by Phoenicians, Romans, Arabs, Normans, the Knights of Malta, France and Britain, the island has been independent since 1964. With the final closing of British naval installations in 1979, Malta's economy has become heavily dependent on tourism to balance its books. Its British connections and widespread use of the English language serve to attract many visitors from Britain, as well as many retirees seeking a longer season in the sun. Agriculture is marginal, and only a modest textile industry and other labor-intensive manufactures produce some export income to balance the needed volume of imports. Nonetheless, the Maltese economy has expanded steadily during the 1990s, and unemployment is low. The country has applied for full membership in the European Union, of which it is now an associate member. Despite its population of only some 368,000 (Valetta, the capital, is a city of about 16,000), Malta's strategic position in the Mediterranean gives it an unusual geopolitical importance.

Figure 21.10 Greece, Malta, Cyprus: Basic data.

Country	Area 1000 km²/mi²	Population million, 1995	GNP $/capita	Capital	Population million, est.
Cyprus	9.3/3.6	0.7	10,380	Nicosia	.18
Greece	131.9/51.0	10.5	7,710	Athens	2.56 (metro)
Malta	.32/.12	0.4	7,970	Valletta	.014

Source: U.S. Department of State, Population Reference Bureau

Italy, Greece, and the Mediterranean Islands 317

Figure 21.11 In mountainous Greece, life changes only slowly: Monastery at Meteora.

To the east, Cyprus lies off the Turkish coast, not far from Syria and Lebanon. A reasonable agricultural base, light industry, a number of small but useful mineral deposits, and the potential for a substantial tourism industry should have laid the foundation for a moderately successful economy when Cyprus became independent in 1960. However, bitter clashes between the Greek majority and Turkish minority populations led to Turkish

Figure 21.12 Greece still appreciates its role as the cradle of western civilization: The antiquities of Delphi.

Figure 21.13   Tourism is a primary support of the Greek economy: Cruise ship at Itea.

military occupation of 40% of the island in 1974, to deployment of a United Nations peacekeeping force between the two sides, and to the unilateral (and internationally unrecognized) declaration of a "Turkish Republic of Northern Cyprus" in 1983.

In these circumstances, it is scarcely surprising that the 660,000 inhabitants of Cyprus (the capital, Nicosia, groups 125,000 of these) have been largely unable to

Figure 21.14   Malta, at the crossroads of the Mediterranean, has a rich history: The harbor of Valletta.

pursue their economic and social development in any meaningful way. Resolution of the impasse depends ultimately on some compromise between Greece and Turkey, whose relations have been embittered for centuries. Turkey, which is an associate member of the European Union and aspires to full membership, has seen its application coldly received by Greece, which views the withdrawal of Turkish forces from Cyprus (also anxious to gain EU membership) as a non-negotiable prerequisite to any regional solution and as a test of Turkey's resolution to move closer to Europe.

ADDITIONAL READING: GREECE AND THE MEDITERRANEAN ISLANDS

HARRINGTON, L., *Greece and the Greeks.* New York: T. Nelson, 1962.

LEONTIDOU, L., *The Mediterranean City in Transition: Social Change and Urban Development.* Cambridge and New York: Cambridge University Press, 1990.

MERRILLEES, R.S., *Alashia Revisited.* Paris: J. Gabalda, 1987. (Cahiers de la Revue Biblique, no. 22).

OGLETHORPE, M.K., *Maltese Development Issues.* Glasgow: Department of Geography, University of Glasgow, (Occasional Papers Series, no. 8), 1982.

TSOUKALIS, L., ed., *Greece and the European Community.* Farnborough: ENG., Saxon House, 1979.

DISCUSSION QUESTIONS

1. Ancient Greece was generally a collection of "city states" with little common cause. What role did the natural environment play in this cultural pattern?

2. Cyprus today is an island divided into hostile Greek and Turkish communities. How did this come about? What future do you see for the country?

3. Greece has been a full member of the European Union since 1981, yet the nation's economy continues to lag behind that of most other members. What problems does Greece face in terms of economic development?

4. Malta relies heavily on tourism to stabilize its economy. How have location and history combined to make the island a favored destination today?

5. Greece faces some of the most severe environmental problems of any European country. What are some of these? Why have they become exaggerated in many sections of the nation?

22
Iberia

At the extreme southwestern corner of the European continent lies the Iberian peninsula, which comprises the countries of Spain and Portugal (Andorra, high in the Pyrenees between France and Spain, is another of Europe's "mini-states"). The peninsula is largely a moderately high plateau, separated and to some extent sheltered from the rest of Europe by the Pyrenees and their westward extension: the Cantabrian mountains. Except in the northwest, Iberia is even more arid than its Mediterranean location would suggest, and this, combined with the long period of Moorish occupation and continued Iberian interest in Africa, creates both physical and cultural landscapes reminiscent of North Africa.

Spain and Portugal were, of course, the first nations to launch Europe into the world in the fifteenth and sixteenth centuries. The great empires they created and the wealth that returned to Europe in their galleons dramatically altered the course of the world's political, cultural, and economic history; but as imperial power gradually eroded away, these countries, with relatively few internal resources of value in the industrial age, entered into a long period of decline, culminating in fascist dictatorships over several decades of the twentieth century. Today, both Spain and Portugal are still in the process of rebuilding democratic institutions, while at the same time attempting to upgrade their economies to competitive European standards. Members since 1986 of the European Union, the Iberian nations are counting on this opportunity to stimulate growth and to some extent overcome the natural isolation that places them on the margins of the continent in both physical and economic senses. Spain, with its larger population and greater resource base, has made substantial economic progress in recent years, while Portugal, although expanding its economy rapidly, remains one of Europe's poorer nations.

PHYSICAL FEATURES

The Iberian plateau, a generally flat surface averaging some 3,000 feet (910 meters) in elevation and dissected here and there by rivers, is higher to the east, lower to the west. This means that nearly all of the important rivers of the peninsula flow across nearly the entire region to empty into the Atlantic (the Guadalquivir, Guadiana, Tagus or Tejo, and Duero or Douro are the best examples). Only the Ebro, draining the south face of the Pyrenees, flows into the Mediterranean.

On the north, the Pyrenees form an abrupt boundary with France, and with their westward extension, the Cantabrians, form a barrier that has extensive climatic, as well as cultural, implications. In the center of the peninsula, near Madrid, such modest mountain ranges as the Sierra de Guadarrama and the Sierra de Gredos, with peaks approaching 8,000 feet (2,450 meters), have climatic and historic significance, and serve to divide Spain into various internal regions. Farther south, the more imposing Sierra Nevada, where the peak of Mulhacen rises to 11,424 feet (3,482 meters), stands near the Mediterranean coast, creating spectacular settings for such cities as Granada and providing extra shelter and sunshine for the booming tourist resorts along the Costa del Sol.

To the west, the plateau is much lower and more dissected in Portugal, and much of the southern part of that country is under 1,000 feet (305 meters) in elevation. While flat land is abundant in Iberia, flat land with abundant water is not, and the few large, permanent rivers and their valleys have accordingly taken on unusual significance, concentrating much of both the traditional and modern Iberian economies.

CLIMATE

Climatically, much of the Iberian peninsula is of course Mediterranean in character, with mild winters and hot, dry summers. However, the physical characteristics of the region, especially the elevation and arrangement of mountain barriers, create some unusual conditions which are found nowhere else in Europe. The height of the plateau results in cooler temperatures, particularly during the winter months, than would be expected at this latitude on Europe's west coast. At the same time, the mountain barriers that run at right angles to the prevailing northwestern storm track produce rain-shadow effects that result in unusually dry climates. South of the Cantabrians, in Old Castile, and along the south coast, in the lee of the Sierra Nevada, even the winter months—so wet across most of Europe—are deficient in rainfall, producing the only steppe or semi-arid climates found west of Ukraine or the Middle East.

Outside the Cantabrian barrier, the region of Galicia enjoys the moderate temperatures and abundant precipitation characteristic of other west-coast locations, and is appropriately known as "green" Spain. Portugal, with its low, open facade to the west, receives substantial winter rainfall, but because of its southerly latitude is very dry during the summer months except in the northern districts, which are much like neighboring Galicia.

Iberia

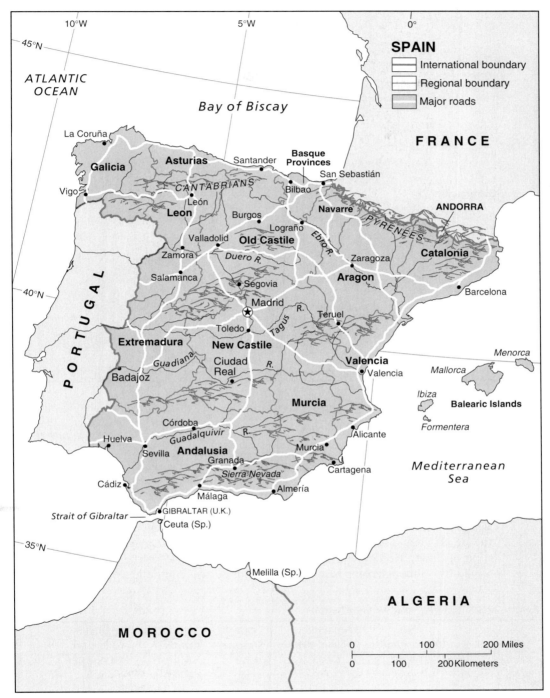

Figure 22.1 Spain.

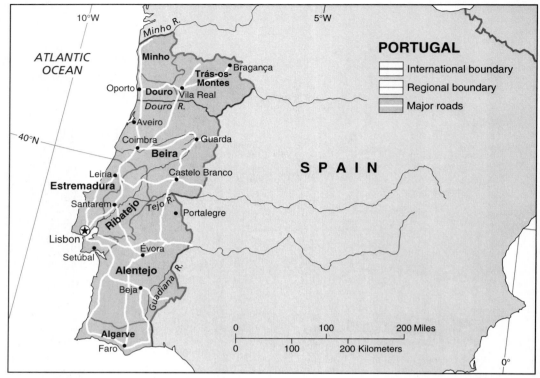

Figure 22.2 Portugal.

CULTURAL BACKGROUND

At the key location where Europe nearly touches Africa and is closest to Latin America, Iberia is the product of the numerous groups that have passed this way—some to stay, others to move on—and was the first region of Europe to move decisively into the "New World." Phoenicians, Greeks, and Carthaginians arrived from the Mediterranean, while Romans came overland in the second century BC and over the next 700 years brought the languages and basic structure of civilization still evident today. As elsewhere in Europe, declining Roman authority was replaced by that of "barbarians"—in this case Visigoths—although some Roman strongholds persisted into the seventh century AD.

Figure 22.3 Spain, Portugal, Andorra: Basic data.

Country	Area 1000 km²/mi²	Population million, 1995	GNP $/capita	Capital	Population million, est.
Portugal	94.3/36.4	9.9	9,370	Lisbon	2.1 (metro.)
Spain	507.6/196.0	39.3	13,280	Madrid	3.5 (metro.)
Andorra	.487/.188	.062	*	Andorra la Vella	.021

*No data calculated.
Source: U.S. Department of State, Population Reference Bureau.

Iberia

Figure 22.4 The interior of Spain is one of Europe's most arid regions: in Extremadura.

A decisive event was the Moorish invasion from North Africa in 711 AD. The Moors brought the Islamic faith (which did not long outlast their stay), numerous place names (which did), magnificent architecture, the techniques of irrigation, and countless other innovations: products of the golden age of Arabic culture. Universities were founded as well, and although the Iberians chafed under "infidel" rule, the judgement of history is that this was indeed a period of enlightenment and progress.

The "reconquest" of Christian Iberia from the Islamic occupation became a powerful cause. Over the next seven centuries, beginning from the northwestern part of the peninsula (which was never occupied by the Moors), the reconquest gradually brought together various kingdoms who often had little in common except their faith and, taking advantage of increasing dissention and weakness among the Moors, pushed steadily southward. The final Moorish stronghold (Granada) fell in 1492, by coincidence the same year as Columbus' first voyage to America. Or was it purely coincidental? Some historians have seen continuity between the fervor of the reconquest and the dramatic thrust into the unknown that launched the "Age of Discovery."

Portugal became independent from Spain in 1140, after a nine-year rebellion. Lisbon was retaken from the Moors in 1147, and by 1249 the present boundaries of the nation had been generally secured. Both countries then began the processes of exploration and exploitation which were to make them the world's first great colonial powers and, during the sixteenth century, Europe's richest nations. Spanish control of the precious metals found in the Americas (notably the silver of the Andes) in fact provided the basis of much of Europe's prosperity, even though Spain itself was continually in debt as a result of nearly continuous wars and the cost of maintaining its vast overseas establishments.

With the rise of the political, military, and economic power of the northern European nations, notably Britain, France, and the Netherlands, the failure of both countries to keep pace with the industrial revolution, and the increasing relevance of their

Figure 22.5 Moorish architecture was a distinctive contribution to the Iberian landscape: Granada, Spain.

geographic isolation on the European fringe, Spain and Portugal entered a long period of decline. This involved the gradual loss of all major overseas colonies and culminated in the political and economic stagnation of long periods of fascist dictatorship (1939–1975 for Spain, 1932–1974 for Portugal). Since the mid-1970s the pace of national life has quickened in both countries, and membership (1986) in the European Union has served as a useful stimulus toward needed economic reforms.

As nations assembled from various groups during periods of strongly centralized authority, both Portugal and (especially) Spain became patchwork countries with substantial regional differences. During a long period of international peace and economic growth, these regionalist distinctions have become more marked, and various forms of separatism, ranging from pressure for greater regional autonomy to calls for outright independence, challenge the national governments. In Portugal, the offshore island groups of Madeira and the Azores have received the status of "autonomous regions." In Spain, such determinedly nationalistic regions as Catalonia, Galicia, and the Basque country, with distinctive linguistic, geographic, and historical personalities, have elected regional parliaments and gained substantial control over local matters. In fact, each of the nation's seventeen regions has some sort of "autonomy statute," as the central government attempts to devolve authority while keeping the national structure intact.

ECONOMIC ACTIVITIES

Economically, Spain and especially Portugal are among the poorer European countries, a result of their lack of internal resources, difficult environment, relative isolation from the economic heart of Europe, and long years of political repression and cultural stagnation. Agriculturally, both nations produce a wide variety of Mediterranean crops, notably wheat, wine, and olives. Portugal is the world's leading producer of cork, which is the protective coating stripped from the trunk of a characteristic species of Mediterranean oak. The large areas of semiarid climate are best suited to grazing animals; thus meat and leather products are also important. Portugal, in fact, has one of the very few European economies maintained largely by the export of agricultural products and other raw materials: a pattern more commonly associated with the less developed countries of other regions of the world.

In the more humid parts of northwestern Spain, conditions are suitable for maize, and this crop has been the food base of Galicia since its original introduction into Europe. Over the rest of the peninsula, permanent sources of water—notably the major river valleys—have long been intensively cultivated. The Islamic invaders of the eighth century brought with them the techniques of irrigation, as well as such now-familiar crops as citrus fruits (Valencia gives its name to one famous variety of orange); and these fertile, well-watered valleys, or *huertas,* became the focal points of urban, cultural, and later industrial growth, a role they still play today.

In terms of resources, neither of the Iberian countries is well provided, even by modest European standards. Portugal is especially poor in minerals, producing only insignificant amounts of coal and iron ore, plus some tin and uranium. Importing virtually all its requirements, particularly including oil, is a major drain on the Portuguese economy. Spain is somewhat better off, having a moderate amount of coal and lignite, but oil production remains negligible, despite offshore exploration in the Mediterranean. Metals are more significant in Spain, however. The country is Europe's leading producer of silver and zinc, and is second in the production of lead and third in that of iron ore. Although in none of these commodities is Spain a leader on the world scale, the famous mercury mines at Almaden remain one of the few sources of that mineral.

Figure 22.6 Portugal remains one of Europe's most traditional nations: Evora.

The major industrial areas of Spain are found in the north. The city of Barcelona, which considers itself Catalan, rather than Spanish, is the country's most industrialized, "Europeanized" and prosperous urban center, with a wide variety of engineering and port-associated industries. On the north coast, at Bilbao and Oviedo, in the Basque region, coal mining has been the basis of the major concentration of heavy industry, including steelmaking. Elsewhere, the rapid growth of Madrid in its role as the national capital has created a large market for many types of "capital city" light manufactured products, while many of the huerta cities have become significant centers for the production of consumer goods. Industry in Portugal is heavily concentrated around the two major urban centers, Lisbon and Oporto, and is mainly concerned with consumer goods for domestic consumption.

No discussion of the economic picture of Iberia is complete without some consideration of the role of tourism. From the late 1960s through the 1980s, the tourism industry grew more rapidly in Spain than in any other European country, and is now the main support of large parts of the nation. In 1989, in fact, Spain at least temporarily surpassed the United States as the nation with the greatest gross income from tourism.

The main attraction is the warmth, color, and beaches of the Mediterranean, and such favored stretches of shoreline as the Costa del Sol have become almost solidly built up with hotels, condominiums, and other tourism infrastructure. The sunshine and relatively low cost of the Spanish coast attract visitors from the cooler, wetter parts of Europe in large numbers, and the contrast between the internationally-flavored resort towns by the sea and the barren landscapes and sleepy villages a few miles inland is striking. Realizing that the profits from mass tourism are often marginal, especially given the environmental and cultural stresses involved, Spain in the 1990s has been increasingly concerned with promoting "upscale" tourism to attract more affluent visitors. Portugal and the interior of Spain share in this tourism boom, but to a more moderate degree; the southern Portuguese region of the Algarve has profited most from these new riches.

Figure 22.7 The Balearic Islands are one of Spain's most important tourism regions: on Menorca.

Iberia

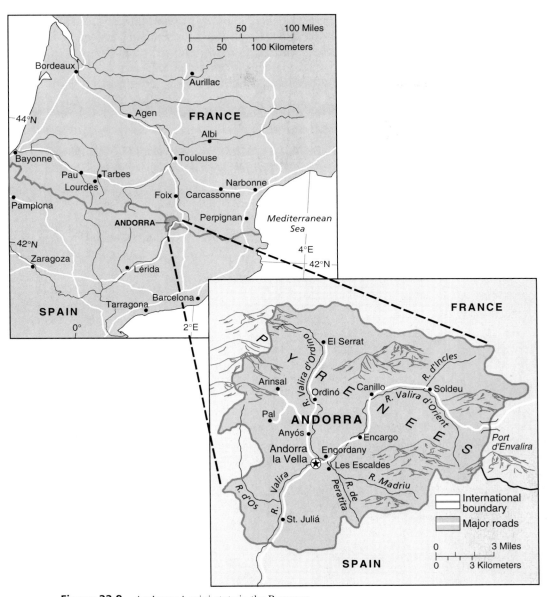

Figure 22.8 Andorra: A mini-state in the Pyrenees.

ANDORRA

The fifth and final European "mini-state," as Liechtenstein a mountainous enclave, Andorra occupies a corner of the high Pyrenees between France and Spain. Its 188 square miles (487 square kilometers) is generally rugged, except for small river valleys draining toward Spain. Governed jointly by the ruling power of France (now the President of the Republic) and the Spanish Bishop of Urguel since 1278, Andorra's

48,000 people exist on an unpromising land base with few resources. The country long had a reputation as a haven for smugglers moving contraband between France and Spain (or vice versa); but today it profits more from tourism, and from the duty-free shopping available to visitors. Andorra is also proud of its recent (1993) acceptance to membership in the United Nations.

ADDITIONAL READING

FISHER, W.B., and H. BOWEN-JONES, *Spain: An Introductory Geography.* New York: Praeger, 1966.

GOMEZ-IBANEZ, D., *The Western Pyrenees: Differential Evolution of the French and Spanish Borderland.* Oxford: Clarendon Press, 1975.

HUDSON, R., and J. LEWIS, *Uneven Development in Southern Europe.* London: Methuen, 1985.

STANISLAWSKI, D., *The Identity of Portugal.* Austin, TX: University of Texas, 1959.

STANISLAWSKI, D., *Portugal's Other Kingdom: The Algarve.* Austin, TX: University of Texas, 1963.

DISCUSSION QUESTIONS

1. The year 1492 AD marks the date both of Columbus' first landfall in the New World and the Spanish capture of the final Moorish stronghold (Granada) in Spain. Is there any connection between the two events?

2. Spain and Portugal became the first great colonial powers, and were among the richest European nations in the sixteenth and seventeenth centuries. Later, however, they entered a period of steep decline. Why were they unable to keep pace with subsequent European development?

3. Spain and Portugal became members of the European Union in 1986, after long and difficult negotiations. Why were the other members reluctant to admit the Iberians?

4. Regional feeling is especially strong in Spain, where several regions have distinctive languages and their own parliaments. Why are centrifugal forces so strong in this country?

5. Spain has become Europe's most important tourism country (based on net income). What special advantages does the country possess for tourism? Why is mass tourism a mixed blessing for Spain as well as other Mediterranean nations?

23
Eastern E[urope]

Of all the regional subdivi[sions...] most difficult to summarize a[nd...] comprise it have little in common, and share only a history of po[litical and] cultural fragmentation, and in most cases economic backwardness as compared to that of the Western European nations.

Eastern Europe is, in fact, what political geographers term a "shatter belt," that is, an area where the interests of strong nations often clash, resulting in political and cultural insecurity and economic uncertainty, particularly for the peoples caught in the middle. Thus, Germanic power to the west, Slavic power to the east, and Islamic power to the south have converged on Eastern Europe for many centuries, with the result that viable independent states have rarely emerged, economic growth has been erratic, and stability and prosperity have been largely unknown. The Middle East and southeast Asia are similar "shatter belts," and it is no coincidence that much of the modern world's turmoil and many of its wars have their origins in these areas, caught as they are in the conflict of big-power interests.

All parts of Eastern Europe have been subjected to these pressures. In the north, Poland has had as troubled a history as any nation, often being partitioned among its neighbors and vanishing from the map for centuries, or seeing its boundaries arbitrarily shifted first eastward, then westward (as was the case during World War II). Nearly all of southeastern Europe was subjugated by the Ottoman Turks from the sixteenth until the nineteenth or even twentieth century, and the economic backwardness that deepened as that empire fell apart created a development gap with the rest of Europe which has still not been closed.

Czechoslovakia and Yugoslavia were patchwork countries created after World War I by the victorious allies, largely from the pieces of the collapsed Austro-Hungarian empire. The hopeful idea was to create viable states that could better resist the unwelcome attentions of their powerful neighbors; but World War II dragged the entire area to disaster again.

... is referred to in terms of its physical geography as the ... important local mountain range), and the term "balkanize" ... European languages. It is a verb implying division into small, ... nits; and it sums up very well the endemic problems faced by the ... gion.

... ose of World War II, most of this area was occupied by the armies of ... nion. To promote expansion of the communist philosophy, to acquire ... arkets and sources of raw materials and industrial products, and above all ... e a buffer zone of friendly nations between the Soviet Union and potential-... eatening countries to the west (from which Russia had been invaded three times ... 130 years, counting only 'major' wars), the Soviets installed puppet governments in all the east European countries they occupied, including their sector of defeated Germany. Moreover, postwar rearrangement of boundaries gave the Soviet Union a common land frontier with Poland, Czechoslovakia, Hungary, and Romania; thus armed force could be used directly to control any threat of political instability, as for example in Hungary in 1956 or Czechoslovakia in 1968.

One of the common features of most modern Eastern European countries, therefore, was until late 1989 a communist government and close political, military, and economic ties to the Soviet Union. East Germany, Poland, Czechoslovakia, Hungary, Romania, and Bulgaria were firmly in the Soviet orbit. Only Yugoslavia, which had evolved its own unique model of socialism and a more open society, and little Albania, which has virtually been cut off from the world until recently by a rigid communist regime, were able to elude the Soviet net. It is undoubtedly significant that neither of these nations had a common frontier with the Soviet Union.

Similarity in political system, however, did not mean that the East European nations were only carbon-copy satellites of the Soviet Union. The east European peoples, with their own languages, cultures, and historical traditions, were not generally fond of the heavy-handed Russians (the Bulgarians were to some extent an exception), and sought in various ways to assert their national personalities and behave as independently as their difficult conditions would allow. It gradually became clear to them, for example, that while the Soviet Union was insistent on political conformity (to the point of enforcing it with armed might), it was much more tolerant of economic "deviation." By the mid-1980s, well before the collapse of their authority in the region, the Soviets appeared to recognize the advantages of these nations maintaining strong economies, which would both enhance political stability and reduce the burden of support on an increasingly stressed Soviet economy.

Accordingly, a wide range of innovative economic practices, some of them suspiciously capitalistic, began tentatively springing up in Eastern Europe. Increasing trade with the west, the spread of decision-making to local regions and even to worker groups, and the encouragement of entreprenurial investment for profit were among the non-Marxist economic strategies which had begun to emerge with the grumbling consent of the Soviet Union by the late 1980s.

While different East European nations innovated in various ways (or in some cases hardly at all), it was probably Hungary that went the farthest toward exploring the limits of Soviet toleration. Since the flowering of renewed independence across

Eastern Europe

Figure 23.1 Poland.

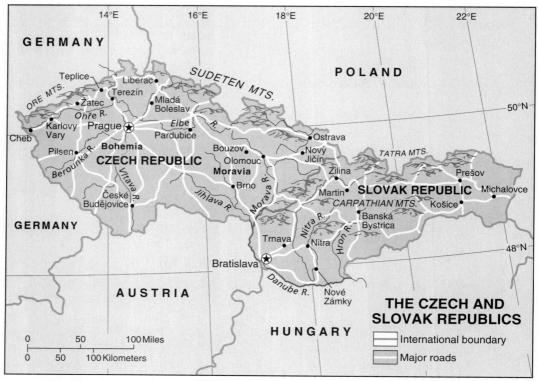

Figure 23.2 The Czech and Slovak republics.

the region, the Hungarians have by most accounts moved as rapidly as any nation to restructure their economic and political framework, and many observers have attributed this to their previous experiences as "reform communists."

In October 1989, the entire political and economic structure of all Eastern Europe began to change with extraordinary speed and dramatic effect. The increasingly evident inability of the centralized planning process to produce a reasonable rate of economic expansion, tangible evidence of an improving quality of life, and a level of competitiveness adequate for the modern, interrelated world began to create heavy pressures for economic reform, particularly in light of surprisingly strong Soviet support for such measures. With its own economy facing similar problems, the USSR was increasingly unwilling and/or unable to prop up its satellites through low-cost sales of energy and other evidences of "socialist solidarity."

Adding to the Eastern European nations' discontent in many cases was a growing awareness that the tightly controlled communist regimes in power since the 1940s had become complacent, corrupt, and generally out of touch with the needs and aspirations of average citizens. Finally, it became evident that the new wind blowing from the Soviet Union had swept away the certainty that "big brother" would intervene militarily to maintain the status quo, as had been the comfortable case in 1956 and 1968. Without at least the implication of forceful imposition of communist hegemony, the governments of Eastern Europe suddenly found themselves reduced to their own resources and accountable to their citizenry.

Figure 23.3 Eastern Europe: Basic data.

Country	Area 1000 km²/mi²	Population million, 1995	GNP $/capita	Capital	Population million, est.
Albania	28.5/11.1	3.3	360	Tirane	.27
Bosnia-Herzegovina	51.1/19.7	3.6	*	Sarajevo	.53
Bulgaria	111.0/44.4	8.4	1,160	Sofia	1.1
Croatia	56.5/21.8	4.4	2,530	Zagreb	1.0
Czech Republic	78.9/30.5	10.3	3,210	Prague	1.2
Hungary	93.0/35.9	10.2	3,840	Budapest	2.0
Macedonia	25.7/9.9	2.1	790	Skopje	.56
Poland	312.6/120.7	38.6	2,470	Warsaw	1.7
Romania	237.5/91.7	22.6	1,230	Bucharest	2.4
Slovakia	49.0/18.9	5.4	2,230	Bratislava	.44
Slovenia	20.3/7.8	2.0	7,140	Ljubljana	.28
Yugoslavia	102.2/39.4	10.2	*	Belgrade	1.5
European nations from the former Soviet Union: (see also Baltic nations, Chapter 15).					
Belarus	207.6/80.2	10.3	2,160	Mensk (Minsk)	1.7
Moldova	33.7/13.0	4.3	870	Chisinau (Kishniev)	.68
Russia	17,075/6,593	147.7	2,650	Moscow	8.8
Ukraine	603.7/233.0	51.1	1,570	Kiev	2.7

*Not calculated as a result of political instability.
Sources: U.S. Department of State, PlanEcon, *The Economist,* Population Reference Bureau.

Many in power appeared to believe that economic reform—partial steps toward a market economy and greater responsibility for their actions on the part of inefficient public enterprises, for example—could be accomplished within the framework of the existing communist political structure (the same belief that later led to the collapse of communist rule in the Soviet Union). However, it soon became evident that economic reform and political reform are two sides of the same coin (not too surprising, perhaps, given the fact that Marxist politics is inevitably founded on Marxist economics) and pressures for political openness and plurality escalated rapidly. Forced by public opinion, international reaction (including "betrayal" by the Soviet Union), mass rallies, strikes, and even violence to abandon the "leading role" written into their post-1945 constitutions, the communist parties of Eastern Europe were largely swept into opposition and forced to appeal to the voters on the basis of their ideas and personal competence, rather than relying on the military hardware of their powerful friends.

Not surprisingly, Hungary, with its long tradition of dogged opposition to Soviet authority and its knack for economic inventiveness, and Poland, whose economy was in total disarray and whose citizens have suffered most from their geopolitical position between Germany and Russia, were the first East European nations to test the limits of Soviet political tolerance. East Germany and later Czechoslovakia followed the process weeks later, and even Bulgaria—traditionally close to Russia—forced out the old guard and began very gradually the process of reformation.

Finally, in the most violent and dangerous confrontation, Romanians, supported by their own disaffected army, overthrew a brutal, corrupt, and repressive regime that had employed an intricate secret police network to maintain order in the best Stalinist style.

At the beginning of 1991, only tiny, isolated Albania retained, at least for the moment, the absolutism of a communist dictatorship; and even here there was abundant evidence of new popular pressure for change that eventually led to political reform.

The last three months of 1989 were without question the most dramatic period in Eastern Europe since World War II and the subsequent establishment of communist regimes. In the excitement of reopening societies and political reform, however, substantial problems remain unsolved and many questions unanswered. Economic difficulties, exacerbated by the heavy hand of centralized planning, but in many cases linked to more fundamental geographic realities (inadequate resources, isolation from the main currents of European growth, weak transportation links and other infrastructural shortcomings), will not miraculously vanish in the dawn of a new political order. Capitalism, like socialism, has its price; and a long period of austerity, unemployment, and slow economic growth may well test the resolve of these populations before a genuine competitiveness can begin to produce living standards rivalling those of the west. In fact, by 1995, "reformed" communist parties were again competing for political respectability and election. This process (also evident in Russia and other former Republics of the Soviet Union) reflects partly nostalgia for the "good old days" of centralized stability; partly the realization that the Communists, as the only political entity for fifty years, at least know how to run governments effectively; and partly a reaction to the economic pain of necessary reforms.

Moreover, the political scene is by no means stabilized, and more serious uncertainties can be seen on the horizon. Foremost among these must be the political evolution of the former Soviet Union. The forces that worked to reshape Eastern Europe so rapidly had their counterparts in most of the Union Republics of the USSR, especially the Baltic Republics (Estonia, Latvia, Lithuania) and those of Transcaucasia (Georgia, Armenia, Azerbaijan). Political pluralism and economic change were embraced by the constituent republics of the Soviet Union in the early 1990s as they had so quickly been in Eastern Europe a few years earlier; but the political future of the region remains clouded (see Conclusion). In any event, the proximity of Russia—still the world's largest and one of its most heavily armed countries—continues to cast a long shadow over Eastern Europe.

Basing their arguments on a clause in the Soviet constitution that guaranteed republics the right to secede from the Union—a purely theoretical idea until 1990—the Baltic republics were in effect the first to declare independence, and are by 1996 clearly restored to the list of European nations. Other republics have made the same claims, and even the giant Russian republic is feeling a new sense of nationalistic pride and asserting its own independence from seventy-five years of Soviet authority. These are deep and dangerous waters that raise profound questions about the future of the region.

Finally, dramatic political reform in Eastern Europe has also had profound implications for the international institutions that have shaped the interrelationships of the continent in recent years and that promise to do so more comprehensively in the future (see discussion of "international institutions," Chapter 14). As the "cold war" has at least moved to the edges of international thinking and peace has come to much of Eastern Europe (with the outstanding exception of ex-Yugoslavia), the role of the North Atlantic Treaty Organization (NATO) has become much discussed. Many Eastern European nations, anxious to ensure their security, would like to join

Eastern Europe

Figure 23.4 Czechs, as other Eastern Europeans, revel in newly discovered freedoms: Prague.

this group, or at least develop closer ties. Russia, however, sees this possibility as a threat, and might use economic or even military pressure to resist such a change.

Similarly, the expanding influence of the European Union (EU) is encountering both problems and opportunities in Eastern Europe as the region seeks more promising economic allegiances and its nations attempt to articulate a new role in international affairs. Again, Russian attitudes toward such expansion are both critical and—given the political uncertainties in Russia itself—difficult to predict.

The most enigmatic, and to many the most unsettling, aspect of recent change in Eastern Europe is the question of Germany's future. If communist rule crumbled in, say, Hungary and Poland, these remained nonetheless distinct nations with unique languages, cultures, and histories. As the same thing happened in East Germany, however, there was really no serious logic by which the state should remain independent from the rest of Germany. The seemingly inevitable tide of German unification, which led to the formal reunification of the nation on October 3, 1990, has profound implications for all of Europe, ranging from the heavy weight of a nation that has by far a larger population and (following the long and expensive task of rebuilding the East German portion) a stronger economy than any other European country including Russia; to the unspoken fears—not least in Russia—of a revival of Germany's military threat. In many ways, the concept of the "shatter belt" remains a valuable way of looking at the hopeful, vulnerable nations that comprise Eastern Europe.

PERCEPTION: THE DANUBE

The history of Eastern Europe—fragmented, bullied by outside powers, always on the fringes of European economic success—can be summarized in the story of its great river: the Danube. Rising in western Germany not far from the Rhine, the Danube is the only major European river flowing east. It passes through or along the border of Austria, Slovakia, Hungary, Serbia, Romania, and Bulgaria, before emptying into the Black Sea in a great delta controlled by Romania, Moldova and Ukraine. With all its potential, the Danube has never developed into an "eastern Rhine" in terms of commercial traffic and industrial development along its banks. Its problems have been both physical and political.

Physically, the Danube has been constricted by its turbulent passage through the "Iron Gate," where it cuts through the Transylvanian Alps between Romania and Serbia. A navigable channel was not opened here until the 1950s.

Figure 23.5 The Danube links many of the nations of Eastern Europe.

Figure 23.6 The Danube River at Budapest, Hungary.

Elsewhere, the Danube is also troublesome in Hungary, where low water during the hot summers is a problem as it flows across the flat landscapes of the Alföld. Finally, flowing into the Black Sea does not open the Danube to the world to nearly the same extent as does the Rhine's arrival at the North Sea.

Politically, development of the Danube has been greatly retarded by fragmentation of the region into several generally hostile states, notably since 1918. Ironically, it was only when the Soviet Union became a riparian power on the Danube (by virtue of its annexation of Moldavia from Romania in 1945) that many of the international conflicts were resolved and full development of the river's possibilities considered. Among the prospects discussed before the Eastern European nations became independent was a canal link between the Danube and the Elbe across Czechoslovakia, thus ensuring waterway transport from the Baltic to the Black Sea, effectively linking all of the East European states. In 1992, however, a more spectacular event occurred when the Danube was linked by canal to the Rhine across southern Germany. Although environmentally contentious, this connection means that the large barges typical of transport on the waterways of western Europe are now able to operate from the North Sea to the Black Sea, theoretically at least increasing the economic impact of the Danube.

The emergence of the East European nations into the modern world raises the prospect of greater cooperation to achieve the potential that the Danube has always promised but never delivered. On the other hand, as the unified direction provided by the heavy-handed planning of the Soviet Union vanishes, it is certainly possible that the various nations will return to their parochial, nationalistic ways. In particular, the tragic events in former Yugoslavia will continue to curtail the river's potential. The fate of the Danube will be a useful indicator of whether Eastern Europe will move toward the future or slip into the divided ways of the past.

Poland. Poland is the largest and, with nearly 38 million inhabitants, the most populous of all the East European states (excluding Ukraine). It is also probably the European nation with the most tragic overall history. Located largely on the flat, featureless, defenseless North European Plain, precisely at the center of virtually continuous conflict among such sometime powers as Prussia (Germany), Russia, Sweden, and Austria, Poland has been shifted east and west with the tides of war, and for centuries at a time has vanished altogether from the map of Europe, literally gobbled up by its greedy neighbors. Only to the north, along the Baltic Sea, and to the south, where the Sudetes and Carpathian mountains form the boundary with the Czech and Slovak republics, is Poland provided with reasonable natural boundaries.

Throughout the setbacks of history, Polish traditions and culture have been kept alive in the nation's heartland, focused on the capital, Warsaw, lovingly rebuilt since its total destruction in World War II; and the ancient city of Krakow to the south. Poland's large agricultural sector is relatively inefficient, even though it was the least collectivized of any East European country under Soviet influence; this is a major focus of concern for the nation in the immediate future.

The modern economy has been founded on the heavy industrial sector made possible by the important Silesian coal fields along the southern border. The industrialized port of Gdansk, on the Baltic, has become known in recent years for the political unrest (the "Solidarity" labor movement) that arose in the shipbuilding industry and proved to be the spearhead of anticommunist political change throughout Eastern Europe. Of all the East European nations, Poland has opted for the most radical program of restructuring its economy along competitive, market-oriented lines. In a very brief time (the "big bang"), socialist protectionist structures were replaced by more open, western-style systems of production, marketing, and accountability. This is a painful process, inevitably involving unemployment, higher prices, and temporary shortages, as the distortions of the previous system are replaced by something more realistic. Poland's experiences will be a good indicator of the problems and potentials facing the other nations of the region. Poland's security concerns also remain acute, since, regardless of hopeful political changes, it remains a vulnerable buffer state between German influence to the west and Russian to the east.

Eastern Europe

The Czech Republic and Slovakia. Bordering Poland and Germany to the south are the successor states to Czechoslovakia, which was created out of the ruins of the Austro-Hungarian Empire after World War I. The idea was to combine one of Europe's most important and advanced industrial regions (Bohemia) with two somewhat less developed areas (Moravia and Slovakia) in the hope that they would blend into a cohesive nation.

The experiment seemed to be going well, but the country was taken over by Nazi Germany in 1938, swept up in World War II, and occupied by Soviet armies. A communist regime was established and the easternmost province of Ruthenia was annexed to the Soviet Union. After the brief uprising of 1968, Czechoslovakia was one of the more tightly controlled East European nations until 1989, when it too was caught up in the profound political and economic changes taking place in the region and moved quickly toward a democratically elected government and a more market-oriented economy.

Despite the relatively peaceful separation from Soviet influence (the "velvet revolution"), the strong forces of nationalism sweeping Eastern Europe made it essentially impossible for a patchwork country to remain unitary. Slovaks, proud of their own traditions and their own capital (Bratislava), became increasingly restive at the domination of the economy and national institutions by the Czechs. Finally, in 1992, the Czech and Slovak Republics (Moravia is part of the "Czech lands") agreed fairly amicably to go their independent ways, although several contentious questions were raised (distribution of industrial resources, for example).

The countries are largely mountainous, with only the valleys of the Elbe to the northwest (in the Czech republic), and the Danube in the south (Slovakia), offering good agricultural possibilities. Elsewhere, the highlands culminate in the Slovakian peaks of the

Figure 23.7 Many parts of Eastern Europe are virtually unknown to western tourists: Telc, Czech Republic.

High Tatra on the Polish border (over 8,000 feet/2,440 meters), where an important winter sports complex has been developed. The major economic sectors are heavy industry based on a share of the important Silesian coal field (shared with Poland), and a varied range of engineering industries around the Czech capital, Prague, and in other cities of Bohemia . Unfortunately, the traditional Czech reputation for quality manufactures (automobiles, armaments, consumer goods) largely vanished under the socialist regime, and will take some time to rebuild, although the Czech Republic has already become one of the region's most successful "new" nations. Slovakia inherited a larger share of the inefficient and heavily indebted heavy industrial structure developed during socialist times, and has been slower to institute radical reforms. Nonetheless, the nation is determined to create an expanding economy and to move closer to western Europe.

While the nations of Eastern Europe already discussed have generally been identified with industrial growth and represent—despite their difficulties—the most economically advanced countries of this group, the states to the south have traditionally been more dependent on agriculture. This reflects both the natural advantages for farming (land base and climate), and the long period of economic stagnation under the colonial rule of either Austria-Hungary or Turkey. Even during the recent socialist period, these nations had considerable difficulty in persuading the Soviet Union and its relevant policy-making bodies that they should also be encouraged to modernize industrially. At the same time, the southern countries were often more innovative in finding ways to expand their economies without unduly ruffling the Soviets.

Hungary. Until the dramatic events of 1989, Hungary was the best example of this innovative trend. The modern Hungarians are descendants of the Magyars, who arrived on the broad plains of the middle Danube (Alföld) at the beginning of the tenth century, and who have been able to maintain their distinctive language, culture, and traditions through all the tribulations that life in the "shatter belt" has brought their way. Fiercely independent, they have treated domination by Austria, Turkey, Germany, and most recently the Soviet Union with a smoldering resignation and a determination to make the very best of the situation.

Following the abortive and brutally crushed revolution of 1956, Hungary took to heart the fundamental lesson that all East European nations were forced to learn with reference to their Soviet neighbors: many things were possible in the cultural and economic areas if political stability was unquestioned. Thus, despite the framework of centralized planning, Hungary was able to build a modestly successful economy through a variety of innovative ideas that looked surprisingly "western" and even "neo-capitalistic" to outside observers, yet which were rarely challenged by the Soviet Union, Hungary remaining politically "reliable."

The country was thus able not only to expand its traditionally strong agricultural sector, exporting high-quality fruits and vegetables throughout Eastern Europe; but also—in the face of much initial Soviet opposition—built a significant industrial base, including iron and steel production and a major line of trucks and busses. Thus, when the winds of "glasnost" and "perestroika" began to blow from the Soviet Union in the late 1980s, Hungary was well placed to take full advantage of the new openness and spirit of economic and especially political pluralism. Not surprisingly, it has been the Hungarians, along with the Poles and Czechs, who have set the pace for the recent dramatic changes in Eastern Europe.

Eastern Europe

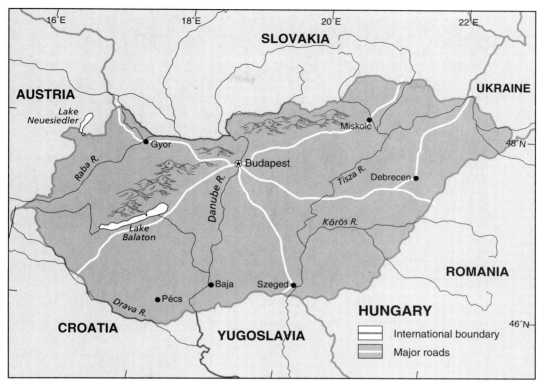

Figure 23.8 Hungary.

Except for modest ranges of hills, notably near the Slovak border, Hungary consists largely of the broad, sometimes poorly drained, Alföld: the Danube flows from north to south directly through the center of the country. The warm, humid summers here, in contrast to regions farther north, create an ideal environment for maize cultivation, and Hungary has been one of Europe's few traditional maize-eating nations since the plant was introduced from the Americas in the sixteenth century. The national capital, Budapest, is similar to Vienna, Austria in concentrating an unusually large share of the national population. In both cases, this reflects their "imperial" status: for half a century Budapest was co-capital of the vast Austro-Hungarian empire, thus growing to its present size. Naturally, it now dominates the relatively small nation it still governs. Southwest of Budapest, Lake Balaton is one of the most important centers for tourism and recreation in Eastern Europe. One of the few large bodies of water over a surprising distance, Balaton attracts visitors from throughout the region, as well as from western European countries.

Romania. Southeast of Hungary and southwest of Ukraine and Moldova, Romania is currently one of the less successful nations of Eastern Europe, a situation which at first glance seems surprising. The interior of the country is largely mountainous, and in the Transylvanian Alps (where peaks exceed 8,000 feet/2,440 meters), modest but—on a European scale—useful deposits of many metallic minerals are found. Productive lowlands are located around the nation's edges: Moldova to the east (although much of this province was incorporated into the Soviet Union after 1945 and is now

Figure 23.9 Lake Balaton is one of Eastern Europe's leading recreational sites: Siofok, Hungary.

an independent nation), extensions of the Alföld along the border with Hungary, and especially the broad lower valley of the Danube to the south, where the capital Bucharest is located.

Long controlled by the ancient Roman empire (Romanian is the only Romance language of Eastern Europe), modern Romania has generally suffered the same fate as its neighbors: domination by Austria, Turkey, Germany, and the Soviet Union. The nation first came to Europe's attention when it was found to have, at Ploesti, the largest oil field ever discovered on the continent (excepting Russia and the modern offshore North Sea fields). Romania briefly led the world in oil production in the mid-nineteenth century, and was a major supplier through the 1950s. The exhaustion of many of the wells at Ploesti, however, has created a considerable irony: Romania is today forced to buy oil from Russia at world market prices and, since funds are lacking, has become one of the most energy-short nations of Europe. Until the revolution of 1989, the average family was restricted to one 40-watt light, and during the bitter winters of 1987–89 central heating was available in Bucharest apartment buildings for only two hours each night!

Romania's recent difficulties have largely resulted from the actions of a particularly repressive Stalinist-style government, supported by a large, well-armed secret police network. More than in any of the other East European nations in the Soviet orbit—where conditions were difficult enough—the personality cult developed by Nicolai Ceausescu admitted no dissent and plunged the country into harsh, unworkable, and almost totally unpopular economic schemes. Paying off the national debt became a governmental obsession, for example, and virtually all the nation's resources, including food, were exported to achieve that doubtful goal. Plans to raze thousands of traditional agricultural villages, replacing them with faceless "agro-cities," were on the verge of implementation. It is little wonder that the sudden overthrow of dictatorial communist rule in Romania was more violent than in other East European countries: several hundred may have died in the worst fighting seen in Europe between the end of World War II and the beginning of the Yugoslav wars in 1991.

Eastern Europe

Figure 23.10 Romania.

Not surprisingly, decline of its energy base and the shortsightedness and inflexibility of government programs have hampered the rational development of the Romanian economy, while nurturing political openness has also proven an uneven and difficult process. Nonetheless, the agricultural base is sound (farm population remains among the largest in Europe), and the long-established trading connections with western nations are well organized and profitable. A potentially flourishing tourism industry along the Black Sea coast and gradually increasing use of the Danube are other economic bright spots.

Bulgaria. Bulgaria lies south of Romania, sharing borders also with Greece, Turkey, Macedonia, and Serbia. The final former Soviet client state of Eastern Europe is also the smallest in population (Albania and the Yugoslav states were not considered within the Soviet orbit), with only 8.5 million inhabitants. The country shares the lower Danubian plain with Romania, but the central and southern regions are mountainous, with peaks in the Balkan and Rhodope ranges exceeding 9,000 feet/2,745 meters. The central Maritsa valley and the Black Sea coast have the major concentrations of population and economic activity.

Figure 23.11 Bulgaria.

A major power in southeast Europe during the middle ages, Bulgaria has since shared the troubled history of the region, gaining partial independence from Turkey only in 1878, reaching its modern dimensions in 1912, participating in both world wars, and falling under Soviet control in 1945. Unlike Hungary and Romania, Bulgarian language and traditions are basically Slavic; thus, although most Bulgarians would doubtless have preferred a greater degree of national independence, they have been able to interact with the Russians more easily than their non-Slavic neighbors. The wave of anticommunist reforming zeal that swept Eastern Europe in late 1989 was more cautious and subdued in Bulgaria; change is underway, but its achievement may be less dramatic than elsewhere.

The Bulgarian economy is essentially agricultural, and its farming had been the most collectivized of any East European country. Although soil and climatic conditions are far from ideal (drought is a persistent problem), Bulgaria has been steadily improving the quality, quantity, and distribution of its products, including such "southern" commodities as tobacco and various soft fruits, which are in great demand farther north. The country has few mineral resources, and its industrial base is comparatively small. As in Romania, tourism along the Black Sea coast and expanded navigation on the Danube offer the possibility of economic growth.

Finally, Eastern or until very recently "socialist" Europe includes two nations which drifted out of the Soviet orbit at a comparatively early date and have moved in vastly different directions: Yugoslavia and Albania.

The Yugoslav States

Yugoslavia, like Czechoslovakia, was a patchwork country pieced together by the victorious allies in the aftermath of World War I. In this case, however, the complexity of the nation was even greater, and its eventual breakup anything but peaceful. Yugoslavs were fond of noting in happier days that their country had two alphabets, three religions, four languages, five nationalities, and six republics! Such diversity, however, does not often result in national cohesion, and the bloody fragmentation of Yugoslavia into its various small components in the 1990s is ample evidence of this principle.

Held together for many years by the forceful personality of its famous World War II partisan hero and subsequent political leader, Josip Tito, the nation suffered weaker leadership following his death in 1980, and many of its regional and ethnic differences quickly became more acute. There was increasing resentment against the dominance of Serbia (and the national capital, Belgrade) in Yugoslav affairs, while the wealthier, more "European" republics of the northwest (Slovenia and Croatia) disliked seeing their prosperity channeled to development of poorer regions in the south.

The fiercely independent Tito was able to break his nation's enforced ties with the Soviet Union (aided certainly by distance, mountainous terrain, and the fact that there is no common frontier), and the economy of communist Yugoslavia accordingly developed its own distinctive pattern, in which innovation and pragmatism often replaced Marxist logic.

Largely a rural country, Yugoslavia worked hard to develop its industrial base, increasing productivity by giving workers a large measure of control over decision-making at local level. While not always an efficient system, such worker participation created a sense of involvement and responsibility. European and American economic assistance and private investment was available and welcomed, tourists flocked to the country (especially to the beautiful coastline of the Adriatic Sea), and over a million Yugoslav workers migrated to temporary employment in western Europe during the 1960s and 1970s, sending or bringing back a substantial amount of money that eventually found its way into economic development.

The sudden transformation of Soviet-dominated Eastern Europe in 1989–90 from a dreary landscape of political conformity (or even repression, as in the case of Romania) and the unproductive economics of centralized planning to a seemingly new world of political pluralism and economic pragmatism left Yugoslavia in an interesting position. From being the most open and adventuresome of Europe's communist nations, it rapidly became one of the more conservative, and the modest economic success it created through western contacts was largely undercut by the new, determined governments emerging elsewhere. Thus, the dramatic events which brought hope and progress to other Eastern European nations helped to precipitate the disaster that has befallen what was in many ways one of Europe's most attractive countries. More conservative, market-oriented governments were elected in Slovenia and Croatia, creating still more friction with Serbia, where a socialist regime continued (and continues) in control. Eventually, these two nations declared total independence from the Yugoslav federation (1992), effectively shattering the myth of Yugoslav unity. As Serbia attempted militarily to restore the traditional Yugoslav order, the ethnic, linguistic, religious, and cultural differences that had been papered over for many years burst into the open, inflamed by strong feelings of historic injustice—especially between Croats and Serbs—and drew virtually all of ex-Yugoslavia into the bitter civil war that still continues. What was a unitary country has now (at least for the moment) divided into five "new" nations.

Yugoslavia. Yugoslavia now consists of two of the six component republics of the original country before its violent breakup: Serbia and Montenegro; as well as the two "autonomous regions" that have been governed and generally dominated by Serbia: Vojvodina and Kosovo. As if this part of Europe needed more political and ethnic complications, both of these regions have considerable interest in greater

Figure 23.12 Ethnic diversity adds both color and stress to the former Yugoslavian states: Islamic elders, Struga, Macedonia.

independence, as well as strong links to outside countries: Vojvodina to Hungary, Kosovo to Albania (see following Perception Box). The heartland of the nation—as in so many Eastern European nations—is the Danube valley, along with those of its major tributaries, the Sava and the Tisza. This area, which forms an extension of the Alföld plain, has been a major part of the region's "breadbasket," and its production of food has permitted Serbia to avoid at least a measure of human suffering following the imposition of an economic blockade by much of the world in protest at Serb responsibility for the current Yugoslav war.

Much of the rest of the nation, including Montenegro and Kosovo, is mountainous, with limited economic prospects. During the federal years, industrial growth and development was particularly strong in the area around Belgrade, and although this sector of the economy has been hard hit by the war and blockade, the potential for future growth remains, should more normal conditions return. Particularly troublesome to Serbia (as indeed to all the ex-Yugoslav states) is the need to import nearly all energy needs, especially oil.

PERCEPTION: KOSOVO—THE NEXT FLASHPOINT?

As the Yugoslav federation began to collapse in the early 1990s, there was widespread international feeling that the most dangerous conflict might be between Serbia and the Autonomous Region of Kosovo. As the subsequent war first involved mainly Serbia and Croatia, later spreading brutally into unfortunate Bosnia-Herzegovina, Kosovo remained in the shadows of the conflict. The problems that make Kosovo a dangerous part of Europe have not been resolved, however, and should the peace process eventually bring more normal conditions to Bosnia, the focus of tension may well shift again, and the "shatter belt" could claim yet more victims.

Two thirds of the population of Kosovo is of Albanian ethnicity, with Turks, Montenegrins, and Serbs composing the rest. Most of the population is Muslim, and links with the adjacent nation of Albania are close. However, Kosovo is administratively controlled by Serbia (now by the reconstituted nation of Yugoslavia), and is in an emotional sense extremely important to the Serbs. It was on the Field of Kosovo, near the modern capital, Pristina, that, in 1389, a Turkish army defeated the Serbians, opening the way for five hundred years of Ottoman rule in the Balkans. Ceding control of this sacred place (memories are very long in the Balkans) to a non-Serbian nation—and an Islamic one at that—would be exceptionally difficult for any modern Serbian government.

Mountainous and poor, with no seacoast and limited infrastructure, Kosovo was the poorest part of federal Yugoslavia. Funds provided from Belgrade represented the only means of agricultural improvement and limited industrial development. As history has repeatedly shown, however, poverty and a sense of disenfranchisement can be powerful revolutionary tools, especially where ethnic/religious differences are also significant. Many citizens of Kosovo would like to see their region incorporated into a "greater Albania;" but any such proposal would certainly elicit a violent reaction from Serbia, as well as from the region's other neighbors: Montenegro and Macedonia. Could Kosovo be the next Bosnia, spilling yet more blood in the Balkans? The world community, as well as most local inhabitants, clearly hope not; yet the potential for another Yugoslav disaster is certainly there.

Slovenia. Slovenia is one of the few parts of the former federal republic of Yugoslavia (Macedonia is another) that was largely able to avoid being engulfed in tragedy as the nation fell apart. One of the primary components of unitary Yugoslavia (the nation was founded in 1918 as the "Kingdom of Serbs, Croats, and Slovenes"), Slovenia has three enormous advantages in this difficult part of Europe. First, the nation is ethnically more homogeneous than any other part of Yugoslavia: over 90% are linguistically and culturally Slovenian, while Roman Catholicism is by far the dominant religion. Second, Slovenia is the only nation in this region to have common boundaries with highly developed European nations (Italy and especially Austria), and to have focused historically on Europe (it was controlled by Austria-Hungary rather than the Ottomans) rather than on the Balkans. Finally, Slovenia is separated from Serbia (generally the aggressor in the Yugoslav wars) by considerable distance, by some rugged topography, and by the nation of Croatia, with which the Slovenes have generally enjoyed harmonious relations.

As a result of its orientation to the west, and its good fortune in escaping the worst excesses of the war, Slovenia is today the most economically advanced of the Yugoslav states. Plains on the floodplains of the upper Sava and Drava rivers provide a solid agricultural base, while the spectacular Alpine scenery—including glacial lakes and a growing winter sports industry—continue to bring in substantial tourism revenues. (Slovenia is the only part of Yugoslavia that has continued to profit from tourism with barely an

Figure 23.13 Slovenia has its Alpine landscapes also: The Soča River near Bovec.

Figure 23.14 The Yugoslav states: Political divisions.

interruption during the 1990s.) Encouraged by relatively low wages and proximity to western European markets, light industry has flourished in the capital, Ljubljana, in Maribor, and in the few other urban centers. Stimulated by its relative success, Slovenia has applied for membership in the European Union, and is very interested in joining NATO or some other international body that might help to guarantee its security.

Croatia. Croatia has been one of the major protagonists in the current Yugoslav wars, and has in fact always played an important role in the troubled history of the region. Uncomfortably caught for centuries at the point of conflict between Austro-Hungarian and Ottoman ambitions, Croatia's development was consistently thwarted by political and military uncertainties. The Croats (largely Roman Catholic in religion, and economically focused to western Europe by way of the Austrian connection) became the major political and cultural antagonists of the Serbs (Orthodox faith and an economic focus toward the Balkan region and Russia). This deep-rooted conflict is at the very heart of the modern troubles, Croats and Serbs being seemingly unable to coexist amicably in the same space. This problem was heightened during World War II, when Croatia became a fascist republic under the Ustasha regime and committed numerous atrocities in Serbia.

Croatia joined Slovenia in declaring independence from the Yugoslav federation in 1991. Unlike the Slovenes, however, the Croats shared a common boundary with the Serbs (through the Autonomous Region of Vojvodina), and the Serb-commanded federal army was able to move against the "rebel" Croats across the lowlands

Figure 23.15 Dubrovnik, Croatia was one of Europe's most attractive cities and a major tourism focus before the tragic civil war.

formed by the Danube and its tributaries. More shocking to the outside world was the wanton destruction of nonmilitary objectives by Serbian forces, particularly along the beautiful Dalmatian coast of the Adriatic Sea, which has long been part of Croatia. The destruction by shellfire of Dubrovnik, a United Nations-designated historic preservation site and surely one of Europe's most beautiful cities, led to the widespread condemnation of Serbian military activity.

In addition to the major tourism attractions of the Dalmatian coast, Croatia has the potential to develop a solid, balanced economy. Agriculture is productive in the eastern part of the country, and the capital, Zagreb (second in size only to Belgrade among Yugoslav cities), has attracted a diverse manufacturing base. As its northern neighbor, Slovenia (with which relations are good), Croatia hopes eventually to become more directed toward western Europe and certainly more integrated into the general European economy. Eventual resolution of the Bosnian problem (which has drained vital resources from all parts of ex-Yugoslavia), is extremely important to Croatia, and should certainly be pursued, despite the ethnic baggage with which the country remains encumbered.

Bosnia-Herzegovina. The desperate situation in Bosnia has been the focus of world attention since 1993. To a considerable extent, the country summarizes the enduring, often intractable, problems of the Balkans. It is certainly a considerable historic irony that the beleaguered national capital, Sarajevo, was the site of the assassination of an Austrian Grand Duke (by a Bosnian Serb) that triggered World War I and dragged nearly all of Europe to disaster. Bosnia epitomizes the ethnic complexity that has been so troublesome throughout the region. Rather than forming a "melting pot," in which various groups might have taken on a specifically "Bosnian" character, the country has remained divided among a number of peoples who have warily coexisted without constructing any durable common institutions. Thus, when the Yugoslav federation collapsed, there was little on which to found a stable nation: there are Bosnian Serbs, Bosnian Croats, and Bosnian Muslims; but too few people who consider themselves simply "Bosnian."

Apart from its cultural and political problems, Bosnia-Herzegovina also faces a challenging economic prospect. Largely mountainous (only in the east and along the valley of the Neretva River is there good quality farm land), nearly landlocked (except for a single small port on the Adriatic), and without substantial mineral resources, this is by nature one of the poorest parts of Europe—a situation that has been exacerbated over the centuries by its tragic history. If, as now seems likely, the country is partitioned into a Bosnian Serb state and a Croat-Muslim federation, it can only be hoped that the larger neighboring powers, Serbia and Croatia, will take an active interest in assuring not only peace, but also a rational program of economic development.

Macedonia. Southernmost of the Yugoslav states is Macedonia which, in addition to the revised version of Yugoslavia, has boundaries with Albania, Greece, and Bulgaria. The country has largely been able thus far to avoid becoming embroiled in the various Yugoslav conflicts; but it has had its own serious problems with Greece. "Macedonia" is a broad and historical term (Alexander the Great was Macedonian) that includes lands in both Bulgaria and Northern Greece. Suspicious of Macedonian territorial ambitions, Greece until 1995 refused to recognize the name "Macedonia" (or the national flag), insisting that the country be called the "Former Macedonian Republic of Yugoslavia." More seriously, Greece obstructed Macedonian exports via the Greek Aegean port of Thessalonika, which was a bitter economic blow to another landlocked

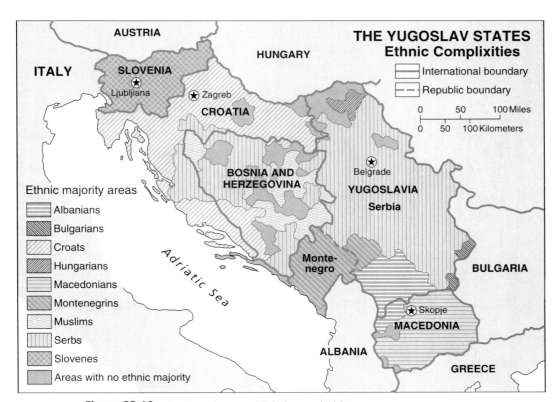

Figure 23.16 The Yugoslav states: Ethnic complexities.

country. By 1996, this problem had been—at least superficially—resolved, largely the result of pressure placed on Greece by its co-members of the European Union; but the persistence of such seemingly petty quarrels certainly casts additional light on the underlying problems of the Balkan region.

The Macedonian economy is, like that of most other southeastern European states, comparatively weak. This reflects lack of a significant resource base, relative isolation in a remote corner of Europe, lack of a coastline and thus independent access to the world, a long history of indifference to development during Ottoman times, and the current violence, which has deprived Macedonia of many of its natural markets. Agriculture is concentrated in the rich Vardar valley, while the national capital, Skopje, has seen the growth of a range of light industry. Absence of an Adriatic or Aegean coastline deprives Macedonia of a significant share of the region's important tourism industry. Should a lasting peace return to the Balkans, however, Macedonia's relatively central location in the region might permit it to play a major role in the rebuilding and eventual coordination of the economies of its various neighbors.

Albania. The final nation of southeast Europe is Albania, which, compared to Yugoslavia, took a precisely opposite course in the years following 1945. This small country, tucked along the Adriatic between Greece, Macedonia and Yugoslavia, became almost totally reclusive following World War II. A conservative, Stalinist, communist leader (Enver Hoxha) severed ties not only with the west, but also with Yugoslavia,

Figure 23.17 Albania.

the Soviet Union, and even China, none of which seemed to him convincingly Marxist. With the passing of Hoxha in 1985, Albania has begun moving very slowly toward some sort of opening to the world, although the heavy weight of one-party rule and still largely communist socioeconomic ideas makes change difficult.

A mountainous nation of marginal agriculture and weakly developed industry, the wealth of Albania resides in its mineral deposits, including oil, copper, nickel, bauxite, and especially chrome, of which it is the world's third leading producer. Closer ties to Yugoslavia seem logical, but the treatment of ethnic Albanians in Kosovo is a major difficulty (see Perception above). Italy, of which Albania was once effectively a colony, is another possible country with which to establish more normal relations. With the sudden liberalization of politics and economics in the other countries of Eastern Europe, Albania stood out conspicuously as the last bastion of traditional orthodox communism on the continent, and its halting modernization in the 1990s suggests that many ghosts have not yet been laid to rest.

Even here, however, pressures for more substantial change are building. Albanians seeking asylum in foreign embassies in the capital, Tirana, and later being evacuated to Italy caught the world's attention in 1990, and led to modest liberalization of domestic life. A modest flow of tourism has also pried open the Albanian door (with a dramatically beautiful coastline on the Adriatic, the potential for tourism development is considerable) and, if the recent experiences of the rest of Eastern Europe are any guide, dramatic changes in the nation's philosophy may eventually permit it to shed the image of being Europe's least known country.

European Countries of the Former Soviet Union

When the Soviet Union—last of the world's great colonial empires—collapsed in 1992 (an event related at least in some measure to the disappearance of inflexible communist authority throughout Eastern Europe), the fifteen "Union Republics" of the Soviet Union became in a short time fully independent nations. Under the Soviet constitution, the Union Republics were defined as having at least one million population, being representative of some distinctive ethnic group, and having an external boundary either with some other nation or with a neutral body of water. This last requirement meant that the Republics were found around the outer margins of the dominant Russian Republic.

Thus, at independence, seven of these new nations suddenly appeared on the eastern borders of Europe (three others constitute the Transcaucasian region; the other five are part of central Asia). Of these, the three Baltic Republics (Estonia, Latvia, Lithuania), with their long history of European orientation, have become fairly easily assimilated into northern Europe (see Chapter 15). Russia—still the world's largest country—is a strong European presence, as it has been for a thousand years; but its interests in Asia and the vastness of its Siberian territory make it impossible to consider as a purely European nation. Pressures for additional fragmentation along ethnic lines remain strong in many parts of Russia, however; and it is by no means excluded that the future may see even more independent states—some of them European—added to the list. The three remaining new nations from the former Soviet Union take their hesitant place in the "shatter belt" between Russia and the west.

Figure 23.18 European nations from the former Soviet Union: Belarus, Moldova, Ukraine.

Ukraine. About the size of Texas, Ukraine is the largest and best known of these states. Its outstanding grassland soils (*Chernozems*) and moderately dry climate have made it one of the world's great wheat growing regions; it was the breadbasket of the Soviet Union, and of imperial Russia before that. Moreover, Ukraine boasts abundant reserves of coal, iron ore, and manganese, and its industrial development, based on these resources, has been historically impressive. The Donets Basin (*Donbas*) was the first heavily industrialized part of Russia in the nineteenth century, and, although saddled with an aging infrastructure and declining productivity, it remains a major asset to the newly independent nation. The national capital, Kiev, a city of some 3 million population, is also a major industrial center and, based on its long and distinguished history and surviving architecture (Kiev was the capital of the first Russian nation from the ninth to the twelfth centuries), a major tourism attraction.

Ukraine is not without its problems, however, and its orientation toward Europe remains shaky at several points. Lack of an indigenous energy base (oil and gas) makes it dependent on Russia or uncertain connections with Transcaucasia or central Asia for these vital commodities. The Chernobyl nuclear disaster (1986) still weighs heavily both physically and psychologically on the nation. Additionally, an authoritarian government of "reformed" communists has been slow to undertake needed (and often painful) economic reforms. Added to these problems is the fact that Ukraine has its own set of ethnic tensions. Twenty-two percent of the population—primarily in the industrialized eastern part of the nation—is Russian, and would politically prefer to be reassimilated into a revived version of either a "Soviet Union" or some form of "Greater Russia." The Crimean peninsula, extending into the Black Sea, is also a contentious territorial issue between Ukraine and Russia. The western region of the country, featuring boundaries with Belarus, Poland, Slovakia, Hungary, Romania, and Moldova, is more "purely" Ukrainian and considerably more attracted to Europe. Will Ukraine effectively join the list of European nations, or will the traditions of the past pull it back into the Russian orbit? The next decade should provide some revealing answers.

Belarus. The second of the "new " European nations derived from the former Soviet Union, Belarus shares in many ways the transitional problems of Ukraine. About the size of Kansas, its location on the western edge of Russia on the flat European plain has often defined its troubled history. Boundaries with Latvia, Lithuania, Poland, and Ukraine (as well as Russia) logically draw the nation toward Europe; but the population is ethnically and linguistically close to Russia, and a very conservative government of "reform" communists seems favorable to some sort of political reunification with the imposing neighbor.

Belarus lacks the resource base that makes Ukraine at least somewhat self-sufficient (it shares a lack of energy resources); on the other hand, it was a favored location for industrial investment under the centralized planning programs of the Soviet Union. This means that the country boasts a relatively sophisticated range of engineering industries—especially around the capital, Minsk—and that it is less dependent than Ukraine on the inefficient heavy industry that was the hallmark of the Soviet Union. In the late 1990s, Belarus has become another country on the edge of Europe. Whether it will open more to the west or slip back under the influence of Russia is a question yet to be decided.

Moldova. Moldova completes the list of "new" European nations. Historically contested between Russia and Romania, "Moldavia" was seized by the Soviet Union following World War II and transformed into the fifteenth "Union Republic." Ethnically, the nation is largely Romanian (some 60% speak that language, and there is a widespread feeling that reunification with Romania may eventually occur. However, there is a strong streak of national pride among Moldovans, and submersion into a greater (and by no means wealthy) Romania is not at all an appealing idea to all citizens. Moreover, there is a serious ethnic conflict in the Trans-Dniestrian region at the extreme eastern edge of the country. Here a substantial Russian minority would prefer to be reunited with Russia (even though the region is separated by Ukraine).

The wealth of Moldova lies in its agriculture. An extension of the rich lands of Ukraine, well watered by the Dniester and its tributaries, and benefiting from the milder (humid subtropical) climate of its more southerly location), this is an outstanding region for the production of maize, fruit crops of many kinds, tobacco, and similar crops in great demand in more northerly areas. Industry is limited to light manufacturing around the capital Kishniev, there are no significant natural resources, and the nation is also landlocked, being separated from the Black Sea by an extension of Ukraine. Whether Moldova can become a successful independent country is yet another of the question marks hanging over southeastern Europe.

ADDITIONAL READING

BATT, J., *Economic Reform and Political Change in Eastern Europe: A Comparison of the Czechoslovak and Hungarian Experiences.* Basingstoke, ENG: Macmillan, 1988.

BENES, V., and N.J.G. POUNDS, *Poland.* New York: Praeger, 1970.

BERNAT, T., ED., *An Economic Geography of Hungary.* tr. L. Veges. Budapest: Akademiai Kiado, 1985.

DEMEK, J., *Geography of Czechoslovakia.* tr. V. Dittrichova and L.G. Winter. Prague: Academia, 1971.

DONIA, R., AND J. FINE, *Bosnia and Hercegovina: A Tradition Betrayed.* New York: Columbia University, 1994.

GELLA, A., *Development of Class Structure in Eastern Europe: Poland and her Southern Neighbors.* Albany, NY: S.U.N.Y. Press, 1989.

JANCAR-WEBSTER, B., ed., *Environmental Action in Eastern Europe: Responses to Crisis.* Armonk, NY: M.E. Sharpe, 1993

KAPLAN, R., *Balkan Ghosts.* New York: St. Martin's, 1993.

LAMPE, J.R., *The Bulgarian Economy in the 20th Century.* London: Croom Helm, 1986.

MORARIU, T., V. CUCU, I. VELCEA, *The Geography of Romania.* 2nd. ed., Bucharest: Meridiane, 1969.

PESCI, M., and B. SARFALVI, *Physical and Economic Geography of Hungary.* Budapest: Corvina Press, 1977.

POLLO, S., and A. PUTO, *The History of Albania: From Its Origins to the Present Day.* tr. C. Wiseman and G. Hole. London and Boston: Routledge and Kegan Paul, 1981.

TEICHOVA, A., *The Czechoslovak Economy, 1918-1980.* London and New York: Routledge, 1988.

TURNOCK, D., *An Economic Geography of Romania.* London: Bell, 1974.

TURNOCK, D., *Eastern Europe (Studies in Industrial Geography).* Boulder, CO: Westview Press, 1978.

DISCUSSION QUESTIONS

1. The Polish economy, among others, has suffered from the distortions of centralized planning. How did this system function? Why did it largely fail to promote economic growth in Eastern Europe?
2. Eastern European nations have generally suffered from the nonconvertibility of their currencies in the larger world. What is a "convertible currency"?
3. Czechoslovakia was a "patchwork" country put together by the victorious European powers following World War 1. What were the component regions of the country? Why has it recently separated into two different nations?
4. In the post-Soviet era, most Eastern European nations have made radical changes in both their political and economic systems with only a minimal amount of conflict. The exception has been Yugoslavia. What are the current "Yugoslav states," and why has Europe's most bitter war since 1945 broken out in this area? What sort of future do you see for this region?
5. Albania has been one of Europe's most reclusive countries. Why is the nation so distinctive, and why has it stayed in the shadows? What role might it play in the tensions developing in the Yugoslav province of Kosovo?
6. With the collapse of the Soviet Union, Ukraine, Belarus, and Moldova (as well as the Baltic Republics and to some extent Russia itself) have become new European nations. Are they really "European"? What developments would you expect to see in this area in the next five years?

Conclusion: Europe and the New Century

Humanity's fascination with symmetry causes great emphasis to be placed on certain dates. Thus, the approach of the year 2000 (by the western calendar) arouses unusual interest. What will the new millenium bring? Rational arguments that any year is much like another, save what we individually or collectively make of it, are not heeded. Celebrations of a new year, with all the implications of a "clean slate" are always enthusiastic; for a new century we will certainly expect something dramatic! Futuristic books and articles focused on the year 2000 have been common currency for at least forty years, and through the 1990s the pace of their appearance has predictably quickened.

In keeping with this tradition, therefore, it is perhaps permissible to speculate on the broad lines of development and change that might mark the European scene as the new century begins to unfold. This is a risky undertaking; the pace of change in recent years has been such as to confound most predictions almost as soon as they are made. Consider, for example, that when the first edition of this book appeared, there were some thirty-three "European" countries (Figure I.3); now the same figure lists forty-four!

In 1977 the geographer Peter Hall edited a volume entitled *Europe 2000*, which reported the findings of a distinguished interdisciplinary committee convoked in 1969 to consider the shape of things to come. With its detailed scenarios of possible political, social, and economic change, the book still makes useful and indeed fascinating reading; but there is at the same time a curiously dated quality about it. Here and there, a widely predicted development never happened. Elsewhere, the potential for dramatic—even revolutionary—changes remained unrecognized or understated. The accelerating march of technology is clearly to blame for much of this; but the pace of change in international and even interpersonal affairs has obviously also been increasing.

If this landmark work shows its age to this extent as its time frame expires, what indeed will the new century bring forth? Other authors are quick to jump into the game of predictions. Masser edited *The Geography of Europe's Futures* in 1992, while Delamaide suggested a radical geopolitical restructuring of the continent (*The New Superregions of Europe*) in 1994. Will these speculations have any relation to reality in, say, 2025? No doubt the unforeseen will figure prominently in Europe's future, as it always has; but there are a number of trends that seem clear, especially since they appear at this time to be gathering strength rather than fading away.

Increasing economic integration at broad scale. The drive toward greater economic integration within Europe seems by now to have an irreversible momentum. The fifteen nations of the European Union seem committed to the creation of a true "common market" by 1999; and although many rough edges will remain, the timetable will probably be generally met. A particularly contentious question involves adoption of a single European currency (the "Euro"), also by 1999, and the economic strength and political will of the member states as they move toward this goal will reveal much about the ultimate dimensions of European integration.

The prospect of exclusion from this large, lucrative market has caused most other western European nations to consider applying to the "club," or at least to explore possible ways of benefitting from association with the Union without accepting all of its binding rules. Farther afield, as the eastern European nations struggle to reform their economies with a view to making them more competitive in European and world markets, they are also seeking some favorable accommodation from the Union. Many, in fact (Poland, Hungary, the Czech Republic, Slovakia, Slovenia), have already submitted applications for membership, and are confident of acceptance by the beginning of the new century. Others will undoubtedly follow as their economies strengthen. Asia, North America, and an increasingly united Europe look set to dominate world trade well into the twenty-first century, and no European country—especially given the small sizes of many of them—wants to be left behind.

The political integration of Europe is and will remain a much more tedious and doubtful process. Most countries have, and will no doubt retain, strong reservations about surrendering many of what are perceived to be their vital national interests to any supranational body, whether this be the European Parliament associated with the Union or some other body. Already, some Union members (notably Britain and Denmark) have obtained "opt-outs" from several of the more contentious proposals. Militarily, NATO (with its strong American participation) now seems to be the only viable security option for most of western Europe, although the Organization for Security and Cooperation in Europe (OSCE) and the Western European Union remain viable options. After much difficulty, NATO has organized a coherent initiative to bring peace to troubled Bosnia (the United Nations was unable to do so); and if this operation meets with any degree of success it will greatly enhance NATO's stature as a defensive shield. Many eastern European countries would like to join NATO, but are constrained by the negative attitude of a suspicious Russia. There are, however, useful proposals for bringing the entire continent under some form of NATO-organized umbrella. Despite some success in the field of defense, however, the concept of a true "United States of Europe" will probably remain as elusive as ever.

The pressures of nationalism. At the same time that Europe becomes more unified, especially in an economic sense, it will also run a serious risk of becoming more fragmented as nationalistic pressures mount. Nearly all European countries contain numerous ethnic and linguistic minority groups, and as the unifying threat of cold war confrontation recedes into the past, these groups will become more assertive in claiming their cultural, political, economic and, in some case, territorial rights.

Western European countries in which these divisions are most strongly marked (Spain, France, Great Britain) have attempted to defuse the nationalistic threat by devolving substantial powers from central to regional governments, and by recognizing the distinctive and deserving character of various minority groups. Whether these steps will be sufficient to allay the fervor of nationalism, or whether they might instead lead to demands for outright independence, will be a major concern into the next century. In any case, many of these groups—Basques, Bretons, Corsicans, Scots, Irish—have a violent fringe that will continue to add an unwelcome element of uncertainty to life in their regions.

As the monolithic political structures of eastern Europe have collapsed in recent years, the strength of many "hidden" nationalistic movements has emerged. Each central government in this region had under its control a number of distinct and vocal minorities, many of which were deprived of their linguistic and cultural identity over long periods of authoritarian rule. These groups are now attempting to use the new democratic freedoms (and often political instability) of the region to assert this identity and state their claims to some sort of official recognition, and the potential for violence as these claims conflict with traditional patterns of authority has unfortunately been shown to be substantial. This pattern has of course been best demonstrated by the tragedy of Yugoslavia, where a unitary country has been transformed into five nations, and will fragment yet again as Bosnia divides into at least two distinct states. The European political map is still volatile: How many Bosnian states will there be? Will Scotland "devolve" from Great Britain? Will Ukraine and especially Belarus be subsumed into a newly expansionist Russia? The European list today stands at forty-four nations: What will be the number in 2010?

Attention to the environment. One of the trends most certain to become dramatically stronger in the years immediately ahead is Europe's awakening concern for its natural environment and its determination to clean up the mess that centuries of neglect and devotion to economic growth have created. Spurred by the new political dimension of "green parties" and by a growing realization of the fragility and vulnerability of many of the continent's most familiar and cherished ecosystems—the Rhine, the Mediterranean, the Black Forest—Europe will be bringing to the task practicality, inventiveness, and an increasing acceptance of the need for many small countries to work together if the slide toward environmental degradation is to be reversed or even halted. Already, countries such as Denmark and Germany (where all packaging and even automobiles must now be recycled) have taken the worldwide lead in implementing aggressive pro-environmental policies.

As eastern Europe emerged from the shadows of secrecy the full extent of environmental mischief throughout a region in which achievement of economic goals was a predominant (and usually failed) objective gradually became known. The challenge to western nations of assisting in the rebuilding of eastern Europe's struggling economies, as well as these nations' own efforts to become internationally competitive in the harsh terms of the twenty-first century, will be compounded by the enormous costs of environmental cleanup, as well as by rebuilding the industrial sector so that renewed pollution becomes less of a probability. In any event, it is possible to predict with confidence that the environment will be one of the most visible causes throughout Europe in the years ahead.

Social change. Other consistent themes in Europe's future will involve a variety of emerging factors and growing pressures relating to social issues. Europe is increasingly an aging continent, and as it becomes "grayer," with increasing numbers at the upper end of the population scale, more demands will inevitably be made on social services and medical care. In Europe as a whole, the fastest growing segment of the population (in percentage terms) consists of those over ninety years of age! This trend is causing (and will continue to cause) considerable reflection across the continent. With an increasing elderly component demanding pensions and health care, and a declining number of workers contributing to social security programs, how will the bills be paid? On the other hand, who will be available to fill the jobs necessary to maintain economic vitality?

A related problem is that of future immigration pressures. Already a cosmopolitan continent, western Europe as a wealthy destination exerts a strong appeal to potential migrants from eastern Europe, from former colonies, and indeed from the entire developing world. Proximity to (and easy access from) the rapidly expanding populations of the relatively poor North African nations puts severe pressure on Europe's Mediterranean nations. Asylum seekers will continue to flock to rich European nations, only to face—in many cases—rejection as these destination countries become increasingly aggressive in closing their borders. Related ethnic tensions will also continue to plague Europe. On the other hand, the skills of the prospective migrants will probably eventually prove to be a necessity in a Europe whose indigenous populations are no longer growing.

Figure C.1 Europe: Continuity and change. Top: Cobblestone street in a timeless French village (Montrichard). Bottom: "Checkpoint Charlie," only vehicle crossing between East and West Berlin during the Cold War, has become since 1991 merely a tourist attraction.

The evolution of Russia and its role in Europe. Finally, looming over the European future is the question of the future shape, direction, and inclinations of Russia. Following the collapse of the Soviet Union and its division into fifteen successor states, several of which have become more "European" (see Chapters 15 and 23), Russia, as did eastern Europe, is currently undergoing dramatic economic and political change. Unlike eastern Europe, however, the ultimate results of this change are by no means clear. Possible scenarios range from gradual democratization and economic reform in a Russia in harmony with its neighbors, to a further—possibly dangerous—fragmentation of Russia into still smaller ethnically defined states (the recent fighting in Chechnya reflects this potential development); to a revolution by the army and other conservative elements seeking to restore the primacy of the communist party and the command economy, and to extend Russian hegemony over many of the newly independent nations of the former Soviet Union.

Whatever happens, all of Europe will be profoundly affected. A Russia evolving further toward a western-style democracy and market economy would offer enormous opportunities of many kinds for Europe's nations, economies, and institutions. In contrast, a Russia bent on a return to the days of the cold war would be a direct threat to the fledgling democracies of eastern Europe and an ominous shadow over the entire continent.

Whatever the impact of these various trends, either singly or in combination, Europe will bring to bear on them the same strengths that have traditionally made the world's smallest major continent (excluding Russia) one of its most significant. An educated, imaginative, and skilled population; a fascination with the possibilities of ever-evolving technological change; and a keen eye for the slight advantages offered by location, resources, and changing circumstances are Europe's greatest strengths. Whatever the exact dimensions of the next century may turn out to be, it is reasonable to expect Europe—collectively and as individual nations—to be at the heart of events.

ADDITIONAL READING

DELAMAIDE, D., *The New Superregions of Europe.* New York: Plume/Penguin, 1994.

HALL, P., ed., *Europe 2000.* New York: Columbia University Press, 1977.

MASSER, I., ed., *The Geography of Europe's Futures.* New York: Bellhaven Press, 1992.

DISCUSSION QUESTIONS

1. Integrating and coordinating the newly liberalizing economies of eastern Europe will be a major concern for all of Europe for many years. Do you think that this process is going smoothly? What are some of the problems that are being encountered?
2. Cleaning up the eastern European environment will be an enormously difficult and expensive process. What caused these problems? How should the costs of cleanup be allocated?
3. Nationalistic pressures are growing rapidly, both in the newly independent states of eastern Europe and in the countries of the west. What dangers does this development pose for the future?
4. What course do you imagine Russia will take over the next decade? What impacts will its political decisions have on the rest of Europe?

Index

Aachen (Germany), 170
Aberdeen (Scotland, Great Britain), 253
Acid rain, 205-08
Adenauer, Konrad, 218, 285
Adriatic Sea, 311, 352-53
Aegean Sea, 314
Age of Discovery, 325
Agriculture:
   extensive model, 139-40
   intensive model, 139
   origins of, 123-15, 139
Agricultural pollution, 210
Airplane, impact on tourism, 195-96
Air pollution, 205-09
Albania, 349, 354-55
   language, 80
   mineral wealth, 355
   Muslim religion, 90
Alföld, 25, 49, 291, 339, 342-43
Algarve (region, Portugal), 328
Almonds, 131
Alpage, 124, 294
Alpine economy, 119-29
Alpine nations, 289-99
Alpine tourism, 197, 296
   See also Switzerland
Alps, 5, 16, 21, 23-24, 119-21, 289-92, 296-97
   Austrian, 291
   Dolomite, 302
   formation of, 21
   glacial features, 23-24
   Italian, 302
   Maritime, 21, 256, 302
   Swiss, 292, 294, 296
Alsace (region, France), 169
Alsacian language, 259
Aluminum, 161, 179
America, agricultural contributions, 144-45
Americas, maize origin, 343
Amoco Cadiz, 212
Andes mountains, 145
Andorra, 321, 329-30
Anglo-Saxons, in England, 250

Animals, as power source, 153-54
Apennine mountains (Italy), 119, 302, 304
Appalachian topography, 289
Apples, 143
Apricots, 131
Apulia (region, Italy), 304
Aquitaine (region, France), 160, 256, 261
Arabian Gulf. See Persian Gulf
Arabic cultures, Iberia, 90
Arabic language, 81
Ardennes mountains (France, Belgium, Luxembourg), 21, 256, 269, 273
Argentina, beef production, 239
Armenia, 336
Athens (Greece), 314
Atlantic ocean, 211, 213, 249, 322
Australia, beef production, 239
Austria, 284, 289, 291-97, 305, 338, 340, 342-43
Austria
   agriculture, 294-95
   climate, 292
   historical development, 292-93
   industry, 295
   photographs, 100, 123, 128, 295
   physical setting, 291-92
   refugees to, 297
   tourism, 191, 296
Austro-Hungarian Empire, 293, 331, 341-43
Auvergne (region, France), 256
Automobile:
   impact on environment, 208-09
   impact on tourism, 195
   invention, 158, 195
Azerbaijan, 336
Azores islands (Portugal), 326

Balaton, lake (Hungary), 343
Balkan mountains, 346
Balkan peninsula, 332
   political conflict, 66, 348-50
Baltic states, 229, 235, 237, 242-43

   See also Estonia, Latvia, Lithuania
Baltic Sea, 25-26, 145, 165-66, 211, 241, 280, 340
Bantry Bay (Ireland), 160
Barcelona (Spain), 134, 328
Bari (Italy), 137, 311
Basel (Switzerland), 295
Basque language, 76, 80, 259
Basque provinces (Spain), 326, 328
Bastille, la, 105
Baux, les (France), 167
Bauxite, 167, 314
   See also aluminum
Bavaria (state, Germany), 280, 286, 288
Beer, 143
Beethoven, Ludwig von, 285
Beets, 142
Belarus, 9-10, 357
Belgium, 269, 272-78, 293
   energy resources, 155, 169-70, 276
   historical development, 274-76
   language complexity, 83-85, 275-76
   photographs, 184, 225, 275
   physical setting, 281-82, 285-86
Belgrade (Yugoslavia), 347
Benelux states, 172, 269-78, 282
   climate, 273
   See also Belgium, Luxembourg, Netherlands
Ben Nevis, mountain (Scotland, Great Britain), 246
Bergslagen (region, Sweden), 240
Berlin (Germany), 282, 285-86
Bern (Switzerland), 83, 292
Beverage crops, role in food economy, 143-44
Bilardello (Italy), 163
Bilbao (Spain), 328
Birmingham (England, Great Britain), 252
Bismarck, Otto von, 284
Black death, 55
Black Forest (Germany), 21, 206, 280
Black Sea, 8-9, 115, 211, 313, 338-40, 345-46

367

Bohemia (region, Czech Republic), 341-42
Bonn (Germany), 285-86
Börde (region, Germany), 48, 187-88, 280, 286, 288
Bordeaux (France), 256, 259
Borinage coalfield, 276
Bosnia-Herzegovina, 349-50, 352-53, 363
Brabant (region, Belgium), 274
Bratislava (Slovakia), 209, 341
Bremen (Germany), 280, 288
Brenner Pass (Austria), 291
British Empire, 316
Brittany (region, France), 21, 144, 249, 256
  language, 76, 80, 259
Broadleaf deciduous forest, 42
Bronze, 164
Bronze Age, 164
Brussels (Belgium), 85, 220, 275, 278
Budapest (Hungary), 343
Bucharest (Romania), 344
Bulgaria, 21, 172, 332, 335, 338, 346-47, 367
  agriculture, 347
  language, 80
Burgenland (province, Austria), 291-92, 295
Burgundy (region, France), 256

Cabbage plants, 142
Caledonian formations, 19-20
California, as Mediterranean environment, 135-36
Camargue, la (France), 46
Cambridge (England, Great Britain), 252
Campine coalfield, 276
Canada, fish exports, 203
Cantabrian mountains (Spain), 21, 321-22
Cap de la Hague (France), 263
Cape of Good Hope (South Africa), 160
Capital, as factor of industrial location, 177-78
Carbon dioxide, 205
Cardiff (Wales, Great Britain), 252
Carinthia (province, Austria), 295
Carpathian mountains (Poland, Slovakia, Romania, Ukraine), 21, 119, 340
Carrara (Italy), 167, 207, 307
Carthaginians, 324
Cassa per il Mezzogiorno, 311
Catalonia (region, Spain), 326, 328
  language, 80

Cattle, domestication of, 116-118
Caucasus mountains (Russia, Trans-Caucasian states), 8
Ceausescu, Nicolae, 344
Celtic languages, 80, 249
  peoples, 61, 283
Central Plateau (region, Switzerland), 290, 294
CERN (European Center for Nuclear Research), 161
Channel tunnel (France, Great Britain), 196, 245-46
Charcoal, 154-55
Charleroi (Belgium), 276
Charles XII (Sweden), 237
Cheese, Swiss, 123-24
Chemnitz (Germany), 288
Cher river (France), 256
Chernobyl (Ukraine), 162, 203-04, 213, 262-63
China, 53-54, 355
Christianity, spread in Europe, 86-87
Church, as urban focus, 101
Cité, Île de la (Paris, France), 105
Citröen automobiles, 177
Citrus fruits, 45, 135, 327
Climate, factors controlling, 30-32
  mountain influences, 13
  types of European, 34-36
Clovis, 274
Coal, 155-58, 163, 187-88
  mining in France, 261
  mining in Germany, 286
  mining in Great Britain, 252
Coastal pollution, 211-12
Colbert, Jean, 264
Cold War, 284, 336
Cologne (Germany), 282, 285
Colonialism, and emigration, 64
COMECON, 223-24,
Common land, 97-98, 187
Common Market, See European Union
Community Workshop industries, 183-85
Como, lake (Italy), 302
Concorde (aircraft), 262
Conservation of energy, 160
Constantinople (Turkey). See Istanbul
Continental climates, 34, 282, 292
Continentality, as climate control, 31
Cook, Thomas, 195, 197
Copenhagen (Denmark), 238, 241
Coriolis effect, 31
Cork oak, 44, 131, 327
Cornwall (region, England, Great Britain), 21, 249
  language, 80

Corsica, island (France), 76, 80, 259
Coruña, la (Spain), 160
Costa del Sol (region, Spain), 136, 322, 328
Cote d'Azur (France). See Riviera
Council of Europe, 225
Council for Mutual Economic Aid See COMECON
Cousteau, Jacques, 212
Coventry (England, Great Britain), 252
Crete (island, Greece), 314
Creys-Malville (France), 263
Croatia, 130, 347-48, 350-53
  photograph, 352
Cunard, Samuel, 64, 195
Cyprus, 316-19
Cyrillic alphabet, 80
Czechoslovakia, 209, 331-32, 335, 341
  See also Czech Republic, Slovakia
Czech Republic, 286, 340-42
  industry, 342
  photographs, 337, 341

Daimler and Benz, automobiles, 195
Dairy products, 143
Danube river (Eastern Europe), 25-27, 49, 145, 308, 312, 338-45, 347, 349
  as diffusion route, 115-16
Defense, as settlement criterion, 99
Defense, la (Paris, France), 108
Delta project (Netherlands), 271-72
Demographic transition, 55-56
Denationalization of industry, 179-80
Denmark, 211, 229-30, 233, 235-241, 243
  agriculture, 147, 238-40
  climate, 230
  energy, 240
  fishery, 150, 238
  historical development, 235-37
  industry, 240-41
  photographs, 163, 239
  physical setting, 229-30, 233, 243
Domestication
  of animals, 116-18
  of plants, 113-15
Donzère-Montdragon project (France), 136
Dresden (Germany), 280, 288
Duero (Douro) river (Spain, Portugal), 322
Dunkerque (France), 261
Durance river (France), 136

Earthquakes, 17-18
Earth rotation, as climatic control, 30-31

Index

East Anglia (region, England, Great Britain), 248, 251
Eastern Europe, 295, 331-48, 362-65
  emigration, 64
  environmental problems, 208-09
  historical development, 331-38
  industrial planning, 179-80
  oil dependency, 160
  political changes, 1-2
  tourism in, 200
  See also individual countries
Ebro river (Spain), 322
Egypt, as tourism destination, 197
Eindhoven (Netherlands), 276
Elbe river (Germany), 339
Enclosure Acts, 187
England, 245
  See also Great Britain
English channel, 245, 256
English language, significance, 74-78
English Midlands (region, England, Great Britain), 246
Environmental determinism, 11
Environmental politics, 213-14
Environmental stress, 203-215, 286-87, 308, 363
Erzgebirge mountains (Germany), 280
Esbjerg (Denmark), 238
Essen (Germany), 188
Estonia, 229, 242, 336
  photograph, 243
  population, 58
Etna, mount (Italy), 17, 304
Eure river (France), 256
Europe, economic model, 6
  emigration from, 61-66
  limits, 8-9
  population change and distribution, 51-59
  resources, 6
  size, 5-6
European Bank for the Reconstruction of Eastern Europe, 298
European Coal and Steel Community, 218-19, 286
European Community, See European Union
European Free Trade Association (EFTA), 221
European Union, 2, 85, 209, 218-23, 242, 253-54, 274-76, 278-79, 311, 316, 319, 321, 326, 337,
  common agricultural policy, 219
  expansion of membership, 220-21
  parliament, 213, 219-20, 362
Europort (Netherlands), 276
Eutrophication, 210

External combustion engine, 155, 158, 195

Fats and oils, role in food economy, 143
Federal Republic of Germany. See Germany
Fenno-Scandian shield, 19-21
Fiat automobiles, 307
Finland, 229-30, 234-38, 240
  agriculture, 237
  climate, 230
  forestry economy, 149
  glacial landscapes, 21-22
  historical development, 235-37
  hydroelectricity, 240
  industry, 241
  language, 81
  minerals, 240
  physical setting, 229-30
Fishing, in the European economy, 149-51
  Mediterranean, 134, 150
Fjeld (Norway), 228
Fjords (Norway), 22-23, 229
Flanders (region, France, Belgium), 256, 272-74
Flemish language, 77, 83-85, 275
Florence (Italy), 301, 305
Föhn (wind), 38
Fontanelli, 302
Ford, Henry, 188, 195
Forest cover, 42-45
Forestry, as an economic support, 147-49
  in the Alpine economy, 124-25
Fos (France), 137, 261
France, 255-267, 305, 329-30
  agriculture, 259-61
  climate, 256, 258
  coal resources, 155, 168-70, 261
  frontiers, 16
  as hexagon, 255
  historical development, 259
  immigration to, 259
  northern, 172
  nuclear electricity, 162, 262-64
  photographs, 89, 107, 133, 143, 144, 157, 175, 193, 200, 208, 214, 258, 261, 262, 265, 364
  physical setting, 255-56
  regional devolution, 259
  southwestern, 145
  wheat production, 142, 260-61
Franco-Prussian war (1870-71), 63 106
Frankfurt-am-Main (Germany), 280, 288
French language, importance of, 78, 82-83

Freshwater pollution, 209-11
Friulian language, 80
Fulton, Robert, 195
Fusion. See nuclear fusion

Galicia (region, Spain), 145, 322, 326-27
Galician language, 80
Gallivare (Sweden). See Kiruna-Gallivare
Garda, lake (Italy), 302
Garonne river (France), 256
Gasoline prices, 160
Gdansk (Poland), 340
Geneva (Switzerland), 161, 296-99
Geneva, lake (Switzerland, France), 144, 150, 203, 292
Genoa (Italy), 305, 307, 311
Georgia, 336
Geothermal energy, 163
German Democratic Republic. See Germany, eastern
Germanic tribes, 283
Germany, 279-88, 340, 344
  agriculture, 286
  air pollution controls, 209
  central, 172
  climate, 282-83
  coal resources, 155, 157, 168-70
  eastern, 10, 285-88, 332
  emigration from, 64
  frontiers, 16
  green parties, 214
  historical development, 283-85
  industry, 287-88
  language, 77-78, 81-85
  lead-free gasoline, 204
  mineral deposits, 286-87
  photographs, 2, 26, 67, 189, 210, 282, 284, 287, 364
  physical setting, 280
  tourism, 191
  unification (1871), 284
  unification (1990), 10, 285, 337-38
  western, 279, 258-86
Ghetto, 89
Gilchrist process, 167
Gironde estuary (France), 256
Glacial landscapes, 229-30
Glaciation, 21-24, 280, 290-91
Glasgow (Scotland, Great Britain), 252
Glitterfinden (mountain, Norway), 229
Goats, domestication of, 116
  in the Mediterranean economy, 132-33
Göteborg (Sweden), 241

Gotha (Germany), 288
Government decision, as factor of industrial location, 179-81
Grain crops, importance to food economies, 140-42
Granada (Spain), 322, 325
Grasslands, 46
  soils of, 49
Gray-brown podzolic soils, 48
Graz (Austria), 291, 295
Great Britain, 16, 245-54. See also Scotland, Wales, Northern Ireland
  agriculture, 250-51
  air pollution, 203
  climate, 246, 248-49
  coal resources, 155
  emigration from, 64
  historical development, 249-50
  industry, 252-53
  North Sea oil production, 159-60
  photographs, 159, 250, 251, 253
  physical setting, 245-46
  tourists from, 136, 197-98
Great European Plain. See North Sea Plain
Greece, 130, 313-19
  agriculture, 314
  climate, 314
  industry, 314
  language, 80-81
  merchant marine, 314
  mountains, 21
  photographs, 44, 131, 162, 317, 318
  physical setting, 314
  tourism, 315-16
Greenhouse effect, 38-41, 205, 314
Greenland, 229, 235-36
Green politics. See Environmental politics
Green Spain, 322
Grenoble (France), 259, 262
Grimaldi family, 266
Grisons (canton, Switzerland), 82
Groningen (Netherlands), 160, 276
Guadalquivir river (Spain), 322
Guadiana river (Spain), 322
Guild system, 183-85
Gulf Stream, 230, 235
Gulf War (1991), impact on tourism, 200

Haarlemmermeer (Netherlands), 270
Habsburg monarchy, 273, 293
Hainault (region, Belgium), 274
Halle (Germany), 288
Hamburg (Germany), 280, 288
Hanseatic league, 273
Hardangerfjord (Norway), 23

Havre, le (France), 256
Herculaneum (Italy), 17
Hercynian formations, 20-21, 273, 280
Highland agriculture, 293-95
Highland climate, 36, 121, 125, 258, 283, 289-92
Highland environment, 11-13, 70-71, 280
High Tatra mountains (Poland, Slovakia), 341-42
Hitler, Adolf, 217, 285, 293
Hohe Tauern mountains (Austria), 291
Holland (region, Netherlands), 269-70
Holy Roman Empire, 284
Homestead acts (United States), 63
Horse, speed of, 194
Hoxha, Enver, 354-55
Huertas, 135, 327
Human factor, in industrial location, 181
Humid subtropical climate, 36, 292, 305
Hungary, 332-35, 337-39, 342-43, 349
  economic innovation, 332, 334, 342
  language, 80-81
  maize cultivation, 145
  photographs, 92, 147, 192, 339, 344
  population, 58
Hunting and gathering economies, 111-13
Hydroelectric energy, 127, 163, 294, 307

Iberian peninsula, 21, 321-30
  Islamic influences, 90-91
Iberian plateau, 21, 322
Iceland, 18, 230, 235, 238
  climate, 235
  fishery, 150, 238
  geothermal energy, 18, 163
  historical development, 235
  physical features, 230
Ijsselmeer lake (Netherlands), 271-72
Île de France (region, France), 255
Immigration, 61, 68-70, 364
Indonesians, in Netherlands, 275
Industrialization, as factor of environmental stress, 204
Industrial location, factors of, 173-81
Industrial revolution, 104, 121, 154-55, 168, 194-96, 238, 250-52, 259-62, 276, 325
Industrial triangle (Italy), 307
Industry, evolution of, 181-87
Inn river (Austria), 291

Internal combustion engine, 158
  contribution to air pollution, 208-09
Internal migration, 70-71
International Atomic Energy Agency, 297
International Labor Organization, 296
International Red Cross, 296
Ionian sea (Greece), 314
Ireland, 172, 246. See also Northern Ireland
  emigration from, 65-66, 254
  language, 80
Ireland, Republic of, 93, 245-54
  agriculture, 251, 254
  climate, 246
  historical development, 249-50
  industry, 254
  physical setting, 20, 245-46, 248
  tourism, 254
Irish question, 250
Irish Sea, 246
Iron Age, 164
Iron curtain, 279
Iron Gate (Romania, Serbia), 338
Iron ore, 164-67, 219, 240
Irrigation, in the Mediterranean economy, 134-36
Islam. See Muslim religion
Isle of Man, 249
Istanbul (Turkey), 9, 87
Italy, 25, 301-13, 355
  agriculture 306-07
  climate, 305, 308
  energy resources, 155, 307
  environmental problems, 308
  historical development, 305
  industry, 307-07
  photographs, 18, 199, 304, 306, 310
  physical setting, 302-04
  southern, 130, 137, 181. See also Mezzogiorno
  unification, 305, 308

Japan, fishing industry, 151
  tourists from, 198
Jena (Germany), 288
Judaism, 89-90
Jura mountains (Switzerland, France), 119, 256, 258, 289-90
Jura (canton, Switzerland), 16, 83, 289-90

Karelia (region, Russia), 80, 242
Kempen. See Campine
Kiruna-Gallivare (Sweden), 165-66, 240

Index

Kola peninsula (Russia), 19, 204, 229
Kombinats, 287
Kosovo (region, Yugoslavia), 80, 90, 348-50
Krakow (Poland), 340
Krupp corporation, 188

Labor, as factor of industrial location, 177
Labor migration, 67-70
Lac Leman. See Geneva, lake
Lacq (France), 160
Lake District (England, Great Britain), 20, 246
Lakes, acidified, 205-06
Languages, factors of competition, 74-77
Languedoc (region, France), 256, 261
Lapps, 112-13
Latin, significance as language, 75
Latitude, as climatic control, 30
  and hours of daylight, 38
Latvia, 229, 242, 336
Lead, as pollutant, 209
Lead-free gasoline, 209
League of Nations, 296
Lebanon, 317
Leeds (England, Great Britain), 252
Leipzig (Germany), 280, 288
Leonardo da Vinci, 305
Lesbos island (Greece), 314
Letzeburgesch language, 276
Liechtenstein, 289, 297-98, 329
  count of, 297
  photograph, 299
Liège (Belgium), 276
Lignite, 208, 286
Lijnbaan, 274
Limburg (region, Belgium, Netherlands), 169, 269, 273, 276
Limon (soil), 48
Limousin (region, France), 256
Lisbon (Portugal), 325, 328
Lithuania, 229, 242, 336
Liverpool (England, Great Britain), 252
Ljubljana (Slovenia), 351
Loess (soil), 46, 187, 280
Loire river (France), 256
Lombardy (region, Italy), 305
London (England, Great Britain), 34, 174, 211, 292-93
Lorraine (region, France), 219, 261, 277
Louis XIV (France), 264
Louvain (or Leuven, Belgium), 84-85

Louvre museum, 105
Lower Austria (province, Austria), 295
Lübeck (Germany), 280
Luleå (Sweden), 165
Luther, Martin, 87
Luxembourg, 269
  banking industry, 277-78
  historical development, 274
  iron and steel industry, 277
  iron ore, 166, 168-70, 277
  photographs, 43, 169, 274
  physical setting, 269, 273
Luxembourg city, 220, 278
Luzern (Switzerland), 296
Lyon (France), 259, 262

Macedonia, 35, 350, 353-54
  photograph, 348
Macchia vegetation, 44
Madeira (island, Portugal), 326
Madrid (Spain), 209, 322, 328
Mafia, 310
Maggiore lake (Italy), 302
Magyars, 25, 342
Main river (Germany), 288
Maize, 113-14, 145, 327, 343
  production, France, 261
Malta, 274, 316
  economy, 316
  language, 81
  photographs, 104, 318
Man, Isle of (Great Britain), 249
  language, 80
Manche, la. See English Channel
Manchester (England, Great Britain), 252
Maquis vegetation, 44
Marble, 167, 307
  destruction by acid rain, 206-08
Maribor (Slovenia), 351
Marine west coast climate, 34, 230, 246, 256, 286, 298
Maritsa river (Bulgaria, Greece, Turkey), 346
Market, as factor of industrial location, 175-76
  as urban focus, 102
Market gardening, 135-36
Marne river (France), 256
Marseille (France), 26, 134, 137, 256, 259, 261
Marshall Plan, 145
Massif Central (France), 21, 256, 261
  uranium mining in, 125
Materials supply, as factor of industrial location, 175-76

Matterhorn (mountain, Switzerland), 23, 197
Mediterranean climate, 34, 37, 39, 129-30, 305, 308, 314, 322
Mediterranean scrub forest, 144-45
Mediterranean Sea, 9, 13, 17-18, 26-27, 129-137, 148, 256, 261-62, 301-02, 305, 307-09, 313-16, 322, 327-28, 363
  agriculture, 327
  as diffusion route, 114-15, 134
  economy, 130-37
  environment, 211-12
  fishery, 134
  irrigation, 134-35
  industrial development, 136-37
  introduced vegetation, 95
  oil exploration, 160
  solar energy, 161-62
  tourism, 136, 191, 197-98, 201
Mediterranean soils, 48
Mercedes-Benz automobile, 288
Mercury, 327
Mersey river (England, Great Britain), 246
Mesopotamia, 113-15, 141
  early cities in, 114
Metallic minerals, 164-70
Mexico, 113, 145
Mezzogiorno (region, Italy), 308-11
  emigration from, 310
  See also Italy, southern
Microclimates, 36, 144
  alpine, 292
Middle Ages, 15, 101, 250, 308, 347
Middle East, 9, 21, 114-15, 139, 143, 164, 263, 305, 313, 316, 322, 331
  oil production in, 159-60
Midlands (region, England, Great Britain), 172, 194
Midnight sun, 38
Migrations, rural to urban, 70-71
  return, 127-28
Milan (Italy), 174, 301, 305, 307
Milford Haven (Wales, Great Britain), 160
Mills. See windmills, water mills
Mining, in alpine economy, 125-26
Mini-states, 265-67
  See also Andorra, Liechtenstein, Monaco, San Marino, Vatican
Mistral (wind), 38
Mittelgebirge (mountains, Germany), 20
Mo-I-Rana (Norway) 181
Moldova, 10, 343, 357-58
Mollet, Guy, 218
Monaco, 266-67
  photograph, 267

Mont Blanc (France), 256
  ascent of, 197
Monte Corno (Italy), 304
Montenegro (Yugoslavia), 348-49
Moors, 325. See also Muslim religion
Moravia (region, Czech Republic), 341
Mosel (Moselle) river (France, Germany, Luxembourg), 256
Mountain barriers, as climatic controls, 32
Muhammad, 90
Mulhacén mountain (Spain), 322
Munich (Germany), 280, 288
Murmansk (Russia), 230
Muslim religion, 90-91, 131, 325
  See also Arabic cultures
Mykonos island (Greece), 314

Nantes (France), 256
Naples (Italy), 17, 304
Napoleon I, 217, 274, 311
Narvik (Norway), 166, 230
National boundaries, relation to resources, 168-70
Nationalistic pressures, 362-63
Natural gas, 160, 276
Needleleaf evergreen forest, 42-43
Netherlands, 269-78
  energy resources, 169-70, 276
  historical development, 273-75
  land reclamation, 269-72
  photograph, 277
  physical setting, 269-73
  population density, 274-75
  windmills, 270
Neuchatel lake (Switzerland), 144
Newcastle (England, Great Britain), 252
New Forest (England, Great Britain), 42
Newfoundland, 150
Nicaragua, population, 53
Nice (France), 198
Nickel, 240
Nicosia (Cyprus), 318
Nitrogen oxides, 208
Nonmetallic minerals, 167
Norman conquest, 250
Normandy (region, France), 21, 256
North Africa, 9, 131, 305, 314, 316, 321, 325
North America, demand for agricultural labor, 63
  and European agriculture, 144-45
  grain production, 238
  tourists from, 198

North Atlantic drift, 32, 230
North Atlantic Treaty Organization (NATO), 85, 224-25, 276, 279, 315, 336, 362
North Rhine–Westphalia (state, Germany), 187
North Sea, 25, 27, 38, 211, 249
  fishery, 150-51
  oil and gas production, 158-60. 240, 253,
  storms, 249, 271-72, 289
North Sea dike (Netherlands), 271
North Sea Plain, 25, 135, 246, 256, 269, 280, 282, 286
Northern Europe, 229-44
  fallout from Chernobyl, 204
Northern Ireland (Ulster), 86, 245, 293
  religion, 93-94
  See also Great Britain
Norway, 211, 229-44
  agriculture, 237
  climate, 13, 230, 235
  economic patterns, 242
  fishery, 150-51, 238
  fjords, 22-23
  historical development, 235-37
  hydroelectric power, 163, 241
  industry, 179, 240
  North Sea oil production, 159-60, 240
  photographs, 24, 112, 149, 178, 240
  physical setting, 229-30
Nuclear electricity, 262-64, 286-87
  fission, 162
  fusion, 162
  pollution, 203-04, 204

Oats, 130, 142
Occitan language, 80
Ocean currents, as climatic controls, 31-32
  See also Gulf Stream; North Atlantic drift
Odille (France), 161
Oil, 158-60, 276
  imports, 159-60
  pollution, 212
Oise river (France), 256
Old Castile (region, Spain), 322
Olives, 130-31, 143, 327
Olympus (mountain, Greece), 314
One-field agricultural system, 97
Oporto (Portugal), 328
Organization for Economic Cooperation and Development (OECD), 226
Organization of Petroleum Exporting Countries (OPEC), 297

Ottoman empire, 9, 90
Oviedo (Spain), 328
Oxford (England, Great Britain), 252

Paid vacation, impact on tourism, 196-97
Papal states, 305
Paris (France), 101, 174, 256, 259, 262, 264, 292-93
  site location, 105
  temperature range, 34
  urban evolution, 104-09
Paris basin, 48, 142, 255-56, 260
Pastoral economies, Alpine, 122-24
  Mediterranean, 132-33
Pays, 258
Peloponnesian peninsula (Greece), 314
Pennine mountains (England, Great Britain), 21, 246, 248, 251-52
Persian (Arabian) Gulf, 159-60
Petroleum. See oil
Petsamo (Finland, Russia), 240
Phoenicians, 324
Pic de Vignemale (mountain, France), 256
Pico de Añeto mountain (Spain), 21
Piedmont (region, Italy), 305
Pindus mountains (Greece), 314
Pipelines, 176-77
Pisa (Italy), 305
Pleistocene glaciation, 21-24
Ploesti (Romania), 158, 344
Po river (Italy), 25, 27, 48, 290, 292, 301-07
Podzolic soil process, 48
Poitiers (France), battle of, 90
Poland, 332, 335, 337, 340-42
  frontiers, 17
Political fragmentation, as factor of environmental stress, 204
Political instability, as emigration pressure, 63
Pollution. See environmental problems
Pompeii (Italy), 17
Population, developed vs. less-developed regions, 51-54
  as emigration pressure, 63
Population density, as factor of environmental stress, 204
Population distribution, 58-59
Population pyramids, 56-57
Portugal, 155, 172, 321-28
  agriculture, 327
  climate, 322
  emigrant workers, 276

historical development, 324-26
photograph, 327
physical setting, 321-22
tourism, 328
Potato, 145, 251
in Ireland, 65
Power and fuel resources, 153-63
costs as factor of industrial location, 178-79
Prague (Czech Republic), 209, 342
Primate city, 259
Primitive household industry, 182
Protestantism, 87-88, 242
Prussia (region, Germany), 284, 340
Public awareness of environmental issues, 204-05, 213-14
Pyrenees mountains (France, Spain, Andorra), 13, 17, 21, 24, 32, 119, 161, 255-56, 321-22, 329

Racial and ethnic stress, 68-69
Rail transportation, 176-77, 194-96
Rance river (France), 163
Reading (England, Great Britain), 252
Reconquest, 325
Reforestation, 42
Refugee movements, 66-67
Regional geography, 3-5
Regional devolution. See individual countries
Reindeer, 203-04
Renaissance, 305, 307-08
Renault automobiles, 180
Rennes (France), 259
Resettlement movements, 66
Resource distribution, 17, 155-56, 168-70
Rhine river (Switzerland, Germany, France, Netherlands), 16, 26-27, 172, 176, 187-88, 256, 269-71, 276, 280, 283-86, 288, 338-39
delta of, 269-72
pollution of, 213
Rhodes island (Greece), 314
Rhodope mountains (Bulgaria), 346
Rhône river (Switzerland, Liechtenstein, France), 26-27, 46, 136, 176, 256, 261-62, 290
Rice, 113-14, 306
Riga (Latvia), 242
Rivers, as settlement sites, 18-19
Riviera, French, 136, 198
Road transport, 176
Roman Catholic church, 242, 305, 312-13
Roman Empire, 217, 301, 305, 324, 344

Romance languages, 78-80
Romania, 209, 335, 338-39, 343-47, 358
language, 344
mineral deposits, 344
oil production, 158, 344
Romans, aqueduct building, 134
conversion to Christianity, 86
in Germany, 283-84
introduction of vine, 143-44
occupation of Britain, 250
Romansh language, 80-83
Rome (Italy), 100, 304, 308, 312
Root crops, importance in food economy, 142-43
Rotterdam (Netherlands), 276
Rouen (France), 259
Ruhr (region, Germany), 187-89, 280, 286-88
coal production, 169, 188
Russia. (See also Soviet Union), 255, 262, 332, 335-38, 340, 344, 347, 355-58, 363, 365
as fishing nation, 151
industrial planning, 180
natural gas reserves, 160
oil production, 158, 160
Rye, 142
Rye-potato belt, 145, 242

Saar coalfield (France, Germany), 169
St. Bernard pass (Switzerland), 294
St. Gallen (Switzerland), 295
St. George's channel (Great Britain, Ireland), 246
St. Kitts and Nevis, 297
St. Petersburg (Russia), 242
St. Vincent and the Grenadines, 297
Salt, 167, 287
Salzburg (Austria), 167, 291, 296
San Marino, 311
Sardinia (island, Italy), 304, 307-08
Sarthe river (France), 256
Savoy (region, France, Italy), 305
Saxons, 250
Scandinavian mountains, 19, 23, 119, 229-30
Scheveningen (Netherlands), 276
Schipol airport (Netherlands), 272
Schuman, Robert, 218
Schwartzwald. See Black Forest
Schwyz (canton, Switzerland), 290
Scilly islands (England, Great Britain), 248
Scotland (region, Great Britain), 245-46, 251. See also Great Britain
economic activity, 250-53

highlands, 20, 119, 246, 249, 251
language, 76, 80
lowlands, 250, 252
photograph, 250
Southern uplands, 20, 246
Seasonal change, as climatic control, 31
Seine river (France), 27, 105, 256
Semitic languages, 81
Separatist movements, and language, 76-77
Serbia (See also Yugoslavia), 347-53
Serbo-Croatian language, 80
Settlement, agglomerated, 99-100
dispersed, 99
Severn river (England, Great Britain), 246
Shannon river (Ireland), 246
Shatter belt concept, 331, 338
Sheep, domestication, 116-17
in the British economy, 251
in the Mediterranean economy, 132-33
Sheffield (England, Great Britain), 252
Shellfish, 150-51
Sicily (island, Italy), 17, 302-04, 308-10
Siena (Italy), 305
Sierra de Gredos mountains (Spain), 322
Sierra de Guadarrama mountains (Spain), 322
Sierra Nevada mountains (Spain), 322
Silesia (region, Czech Republic, Slovakia, Poland), 286
coalfields, 185, 340, 342
Simple powered household industry, 182-83
Sirocco wind, 38
Slavonic languages, 80
Slovakia, 220, 341-42, 362
Slovenia, 220, 348, 350-52, 362
photograph, 350
Sognefjord (Norway), 23
Soil classification, 46-49
Solar energy, 161-62
Solidarity labor movement, 340
Solid waste pollution, 212-13
Southeast Asia, 331
Southeastern Europe, 332. See also Balkans
Southern Europe, emigration from, 64
Sovereign Military Order of Malta, 312-13

Soviet Union (former), See also Russia. 1, 9, 284-85, 287, 293, 332-36, 339-44, 346-48, 355-58, 365
  Baltic republics. See Estonia, Latvia, Lithuania
  relations with COMECON, 223-24
  relations with Finland, 237-38
  Transcaucasian republics, 336
Spaak, Paul-Henri, 218
Space requirements, as factor of industrial location, 179
Spain, 321-30
  agriculture, 327
  civil war, 67
  climate, 322
  fishery, 150
  frontiers, 17
  historical development, 324-26
  industry, 328
  minerals, 327
  photographs, 77, 105, 137, 151, 325, 326, 328
  physical setting, 322
  as pollution importer, 209, 213
  southern, 130
  tourism, 191, 328
Steam engine. See external combustion engine
Steamship, 195
  role in emigration, 64
Steelmaking, 188
  in France, 261
Steppe climate, 36, 322
  vegetation, 46
Stockholm (Sweden), 241
Stockton and Darlington railway, 194
Strasbourg (France), 220, 259, 278
Stuttgart (Germany), 288
Sudetes (mountains, Czech Republic, Poland), 340
Sulphur, 307
  dioxide, 205
Sunflowers, 143
Svalbard island (Norway), 240
Swansea (Wales, Great Britain), 252
Sweden, 229-38, 240-43
  agriculture, 237-38
  air pollution, 203
  climate, 13, 230, 235
  coal deficiency, 155
  economic patterns, 237-38, 240-41
  glacial landscapes, 21-22, 25
  historical development, 235-37
  industry, 240-41
  iron ore, 165-66, 240-41
  photograph, 241
  physical setting, 229-30
Switzerland, 289-97
  agriculture, 294-95
  alpine boundary, 16
  cantons, 17, 86, 292
  climate, 292
  confederation, 292-93
  economy, 124
  financial industry, 297
  historical development, 292-93
  hydroelectric power, 127, 163
  industry, 295
  language complexity, 81-83, 293
  neutrality, 289, 296-97
  photographs, 15, 35, 69, 101, 126, 137, 201, 294, 296
  physical setting, 289-90
  religions, 293
  tourism, 197, 201-02, 296
Syria, 317
Systematic geography, 3-5

Tagus (Tejo) river (Spain, Portugal), 322
Tallinn (Estonia), 242
Taranto (Italy), 137, 311
Taunus mountains (Germany), 280
Textiles, 252
TGV (high-speed train), 196, 262
Thames river (England, Great Britain), 27, 246
Thrace (region, Greece, Turkey), 9
Three-field agricultural system, 97-98
Thuringerwald (region, Germany), 280
Thyssen industrial group, 188
Ticino (canton, Switzerland), 290, 292, 295
Tidal power, 163
Tirane (Albania), 355
Titano (mountain, San Marino), 311
Tito, Josip Broz, 347
Torrey Canyon (tanker), 212
Tourism, 191-202. See also individual countries.
  early European, 197-98
  in the Alpine economy, 127
  infrastructure, 198-99
  in the Mediterranean economy, 136
  modern European, 198-201
  psychology of, 199-200
  time and space concentration of, 200
  visitors from North America, 254
Transhumance, 122-24, 133
Transportation, as factor of industrial location, 176-77
Transylvanian Alps (Romania), 338, 343
Travelers, early, 192-93
Treaty of Rome (1957), 219
Tuberculosis, 197, 205
Tundra climate, 36, 248
  vegetation, 45
Turin (Italy), 302, 307
Turkey, 10, 313-19, 331, 342, 344, 346-47, 349
  language, 81
Turkish Republic of Northern Cyprus, 318
Turks, in Bulgaria, 66
Tuscany (region, Italy), 304
Tyne river (England, Great Britain), 246
Tyneside (region, England, Great Britain), 252

Ukraine, 10, 322, 340, 343, 357-58
Ulster. See Northern Ireland
Unionization, 196
United Kingdom. See Great Britain, Northern Ireland
United Nations, 297-98, 318
United States, industrial planning, 179
  oil production, 158
  tourism, 328
Ural mountains (Russia), 8
  river (Russia), 8
Ural-Altaic languages, 80-81
Urbanization, 70, 100-08, 114
  and industry, 173-75
Urgel, Bishop of (Spain), 329
USSR. See Russia, Soviet Union

Vaduz (Liechtenstein), 297
Valencia (Spain), 45, 135, 327
  language, 80
Valletta (Malta), 316
Vatican City, 265, 312
Venice (Italy), 191, 301, 305
Vertical integration of industry, 188
Vesuvius mountain (Italy), 17, 304
Vienna (Austria), 291, 293, 296-97, 343
Vienne river (France), 256
Vikings, 23, 236, 250
Vilnius (Lithuania), 242
Vine, as beverage crop, 143-44
  extent of, 125
  in the Mediterranean economy, 131, 143
Violent weather, 38
Visigoths, 324
Volcanic activity, 17-18
Vosges mountains (France), 21, 256

Wales (region, Great Britain), 20, 245-46, 248, 251. See also Great Britain

Index

language, 76
photograph, 251
South, 252
Walloons, 83-85, 275
Walls, as urban features, 103-07
of Paris, 105-07
Warsaw (Poland), 340
Warsaw Pact, 224
Water, as power source, 154
mills, 154
as transportation, 176-77, 211
Water power in the Alpine economy, 126
Waterways, locational advantages of, 99-100
Watt, James, 155
Western European Union, 225
Wheat, 113-14, 140-42, 327
domestication, 113-14

growing requirements, 142
in the Mediterranean economy, 130, 141
production, France, 260-61
Wiener Neustadt (Austria), 295
William the Conqueror, 245
Wind, as power source, 154, 163
mills (Netherlands), 154
Wine, 306, 327
production, France, 261
Winterthur (Switzerland), 295
Wood, as fuel, 154-55
changing uses of, 148-49
See also forestry
Woolen textiles, 251-52
World War I (1914-18), 63, 104, 197, 293, 298, 302, 331, 347, 352
World War II (1939-45), 66, 68, 158, 197, 240, 284, 288, 293, 331-32, 340-41, 347, 351, 354
Wright brothers, 195

Yorkshire (region, England, Great Britain), 251-52
Yugoslavia, 200, 331-32, 336, 347-52, 363. See also Bosnia, Croatia, Macedonia, Montenegro, Serbia, Slovenia
landforms, 21
migrant labor from, 348
tourism, 200

Zagreb (Croatia), 352
Zeeland (region, Netherlands), 270
Zeiss optical company, 288
Zinc, 240
Zuider Zee (Netherlands), 271
Zurich (Switzerland), 292, 296